T0276078

NON-EXHAUST EMISSIONS

NON-EXHAUST EMISSIONS

AN URBAN AIR QUALITY PROBLEM FOR PUBLIC HEALTH

IMPACT AND MITIGATION MEASURES

Edited by

FULVIO AMATO

Institute of Environmental Assessment and Water Research (IDÆA)
Spanish National Research Council (CSIC), Barcelona, Spain

ACADEMIC PRESS

An imprint of Elsevier

Academic Press is an imprint of Elsevier
125 London Wall, London EC2Y 5AS, United Kingdom
525 B Street, Suite 1800, San Diego, CA 92101-4495, United States
50 Hampshire Street, 5th Floor, Cambridge, MA 02139, United States
The Boulevard, Langford Lane, Kidlington, Oxford OX5 1GB, United Kingdom

Library of Congress Cataloging-in-Publication Data
A catalog record for this book is available from the Library of Congress

British Library Cataloguing-in-Publication Data
A catalogue record for this book is available from the British Library

ISBN: 978-0-12-811770-5

For information on all Academic Press publications visit our website at
https://www.elsevier.com/books-and-journals

Working together
to grow libraries in
developing countries

www.elsevier.com • www.bookaid.org

Publisher: John Fedor
Acquisition Editor: Erin Hill-Parks/Kattie Washington
Editorial Project Manager: Kristi Anderson
Production Project Manager: Anusha Sambamoorthy
Designer: Victoria Pearson

Typeset by TNQ Books and Journals

Contents

List of Contributors vii

Preface ix

Acknowledgments xi

1. Current State of Particulate Air Quality
PHILIP K. HOPKE, ROY M. HARRISON,
FRANK DE LEEUW, XAVIER QUEROL

1. Introduction 1
2. Air Quality Guidelines 2
3. Air Quality Data 5
References 18

2. Vehicle Non-Exhaust Emissions: Impact on Air Quality
ELIO PADOAN, FULVIO AMATO

1. Introduction 21
2. Review Method and Limitations 24
3. Results 24
4. Conclusions 55
References 57

3. Impact on Public Health—Epidemiological Studies
MASSIMO STAFOGGIA, ANNUNZIATA FAUSTINI

1. Introduction 67
2. Non-Exhaust Particles: A Brief Overview 68
3. Health Effects of Non-Exhaust Particles: Epidemiological Evidence 71
Epidemiological Study Designs for the Analysis of Air Pollution and Health 81
References 83

4. Regulation on Brake/Tire Composition
THEODOROS GRIGORATOS

1. Brakes Regulation 89
2. Tires Regulation 97

3. Summary 99
References 100

5. European Emission Inventories and Projections for Road Transport Non-Exhaust Emissions
HUGO DENIER VAN DER GON, JAN HULSKOTTE,
MAGDALENA JOZWICKA,
RICHARD KRANENBURG, JEROEN KUENEN,
ANTOON VISSCHEDIJK

1. Introduction 102
2. Trends in Road Transport Exhaust and Wear Emissions 104
3. Emission Reporting in Europe for Brake and Tire Wear and Road Abrasion 108
4. Analysis and Comparison of Official Reported Emissions Data 109
5. Resuspension due to Road Transport 114
6. Tracers to Verify Brake, Tire, and Road Wear 116
7. Conclusions 118
References 120

6. Review of Brake Wear Emissions
JANA KUKUTSCHOVÁ, PETER FILIP

1. Introduction 123
2. Automotive Friction Brake Materials 125
3. Transformations of Friction Brake Materials and Release of Wear Emissions 126
4. Brake Wear Emission Studies 129
5. Identification of Gaps 141
6. Future Trends in Development of Materials for Automotive Brakes 142
7. Summary 143
Acknowledgments 143
References 144

7. Review of Tire Wear Emissions
JULIE PANKO, MARISA KREIDER, KENNETH UNICE

1. Introduction   147
2. Characteristics of Tire Wear Particles   148
3. Tire Wear Studies   153
4. Data Gaps and Needs   156
5. Conclusions   156
References   157

8. Review of Road Wear Emissions
MATS GUSTAFSSON

1. Background   161
2. Road Wear Particle Emissions From Nonstudded Tires   162
3. Road Wear Particle Emissions From Studded Tires   167
4. Road Wear Emission Factors   172
5. Road Wear Particle Properties   173
6. Discussion and Identification of Gaps and Future Needs   177
References   179

9. Review of Road Dust Emissions
BRUCE R. DENBY, KAARLE J. KUPIAINEN, MATS GUSTAFSSON

1. Introduction   184
2. Overview of Sources and Processes Affecting Road Dust Loading and Emissions   185
3. Sampling of Road Dust Loading   191
4. Measurement of Road Dust Emissions   194
5. Process-Based Modeling of Road Dust Loading and Emissions   196
6. Conclusions and Future Research Needs   199
References   200

10. Technological Measures for Brake Wear Emission Reduction
SEBASTIAN GRAMSTAT

1. Introduction   205
2. Reduction Measures of Emitted Brake Particles   206
3. Collection of Emitted Brake Particles   218
4. Prevention of Brake Particle Generation   220
5. Summary   225
Glossary   225
References   226

11. Non-technological Measures on Road Traffic to Abate Urban Air Pollution
XAVIER QUEROL, FULVIO AMATO, FRANCESC ROBUSTÉ, CLAIRE HOLMAN, ROY M. HARRISON

1. Introduction   230
2. Non-technological Road Traffic Measures in Urban Areas   234
Acknowledgments   255
References   255

12. Non-Exhaust PM Emissions From Battery Electric Vehicles
VICTOR R.J.H. TIMMERS, PETER A.J. ACHTEN

1. Introduction   262
2. Weight and Non-Exhaust Emissions   264
3. Non-Exhaust Emissions of Electric Vehicles Compared With Internal Combustion Engine Vehicles: A Review of Previous Studies   269
4. Discussion   278
5. Conclusions   281
References   282

13. Air Quality in Subway Systems
TERESA MORENO, VÂNIA MARTINS, CRISTINA RECHE, MARIA CRUZ MINGUILLÓN, ELADIO DE MIGUEL, XAVIER QUEROL

1. Particulate Matter Concentrations and Size Ranges   301
2. Airborne Particle Characteristics, Sources, and Chemical Composition   308
3. Bioaerosols in the Subway System   312
4. Health Effects, Mitigation, and Future Trends   313
Acknowledgments   316
References   316

Index   323

List of Contributors

Peter A.J. Achten INNAS BV, Breda, The Netherlands

Fulvio Amato Institute of Environmental Assessment and Water Research (IDÆA), Spanish National Research Council (CSIC), Barcelona, Spain

Frank de Leeuw National Institute for Public Health and the Environment (RIVM), Centre for Environmental Quality (MIL), Bilthoven, The Netherlands

Eladio de Miguel Transports Metropolitans de Barcelona, TMB Santa Eulalia, L'Hospitalet de Llobregat, Spain

Bruce R. Denby The Norwegian Meteorological Institute (MET), Norway

Hugo Denier van der Gon TNO, Department Climate, Air and Sustainability, Utrecht, The Netherlands

Annunziata Faustini Department of Epidemiology, Lazio Region Health Service/ASL Roma 1, Rome, Italy

Peter Filip VSB — Technical University of Ostrava, Ostrava, Czech Republic; Southern Illinois University Carbondale, Carbondale, United States

Sebastian Gramstat Development Foundation Brake, AUDI AG, Ingolstadt, Germany

Theodoros Grigoratos Environmental Chemist, Ranco (VA), Italy

Mats Gustafsson Swedish National Road and Transport Research Institute, Linköping, Sweden

Roy M. Harrison Department of Environmental Sciences/Centre for Excellence in Environmental Studies, King Abdulaziz University, Jeddah, Saudi Arabia; School of Geography, Earth & Environmental Sciences, University of Birmingham, Birmingham, United Kingdom

Claire Holman Brook Cottage Consultants Ltd, Bristol, United Kingdom

Philip K. Hopke University of Rochester School of Medicine and Dentistry, Rochester, NY, United States

Jan Hulskotte TNO, Department Climate, Air and Sustainability, Utrecht, The Netherlands

Magdalena Jozwicka TNO, Department Climate, Air and Sustainability, Utrecht, The Netherlands; European Environment Agency, Copenhagen, Denmark

Richard Kranenburg TNO, Department Climate, Air and Sustainability, Utrecht, The Netherlands

Marisa Kreider Cardno ChemRisk, LLC, Pittsburgh, PA, United States

Jeroen Kuenen TNO, Department Climate, Air and Sustainability, Utrecht, The Netherlands

Jana Kukutschová VSB — Technical University of Ostrava, Ostrava, Czech Republic

Kaarle J. Kupiainen Nordic Envicon Oy, Helsinki, Finland

Vânia Martins Universidade de Lisboa, Portugal

Maria Cruz Minguillón Institute of Environmental Assessment and Water Research (IDÆA), Spanish National Research Council (CSIC), Barcelona, Spain

Teresa Moreno Institute of Environmental Assessment and Water Research (IDÆA), Spanish National Research Council (CSIC), Barcelona, Spain

Elio Padoan Institute of Environmental Assessment and Water Research (IDÆA), Spanish National Research Council (CSIC), Barcelona, Spain; Università degli Studi di Torino, DiSAFA - Chimica agraria, Grugliasco (TO), Italy

Julie Panko Cardno ChemRisk, LLC, Pittsburgh, PA, United States

Xavier Querol Institute of Environmental Assessment and Water Research (IDÆA), Spanish National Research Council (CSIC), Barcelona, Spain

Cristina Reche Institute of Environmental Assessment and Water Research (IDÆA), Spanish National Research Council (CSIC), Barcelona, Spain

Francesc Robusté Civil and Environmental Engineering, ETESCCP, Universitat Politècnica de Catalunya, Barcelona, Spain

Massimo Stafoggia Department of Epidemiology, Lazio Region Health Service/ASL Roma 1, Rome, Italy

Victor R.J.H. Timmers INNAS BV, Breda, The Netherlands

Kenneth Unice Cardno ChemRisk, LLC, Pittsburgh, PA, United States

Antoon Visschedijk TNO, Department Climate, Air and Sustainability, Utrecht, The Netherlands

Preface

Air pollution, especially from particulate matter (PM), is widely recognized as one of the main risk factors for premature deaths and hospital admissions worldwide. PM is a heterogeneous mix of chemical elements and origins, with road traffic being the major source of PM and related toxicity in cities. The increasing urbanization in developing countries and aging of population in developed countries enhance considerably the risk for the future. Over the last two decades most of the attention regarding this risk has been focused on vehicle exhaust emissions and their successful remediation and regulation. Consequently, nowadays about half of traffic emissions derive from non-exhaust processes such as direct emissions from the wear of brakes, tires, and road surfaces, and the suspension of particles previously deposited on road surfaces (road dust suspension) due to vehicle-induced turbulence and wind. Non-exhaust emissions are projected to dominate traffic emissions within a few years, contributing heavily to degraded air quality and public health. As the relative contribution from non-exhaust emissions to total urban PM increases, so should our understanding of their health impact and abatement. Nevertheless, the current knowledge on this topic is quite limited, mostly concerning effective mitigation. Non-exhaust emissions also produce high levels of pollution in underground urban transport systems (up to seven times higher than outdoor air), where millions of passengers are exposed during their daily commuting.

This book includes 13 independent chapters focusing on non-exhaust PM grouped under the following headings:
- impact on health
- impact on air quality in relation to national and WHO standards
- emission inventorying and modeling
- technological measures
- regulation and policy measures
- role in subway air quality

This book gathers the most up-to-date knowledge on non-exhaust emissions, identifying policy and research needs to support national and local authorities in the challenging task of meeting air quality guidelines and thus improving public health. I am thankful to the experts in epidemiology, air quality, vehicle and road engineering, and policy for contributing chapters that make this book a reference document for:
- Researchers and students of a wide range of disciplines (environmental sciences, health sciences, air quality, atmospheric chemistry and physics, geochemistry, toxicology, epidemiology, built environment, road and vehicle engineering, city planning).
- Society: nongovernmental organizations, citizens associations, among others.
- Industry: including vehicle and brake manufacturers and those involved in road and subway construction and maintenance.
- Governance: authorities responsible for air quality and public health.

Fulvio Amato
Editor

Acknowledgments

The Editor would like to thank the authors for the big effort and several revisions they have made into this book. Support from AXA Research Fund and BBVA Foundation is also acknowledged. The Editor is thankful to Xavier Querol (CSIC), Teresa Moreno (CSIC), Wes Gibbons, Franco Lucarelli (University of Florence), and Mar Viana (CSIC) for their advices and friendship and finally to his family (Norma, Abril, and Gerard) for their love.

CHAPTER

1

Current State of Particulate Air Quality

Philip K. Hopke[1], Roy M. Harrison[2,3], Frank de Leeuw[4], Xavier Querol[5]

[1]University of Rochester School of Medicine and Dentistry, Rochester, NY, United States;
[2]Department of Environmental Sciences/Centre for Excellence in Environmental Studies, King
Abdulaziz University, Jeddah, Saudi Arabia; [3]School of Geography, Earth & Environmental
Sciences, University of Birmingham, Birmingham, United Kingdom; [4]National Institute for
Public Health and the Environment (RIVM), Centre for Environmental Quality (MIL),
Bilthoven, The Netherlands; [5]Institute of Environmental Assessment and Water Research
(IDÆA), Spanish National Research Council (CSIC), Barcelona, Spain

OUTLINE

1. Introduction	2	3.1.3 Trends of Average PM_{10}, $PM_{2.5}$, and $PM_{2.5-10}$	7
2. Air Quality Guidelines	2	3.1.4 Trends of Exceedances of the Air Quality Standards	10
3. Air Quality Data	5	3.2 United States	12
3.1 Europe	5	3.3 South and Southeastern Asia	14
3.1.1 European Union Limit Values for Particulate Matter	5	References	18
3.1.2 Proportion of the Coarse Particles in PM_{10}	6		

Non-Exhaust Emissions
https://doi.org/10.1016/B978-0-12-811770-5.00001-7

1. INTRODUCTION

Non-exhaust emissions make significant contributions to the concentration of airborne particulate matter (PM), particularly to coarse mode particles. Thus, these emissions play an important role in degrading air quality such that they fail to achieve health-based regulatory levels or global guidelines. These emissions include brake and tire wear particles and particles of suspended road dust. Road dust includes soil that has been deposited onto paved roads as well as the unpaved surfaces, including parking lots, and work areas that have been intermixed with wear particles, leaking oil, and tailpipe emissions. The major impact of non-exhaust emissions in terms of mass contributions is on the concentrations of supermicron particles in PM_{10} or $PM_{2.5-10}$. The purpose of this chapter is to review measured concentrations of PM_{10} or $PM_{2.5-10}$, the trends over time at multiple locations around the world, and the relationship of those concentrations with health-based guidance developed by the World Health Organization (WHO) and regulatory standards promulgated by the United States (US) Environmental Protection Agency and the European Union (EU).

2. AIR QUALITY GUIDELINES

The WHO first published guidelines under the title of "Air Quality Guidelines for Europe" in 1987. Those guidelines covered the effects of a wide range of organic compounds and inorganic substances on human health, as well as effects of inorganic substances on vegetation. The guidelines were revised as a second edition, which was published in 2000, again covering a wide range of substances listed, together with the guideline values shown in Table 1.1. At a

TABLE 1.1   World Health Organization Air Quality Guidelines: Individual Substances Based on Effects Other Than Cancer or Odor/Annoyance

Substance	Time-Weighted[a] Average	Averaging[a] Time	Time-Weighted[b] Average	Averaging[b] Time
Cadmium	$5\,ng/m^{3c}$	annual		
Carbon disulfide[d]	$100\,\mu g/m^3$	24 h		
Carbon monoxide	$100\,mg/m^{3e}$	15 min		
	$60\,mg/m^{3e}$	30 min		
	$30\,mg/m^{3e}$	1 h		
	$10\,mg/m^3$	8 h		
1,2-Dichloroethane	$0.7\,mg/m^3$	24 h		
Dichloromethane	$3\,mg/m^3$	24 h		
	$0.45\,mg/m^3$	1 week		
Fluoride[f]	—	—		
Formaldehyde	$0.1\,mg/m^3$	30 min		
Hydrogen sulfide[d]	$150\,\mu g/m^3$	24 h		
Lead	$0.5\,\mu g/m^3$	annual		
Manganese	$0.15\,\mu g/m^3$	annual		

TABLE 1.1 World Health Organization Air Quality Guidelines: Individual Substances Based on Effects Other Than Cancer or Odor/Annoyance—cont'd

Substance	Time-Weighted[a] Average	Averaging[a] Time	Time-Weighted[b] Average	Averaging[b] Time
Mercury	$1 \, \mu g/m^3$	annual		
Nitrogen dioxide	$200 \, \mu g/m^3$	1 h	$200 \, \mu g/m^3$	1 h
	$40 \, \mu g/m^3$	annual	$40 \, \mu g/m^3$	1 year
Ozone	$120 \, \mu g/m^3$	8 h	$100 \, \mu g/m^3$	8 h
Particulate matter[g]	Dose–response	—	$20 \, \mu g/m^3$ (PM$_{10}$)	1 year
			$10 \, \mu g/m^3$ (PM$_{2.5}$)	1 year
			$50 \, \mu g/m^3$ (PM$_{10}$)	24 h
			$25 \, \mu g/m^3$ (PM$_{2.5}$)	24 h
Platinum[h]	—	—		
Polychlorinated biphenyls (PCBs)[i]	—	—		
Polychlorinated dibenzo-p-dioxins (PCDDS)/ polychlorinated dibenzofurans (PCDFs)[j]	—	—		
Styrene	$0.26 \, mg/m^3$	1 week		
Sulfur dioxide	$500 \, \mu g/m^3$	10 min	$500 \, \mu g/m^3$	10 min
	$125 \, \mu g/m^3$	24 h	$20 \, \mu g/m^3$	24 h
	$50 \, \mu g/m^3$	annual		
Tetrachloroethylene	$0.25 \, mg/m^3$	annual		
Toluene	$0.26 \, mg/m^3$	1 week		
Vanadium[d]	$1 \, \mu g/m^3$	24 h		

[a]From Air Quality Guidelines for Europe, Second Edition (WHO, 2000).

[b]From Air Quality Guidelines: Global Update 2005 (WHO, 2006).

[c]The guideline value is based on the prevention of a further increase of cadmium in agricultural soils, which is likely to increase the dietary intake.

[d]Not reevaluated for the second edition of the guidelines.

[e]Exposure at these concentrations should be for no longer than the indicated times and should not be repeated within 8 h.

[f]Because there is no evidence that atmospheric deposition of fluorides results in significant exposure through other routes than air, it was recognized that levels below $1 \, \mu g/m^3$, which is needed to protect plants and livestock, will also sufficiently protect human health.

[g]The available information for short- and long-term exposure to PM$_{10}$ and PM$_{2.5}$ does not allow a judgment to be made regarding concentrations below which no effects would be expected. For this reason no guideline values have been recommended, but instead risk estimate have been provided (see Chapter 7, Part 3).

[h]It is unlikely that the general population, exposed to platinum concentrations in ambient air at least three orders of magnitude below occupational levels where effects were seen, may develop similar effects. No specific guideline value has therefore been recommended.

[i]No guideline value has been recommended for PCBs because inhalation constitutes only a small proportion (about 1%–2%) of the daily intake from food.

[j]No guideline value has been recommended for PCDDs/PCDFs because inhalation constitutes only a small proportion (generally less than 5%) of the daily intake from food.

meeting in 2005, work was carried out for a global update, ultimately published in 2006. The update carried out a reassessment of just four pollutants, i.e., PM, ozone, nitrogen dioxide, and sulfur dioxide. The revised guidelines appear for these pollutants in Table 1.1. This was followed by a consideration of guidelines for indoor air quality that culminated in a 2010 publication. Those guidelines are less relevant to the purpose of this book and are not included in the tabulation. Other than changes to some of the numerical values, there were two major differences between the recommendations of 2005 and those from 2000. Firstly, the WHO (2006) proposed not only guidelines but also interim targets of three higher concentration levels to be used as targets for countries with very high air pollutant levels, which provide a lower level of protection than the WHO Air Quality Guideline itself. Secondly, the 2000 publication treated PM as a nonthreshold pollutant providing only a dose–response relationship, whereas the 2006 publication, while still acknowledging the nonthreshold behavior of PM, recommended explicit numerical values for the guideline. Air Quality Guidelines for Europe (2000) also recommended unit risk factors for use in assessing chemical carcinogens and these values are listed in Table 1.2, although few of these are directly relevant to non-exhaust emissions from road vehicles.

The International Agency for Research on Cancer has classified air pollution, in general, as well as PM as a separate component of air pollution mixtures, as carcinogenic (Loomis et al., 2013).

TABLE 1.2 World Health Organization Air Quality Guidelines: Carcinogenic Risk Estimates Based on Human Studies[a]

Substance	IARC Group	Unit Risk[b]	Site of Tumor
Acrylonitrile[c]	2A	2×10^{-5}	Lung
Arsenic	1	1.5×10^{-3}	Lung
Benzene	1	6×10^{-6}	Blood (leukemia)
Butadiene	2A	–	Multisite
Chromium (VI)	1	4×10^{-2}	Lung
Nickel compounds	1	4×10^{-4}	Lung
Polycyclic aromatic hydrocarbons (BaP)[d]	–	9×10^{-2}	Lung
Refractory ceramic fibers	2B	1×10^{-6} (fiber/L)$^{-1}$	Lung
Trichloroethylene	2A	4.3×10^{-7}	Lung, testis
Vinyl chloride[c]	1	1×10^{-6}	Liver and other sites

IARC, International Agency for Research on Cancer.
[a]Calculated with average relative risk model.
[b]Cancer risk estimates for lifetime exposure to a concentration of 1 µg/m³.
[c]Not reevaluated for the second edition of the guidelines.
[d]Expressed as benzo(a)pyrene (based on a benzo(a)pyrene concentration of 1 µg/m³ in air as a component of benzene-soluble coke-oven emissions).

3. AIR QUALITY DATA

3.1 Europe

3.1.1 European Union Limit Values for Particulate Matter

The EU sets legally enforceable limit values covering a wide range of air pollutants. Most relevant to this chapter are those for PM. These are as follows:

PM_{10}	Annual mean	$40\ \mu g/m^3$
PM_{10}	24-h mean	$50\ \mu g/m^3$ with 35 permitted exceedances per year
$PM_{2.5}$	Annual mean	$25\ \mu g/m^3$

For $PM_{2.5}$, there are also exposure objectives, based on the average exposure indicator (AEI). The AEI is defined as "a 3-year running annual mean $PM_{2.5}$ concentration averaged over the selected monitoring stations in agglomerations and large urban areas, set in urban background locations to best assess the $PM_{2.5}$ exposure of the general population." The $PM_{2.5}$ exposure concentration obligation is set at $20\ \mu g/m^3$ (AEI) based on the 3-year average for 2013−15. The $PM_{2.5}$ exposure reduction target describes the reduction required by 2020 based on the AEI in 2010. Exposure reduction targets may be 0%, 10%, 15%, or 20%, and if the AEI exceeds $22\ \mu g/m^3$ in 2010, "all appropriate measures need to be taken to achieve $18\ \mu g/m^3$ by 2020."

Although the PM_{10} limit values had to be met in 2010, 2015 monitoring data still showed frequent exceedances of the daily limit value in large parts of Europe (AirBase-EEA, 2017). Exceedances are mostly observed in urban and suburban areas. The annual limit value is exceeded at about 4% of the monitoring stations, mainly located in Bulgaria and Poland. The $PM_{2.5}$ annual limit value had to be met in 2015. In 2014 (the most recent data available currently at Airbase (EEA, 2017), the level of $25\ \mu g/m^3$ was exceeded in 4 of the 28 EU member states. It is estimated (EEA, 2017) that 8%−12% of the urban population in the EU is exposed to concentrations above the $PM_{2.5}$ limit value; a larger fraction (16%−21%) is exposed to levels above the PM_{10} daily limit value.

A more detailed analysis of selected sites was performed to assess the concentrations and trends across Europe. From the EEA (2016) Air Quality (AQ) e-Reporting Database (http://www.eea.europa.eu/data-and-maps/data/aqereporting-1), 176 AQ monitoring sites were selected for this study based on the simultaneous PM_{10} and $PM_{2.5}$ data availability of 10 years over the 15 study years (2000−14) and >25% of annual data availability. In the 2000−03, only 11, 19, 40, and 60 stations meet these criteria, whereas in the 2004−14, from 94 to 118 stations/year met them.

According to this data availability, trend analyses focus only on 2004−14 for stations having valid annual data during 73% of the period (8 years for the 11-year period 2004−14), although data from 2000 to 2003 is used sporadically for other purposes in a normalized way. With the data arising from these collocated PM_{10} and $PM_{2.5}$ measurements, we calculated the coarse fraction ($PM_{2.5-10}$) and we performed the trend analysis of the three fractions individually using the Mann−Kendall (MK) method as applied also in EEA (2017) and EMEP (2016) air quality trend reports. Mean trends and 95% confidence limits ($\mu g/m^3$/year or percentage increase or decrease/year) by country and by station type were calculated. National trends are calculated by averaging the trends estimated at individual stations. Only countries included in the consistent set have been included in the analysis. The 95% confidence interval ($\pm 2\sigma_R$) of a country-averaged trend is calculated through error propagation from the 95% confidence intervals of slopes estimated at all N_R individual stations ($\pm 2\sigma_i$) in country R, see Annex 1 for further details. The star ranking refers to the statistical significance, from the highest to the lowest significance: ***$\alpha < 0.001$; **$\alpha < 0.01$; *$\alpha < 0.05$; +$\alpha < 0.1$; blank, no significance. In this chapter, only data from colocated PM_{10}/$PM_{2.5}$ stations were employed because $PM_{2.5-10}$ could then be calculated. In the EEA (2016) report on air quality in Europe, also PM_{10} and $PM_{2.5}$ trends were analyzed separately with a much higher number of stations, and this yielded to different results than reported in EEA (2016).

3.1.2 Proportion of the Coarse Particles in PM₁₀

With the collocated $PM_{2.5}/PM_{10}$ data from 2004 to 2014, the coarse fraction ($PM_{2.5-10}$) accounts 37 ± 2, 34 ± 10, and $34 \pm 6\%$ of the averaged PM_{10} concentrations from the rural, suburban, and urban background sites, respectively, and 43 ± 3 and $30 \pm 4\%$ for the traffic and industrial sites, respectively. Thus, as expected, the traffic sites are characterized by the coarsest PM_{10} (Fig. 1.1). The finest PM_{10} was recorded at the industrial sites ($70 \pm 4\%$ of PM_{10} being made of $PM_{2.5}$) and the urban and suburban background sites (66 ± 6 and $66 \pm 10\%$) (Fig. 1.1). These values refer to mean ratios, but there are very large variations of the $PM_{2.5-10}/PM_{10}$ proportions across EU state members. Thus, in specific Portuguese stations, 65%–80% of the annual PM_{10} mean is accounted for the coarse fraction, probably due to high sea salt contributions. This proportion reaches 65%–72% of $PM_{2.5-10}$ in PM_{10} measured at Spanish sites in the Canary Islands due to the contribution of African dust. In addition, 50%–66% values were recorded in specific traffic sites in Norway, Finland, Sweden, Denmark, Czech Republic, and United Kingdom, probably due to large contributions from winter road sanding and road dust emissions enhanced by the use of studded tires in

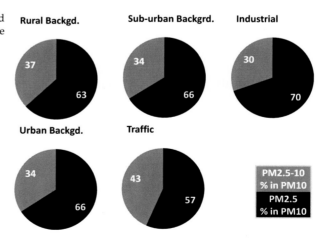

FIGURE 1.1   2004–14 averaged $PM_{2.5-10}$ and $PM_{2.5}$ mass contributions (in %) to PM_{10} for the selection of sites analyzed in the present study.

TABLE 1.3 Average Percentages of $PM_{2.5-10}$ in PM_{10} for Milano (ML), Porto (PO), Florence (FL), Barcelona (BCN), and Athens (AT) (UB, Urban Background; SUB, Suburban Background; TR, Traffic Sites) as a Function of the Daily PM_{10} Concentrations

%	PM_{10} μg/m³					
	≤40	40–45	45–50	50–55	55–60	>60
ML-UB	20	14	24	18	21	22
PO-TR	34	21	28	18	4	12
FL-UB	30		13		14	6
BCN-UB	29	42	21			
AT-SUB	33	60		62		63
AT-TR	40	35	33	40		

some countries. Specific rural and urban background sites from United Kingdom, France, Italy, Czech Republic, Germany, and Northwest Spain reach values of 7%—25% for the coarse fractions, due probably to the high prevalence of secondary PM in PM_{10}.

The AIRUSE-LIFE+ (2015) and Amato et al. (2016) reported source apportionment results for PM_{10} and $PM_{2.5}$, both for annual (2013) averages and high PM episodes of five Southern European cities. Table 1.3 presents the proportions of the coarse particle fractions in PM_{10} as a function of the increase of daily PM_{10} concentrations for these five cities. The results do not show a relation of the increase of PM_{10} concentrations with its coarser or finer proportions. However, a clear differentiation is evidenced for the type of sites and the climate region, with higher coarse proportion being recorded closer to road traffic and/or in drier climates.

3.1.3 Trends of Average PM_{10}, $PM_{2.5}$, and $PM_{2.5-10}$

Based on time trends for measured PM_{10}, $PM_{2.5}$, and $PM_{2.5-10}$, a number of studies have shown large improvements in ambient air PM_{10} in Europe (Barmpadimos, Keller, Oderbolz, Hueglin, & Prevot, 2012; EEA, 2016b; EMEP, 2016; Guerreiro, Foltescu, & de Leeuw, 2014; Querol et al., 2014). According to the most recent evaluation (EEA, 2016), there is a dominant, statistically significant, decreasing trend from 2000 to 2014 in the annual mean PM_{10} values with rates reaching averages of −0.6 and −0.9 μg/m³PM_{10}/year (the uncertainty [given as 2σ value] in the slope values is 2σ = 0.03 and 0.05 μg/m³PM_{10}/year) in urban background and road traffic sites in the 28 EU Member States (EU-28), respectively. Even more pronounced reductions were observed in some regions of Southern Europe (−1.0 and −1.5 μg/m³PM_{10}/year in Spain, 2σ = 0.14 and 0.19 μg/m³PM_{10}/year). According to EEA (2016), most European stations do not register a $PM_{2.5}$ statistically significant trend over the period 2006—14, although there is a clearly decreasing trend in many state members.

For analyzing the possible trends in observed time series, the MK method for identifying significant trends combined with the Sen's slope method for estimating slopes and confidence interval has been applied. The nonparametric MK test is particularly useful because missing values are allowed and the data need not to conform to any particular distribution. Moreover,

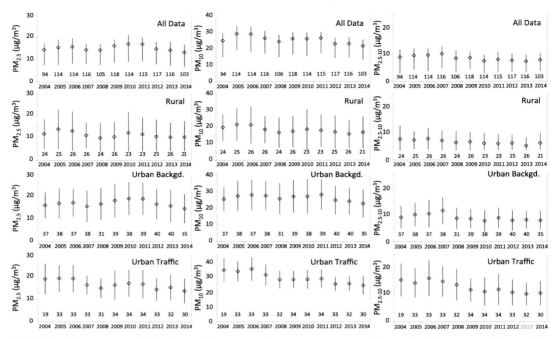

FIGURE 1.2   2004–14 $PM_{2.5}$, PM_{10}, and $PM_{2.5-10}$ median concentrations ± the standard deviation and the number of stations each year.

as only the relative magnitudes of the data rather than their actual measured values are used, this test is less sensitive toward incomplete data capture and/or special meteorological conditions, leading to extreme values. A detailed description of the MK test is given by Gilbert (1987).

Fig. 1.2 shows the average annual concentrations ± standard deviation for $PM_{2.5}$, PM_{10}, and $PM_{2.5-10}$ at 176 selected sites, and for urban and rural background and traffic sites, showing that the decreasing trends are also larger than those for PM_{10}. Below presented are the trend analyses for each of these three PM fractions.

3.1.3.1 $PM_{2.5}$

A statistically significant 2004–14 decreasing trend for $PM_{2.5}$, with a slope of -0.37 $\mu g/m^3/year$, $2\sigma = 0.07$ $\mu g/m^3/year$ (equivalent to $-2.8\%/year$, $2\sigma = 0.57\%/year$), was obtained when all the sites are included in the MK analysis. Negative slopes were obtained for 77% of the individual trends, with 41% showing statistical significance (+, *, **, and *** ranks, with a mean slope of -0.81 $\mu g/m^3/year$ for these sites). However, 4% of the sites exhibited a statistically significant increasing trend (specific sites from Czech Republic, France, and Lithuania). The largest decreasing trends were obtained for Southern Europe, with high significance ($2\sigma = 0.30-0.61$ $\mu g/m^3/year$) from -0.74 to -1.64 $\mu g/m^3/year$ for Greece, Italy, Spain, and Portugal. In addition, Austria and Bulgaria had significant decreasing trends of -0.77 and $-1.06\,\mu g/m^3/year$ ($2\sigma = 0.26$ and $0.95\,\mu g/m^3/year$). Conversely, Lithuania's (all traffic) sites recorded a highly statistically significant increasing trend of $+1.19\,\mu g/m^3/year$ ($2\sigma = 0.61\,\mu g/m^3/year$), equivalent to an increase of $8.57\%/year$ ($2\sigma = 4.22\,\mu g/m^3/year$). These sampling locations were all traffic sites. Poland also recorded a similar slope but without reaching statistical significance. When distinguishing between types of environments, the highest decreases were recorded for industrial sites, followed by high traffic locations. Thus, slopes of -0.24 ($2\sigma = 0.16$), -0.26 ($2\sigma = 0.10$), -0.50 ($2\sigma = 0.13$), and -0.79 ($2\sigma = 0.29$) $\mu g/m^3/year$ (equivalent to -3.25%, -1.57%, -4.03%, and $-5.59\%/year$, referred to the initial year) were obtained for regional and urban background, high traffic, and industrial sites, respectively. The fact that the recent EEA (2016) report did find an averaged decreasing trend for $PM_{2.5}$, but without statistical significance, is probably due to the fact that the number of sites included in this report was much higher because they did not apply the restriction of the collocated, simultaneous PM_{10} and $PM_{2.5}$ measurements.

3.1.3.2 PM_{10}

For PM_{10} a statistically significant decreasing trend, with a slope of $-0.69\,\mu g/m^3/year$, $2\sigma = 0.09$ (equivalent to $-2.6\%/year$, $2\sigma = 0.32$), was obtained when all sites were included. Among these sites, 82% of the trends have a negative slope, and 52% were statistically significant (with a mean slope of $-1.18\,\mu g/m^3/year$ for these sites). Alternatively, 2% of these exhibit a statistically significant increasing trend (specific sites from France, Lithuania, and Spain). The largest decreasing trends were obtained for Southern Europe, with highly significant trends from -1.16 to $-2.14\,\mu g/m^3/year$ ($2\sigma = 0.36-1.61$) for Spain, Greece, Portugal, and Italy. For $PM_{2.5}$, Lithuania's traffic sites recorded a highly statistically significant increasing trend of $+1.40\,\mu g/m^3/year$ ($2\sigma = 0.83$), equivalent to an increase of $4.75\%/year$ ($2\sigma = 2.76$). Poland produced a similar slope but without reaching statistical significance. When distinguishing between types of environments, the highest decrease was recorded similarly at traffic and industrial sites. Thus, slopes of -0.48 ($2\sigma = 0.20$), -0.57 ($2\sigma = 0.14$), -0.88 ($2\sigma = 0.14$), and -0.98 ($2\sigma = 0.46$) $\mu g/m^3/year$ (equivalent to -2.50, -2.11, -3.16%, and $-3.04\%/year$) were obtained for regional and urban background, traffic, and industrial sites, respectively. The causes for these decreasing trends are unknown. The most notable decreasing trends were observed not only in South European countries but also in some cold climate countries. Trends in African dust outbreaks, measures related to the use of sanding and studded tires to abate resuspension, and the economic recession in some countries are likely to have influenced these trends, as well as the implementation of EU legislation on emissions and the national and local air quality plans.

3.1.3.3 $PM_{2.5-10}$

For the coarse PM fraction, we also obtained a statistically significant decreasing trend with a slope of $-0.30\,\mu g/m^3/year$, $2\sigma = 0.07$ (equivalent to $-0.2\%/year$, $2\sigma = 0.73$) when all the sites were included. From these, 73% trends have a negative slope with only 29% being statistically significant (with a mean slope of $-0.53\,\mu g/m^3/year$ for these sites). However, 4% of these exhibit a statistically significant increasing trend (specific sites from the Czech Republic, Finland, Luxemburg, Malta, Portugal, Spain, and Sweden). The largest decreasing trends were highly significant from -0.60 to $-1.63\,\mu g/m^3/year$ ($2\sigma = 0.40$ to 0.80) for Bulgaria,

Denmark, Italy, Latvia, and Spain. Only Luxembourg (with only one industrial station included in our study) exhibited a highly statistically significant increasing trend of $+0.78\ \mu g/m^3/year$ ($2\sigma = 0.46$), equivalent to an increase of 9.26%/year ($2\sigma = 5.45\%$). When distinguishing between types of environments, trends were not significant for regional background and industrial sites but were significant for the other types. Thus, slopes of -0.30 ($2\sigma = 0.9$) and -0.37 ($2\sigma = 0.14$) $\mu g/m^3/year$ (equivalent to -3.46 and $-3.71\%/year$) were obtained for urban background and traffic sites, respectively. Similar to the results for PM_{10} and $PM_{2.5}$, the causes of these decreasing trends cannot be definitively identified, but the fact that the most marked decreasing trends are obtained at traffic sites points to a likely decrease of traffic, in part due to the economic recession and to control measures taken. The impact of the economic recession is further supported by the fact that the $PM_{2.5-10}$ steeply decreased in 2008−09 and remained constantly lower from 2009 to 2014 (see Fig. 1.2).

3.1.4 Trends of Exceedances of the Air Quality Standards

According to the selection of sites with colocated PM_{10} and $PM_{2.5}$ monitors, 1656 years-stations (number of years available at each site multiplied by the number of sites) were available for the whole period, and from these 153 (9%) exceeded the European PM_{10} annual limit value of 40 $\mu g/m^3$. This proportion reached 8%, 10%, 7%, and 21% (only five stations) of the rural and urban background, traffic, and industrial sites, respectively. Starting from 2004 to 2006, 10−27 stations/year exceeded the annual PM_{10} limit value. Subsequently, a clear and progressive decrease is evidenced until 2014, where only six sites recorded exceedances. From 2004 to 2007, 11%−19% of the sites exceeded this limit value, whereas from 2007 to 2014, these proportions ranged from 4% (2014) to 9% (2007). Focusing on rural and urban background and traffic sites, data from a relatively high number of sites reporting PM_{10} levels (13−24, 45−80, 22−61 sites/year, respectively) between 2004 and 2014 were available for analysis. A high statistically significant (***) decreasing trend of $-1.6\%/year$ of the percentage of sites exceeding the annual PM_{10} limit value at traffic sites was found using a MK analysis (Fig. 1.3). However, this decline was not the case for the urban and rural background sites, where the trends were not statistically significant, although with a slope of $-1.1\%/$ year was calculated in the case of the urban background sites (Fig. 1.3).

For the PM_{10} daily values (not exceeding 50 $\mu g/m^3$, as the percentile 90.4 of the daily concentrations), 541 (33%) years-stations exceeded this European PM_{10} standard. This proportion reached 19%, 30%, and 38% of the rural and urban background and traffic sites, respectively. From 2004 to 2012, 40−61 (24%−51%) stations/year exceeded the daily PM_{10} limit value, and in 2013 and 2014, only 29 (19%) and 23 (15%) sites recorded exceedances. Focusing on the type of site, marked decreasing trends of -2.2 and $-5.9\%/year$ of the percentage of sites exceeding the daily PM_{10} limit value, with low statistical significance (*), were obtained at urban background and traffic sites, respectively, but there was no significance at rural sites (Fig. 1.3).

For $PM_{2.5}$, 185 (11%) accumulated years-sites exceeded the European annual limit value of 25 $\mu g/m^3$. This proportion reached 9%, 16%, 5%, and 23% (only five stations) of the rural and urban background, traffic, and industrial sites, respectively. Decreasing trends were evidenced for the rural, urban, and traffic sites, but only that of the traffic sites reached a (low) statistical significance (*, $-0.9\%/year$, Fig. 1.3). For these, a sharp decreasing trend of exceedances were recorded in 2004−06, but not so marked in 2006−14.

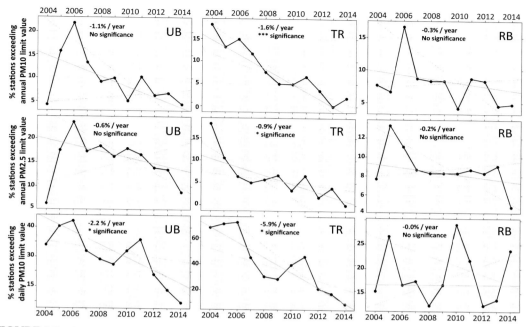

FIGURE 1.3 2004–14 trends of percentage of monitoring sites exceeding the EU PM_{10} and $PM_{2.5}$ annual and daily air quality standards.

Concerning the spatial distribution of the exceedance to the annual PM_{10} limit value, in the Czech Republic nine stations exceed this standard, with 43 accumulated years of exceedances (up to 11 days of exceedance/site); three sites in Portugal, with 10 accumulated years (up to 5 years/site); two sites in Poland and United Kingdom, with up to 10 and 8 years/site (11 and 9 accumulated years), respectively; two sites in Belgium, with 5 accumulated years (up to 4 years/site); two sites in Germany with 1 year each; one site in Italy with 7 years of exceedance; and one site in Slovakia, Denmark, Spain, France, and Norway, with 3 years in the first two sites and 1 year in the last three sites. For the annual $PM_{2.5}$ limit value, in the Czech Republic 10 stations exceed this standard, with 64 years of exceedances (up to 11 years/site); 3 sites in Portugal, with 4 years of exceedance; 2 sites in Poland and France with 19 and 3 accumulated years of exceedances; 1 site in Italy, Slovakia, Switzerland, Austria, United Kingdom, and Spain, with 10, 4, 2, 2, 1, and 1 years of exceedances, respectively.

Sites in Spain, Norway, Germany, Denmark, United Kingdom, Portugal, and Belgium exceeded the PM_{10} annual limit value, but did not exceed the $PM_{2.5}$ limits, with these locations being mostly traffic sites. This caused the decrease from 10% of years-traffic sites exceeding the PM_{10} annual limit value to 6% for $PM_{2.5}$. This shows probably the effect of the lack of measures for non-exhaust emissions. In other urban background sites of this group of stations, this difference was probably attributable to the impact of African dust and or sea salt. There are also a number of stations from Austria, Switzerland, Italy, or Poland that increased the number of exceedances in $PM_{2.5}$ compared with PM_{10}, or that exceeded the

$PM_{2.5}$ and not the PM_{10} limit value (Austria and Switzerland). This difference might be the result of the influence of the use of solid fuels, particularly wood for domestic, commercial, and institutional heating in these two countries.

The percentage of sites exceeding the WHO PM_{10} annual guideline ranged from 61% (2014) to 93% (2006), with an average of 73% for 2004—14, and 22%, 83%, 70%, 87%, and 89% for the rural, suburban, urban background, traffic, and industrial sites, respectively. For the annual WHO $PM_{2.5}$ annual guideline, the percentage of sites ranged from 68% (2014) to 89% (2006), with an average of 78% for 2004—14, and 46%, 95%, 89%, 83%, and 94% for the rural, suburban, urban background, traffic, and industrial sites, respectively. According to EEA (2016b), 8%—12% and 16%—21% of the EU-28 population are exposed to levels that exceed the annual $PM_{2.5}$ limit value (25 µg/m^3) and the PM_{10} levels daily limit value (50 µg/m^3 PM_{10} as 90.4 percentile of the daily concentrations in a year). If we use the WHO guidelines, these increase to 85%—91% for the WHO $PM_{2.5}$ annual guideline and 50%—63% for the daily PM_{10} threshold.

The PM_{10} and $PM_{2.5}$ annual limit values from the EU standard (2008/50/CE) are higher by ×2 and ×2.5 than the respective WHO health protection guidelines (WHO, 2006). During the elaboration of the 1999/30/CE directive (first EU daughter air quality directive), working groups proposed several PM_{10} limit values, based on integrated modeling assessment, ranging from the WHO guideline to different targets exceeding this guideline by several fold. However, it was stated clearly that these target values were interim values to reach the WHO target in forthcoming years. The 1999/30/CE was finally set up with two stages for the PM_{10} limit values: stage I, with an implementation from 2005 to 2010 with the annual limit value of 40 µg/m^3, and stage II, with an implementation from 2010 with the annual limit value of 20 µg/m^3 (coinciding with the WHO guideline). During the 2007 revision of this directive, the new $PM_{2.5}$ annual limit value that was introduced exceeded the US equivalent standard by a factor of 2 and the WHO guideline by 2.5. This directive became mandatory in 2015, and the PM_{10} limit value established by the 1999 directive for Stage I was confirmed. This reaffirmation was set by the new air quality directive 2008/50/CE. It also was stated that a revision of these would take place in 2013. In 2013—14 the Clean Air for Europe Legislation Package announced that the revision of the 2008/50/CE, containing the PM_{10} limit value (fixed with criteria of 1999), will continue in force until 2020, when the new revision of the directive will take place. Consequently, the fact that 10% and 20% of the European population in 2014 were exposed to levels higher to the ones fixed by the EU standards is not very meaningful because (1) these standards are old and not revised simultaneously with technological and economic developments and (2) these are not thresholds for health protection but more than double those from WHO. We estimated that meeting the annual WHO $PM_{2.5}$ current guideline (10 µg/m^3) throughout the EU-28 would lead to a reduction of 123,000 in the number of premature death compared with the 2014 situation.

3.2 United States

The US instituted a PM_{10} National Ambient Air Quality Standard (NAAQS) in 1987 with an annual average standard level of 50 µg/m^3 and a 24-h standard of 150 µg/m^3 not expected to be exceeded more than one time per year in any given 3-year period. In 2006, the 24-h standard was discontinued and only the annual average standard remains in effect.

The promulgation of this resulted in many US counties that included major urban areas initially being in nonattainment of this NAAQS. Fig. 1.4 presents the US national PM_{10} trend from 1990 to 2014 where the error bars represent the 90th and 10th percentile values. In the first years of this standard, the country was on average out of attainment with mean values in excess of $50\ \mu g/m^3$. However, over the ensuing decade, cities instituted effective mitigation strategies mostly based on effective street sweeping that has resulted in substantial reductions in urban PM_{10}. By 2014, the national mean value declined to $41.93\ \mu g/m^3$ with a substantially reduced 90th percentile value of $61\ \mu g/m^3$. There are now only 37 areas in 31 counties in the US (total 2010 population of 9,135,291) that are in nonattainment of the standard (https://www3.epa.gov/airquality/greenbook/pbtc.html). Most of the remaining areas are rural areas with extensive agricultural activity or are desert as shown in Fig. 1.5

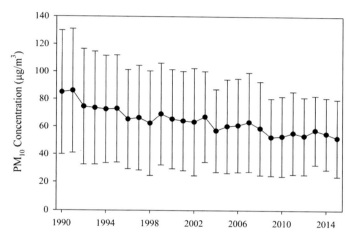

FIGURE 1.4 US national mean PM_{10} values from 1990 to 2015. Error bars represent the 90th and 10th percentile values of annual average PM_{10}.

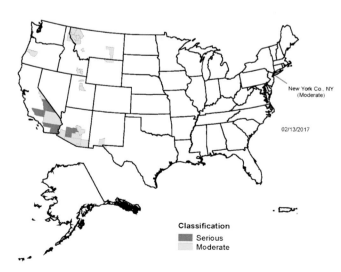

FIGURE 1.5 Counties designated nonattainment for PM_{10}. Classification colors are shown for whole counties and denote the highest area classification for that county. Map taken from https://www3.epa.gov/airquality/greenbook/map/mappm10.pdf.

(https://www3.epa.gov/airquality/greenbook/map/mappm10.pdf). New York City is the only eastern location that remains in nonattainment of the PM NAAQS. However, although measures taken, such as street sweeping, have generally allow attainment of the PM_{10} NAAQS standard, the average value in the US remains well above the WHO guidance value.

Over the same period, vehicular miles traveled in the US increased by 41% from 2.1 to 3.0 million miles per year (https://www.rita.dot.gov/bts/sites/rita.dot.gov.bts/files/publications/national_transportation_statistics/html/table_01_35.html). Thus, it was possible to reduce the coarse particle, road dust emissions even with a substantial increase in road traffic. Thus, road dust resuspension appear to generally be under reasonable control.

The $PM_{2.5}$ NAAQS was instituted in 1997 and monitoring data are available from 2000 onward. The current standard is an annual average of 12 $\mu g/m^3$ and a 24-h standard of 35 $\mu g/m^3$ as the 98th percentile value in the distribution (Salako & Hopke, 2012). Fig. 1.6 presents the national trend for $PM_{2.5}$. A clear decrease in the $PM_{2.5}$ concentrations can be observed over the 16 years that the standard has been enforceable. Enforcement requires 3 years of valid data at any given location.

There are few locations in the US that are now in nonattainment of the 2012 $PM_{2.5}$ NAAQS as shown in Fig. 1.7. The areas that continue to have problems are concentrated in southern California and its central valley region.

3.3 South and Southeastern Asia

There are no publicly available, long-term data collected for regulatory purposes across south and southeastern Asia. However, there has been an ongoing airborne particle monitoring network supported by the International Atomic Energy Agency (IAEA) (https://www.iaea.org/newscenter/news/past-future-measuring-air-pollution-asia-and-pacific-using-nuclear-techniques). Participating countries include Australia, Bangladesh, China, India,

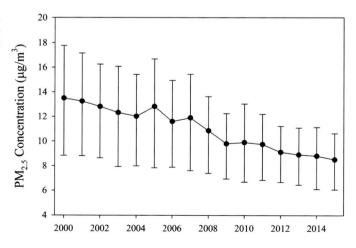

FIGURE 1.6 US national mean $PM_{2.5}$ values from 2000 to 2015. Error bars represent the 90th and 10th percentile values of annual average $PM_{2.5}$.

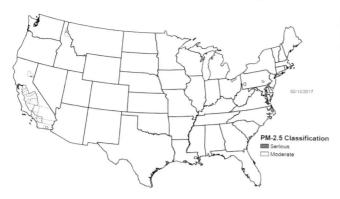

FIGURE 1.7 Counties designated non-attainment for $PM_{2.5}$. Map taken from https://www3.epa.gov/airquality/greenbook/map/mappm25_2012.pdf.

Indonesia, Malaysia, Mongolia, Myanmar, New Zealand, Pakistan, Philippines, Republic of Korea, Sri Lanka, and Vietnam. Initial data from this IAEA program have been previously reported by Hopke et al. (2008). Sampling was performed using a Gent stacked filter unit sampler that provides samples of coarse ($PM_{2.5-10}$) and fine ($PM_{2.5}$) particles (Hopke et al., 1997). For many of these countries, data are available up to 2014 and can be obtained from http://www.ansto.gov.au/ResearchHub/OurResearch/IER/Capabilities/ASP/FineParticle Databases/index.htm. The annual mean $PM_{2.5-10}$ values and their minimum and maximum values (error bars) for 14 countries (Australia, Bangladesh, China, India, Indonesia, Korea, Malaysia, Mongolia, New Zealand, Pakistan, Philippines, Sri Lanka, Thailand, and Vietnam) are presented in Figs. 1.8 and 1.9. For this monitoring program, the coarse particle concentrations are available and provide a more direct indication of the potential for road dust and, thus, non-exhaust emissions.

In Sydney, Australia, particulate concentrations are generally very low. Liverpool is an urban site, whereas Lucas Heights is a suburban location. It can be seen that the urban site has consistently higher than the suburban site. There were anomalously higher concentrations in 2009 because of substantial bush fires producing unusually high PM concentrations. In Dhaka, Bangladesh, concentrations of $PM_{2.5-10}$ increased slightly over this period. At the same time, $PM_{2.5}$ remained relatively constant even with increasing traffic because of the transition from gasoline to CNG as the primary motor vehicle fuel (Begum, Hopke, & Markwitz, 2013). Thus, increased traffic was likely to be driving increased road dust, contributing to high concentrations of coarse airborne PM. With typical $PM_{2.5}$ concentrations, the resulting PM_{10} values would be in excess of existing health-based guidelines and standards.

Among the other locations, the sites in Indonesia, Korea, and New Zealand had low concentrations, whereas the other locations had high coarse particle concentrations with Mongolia having very high concentrations largely as a result of residential coal combustion. Source apportionment studies have been performed in a number of these locations (Begum, Biswas, Markwitz, & Hopke, 2010; Begum, Nasiruddin, Randal, Sivertsen, & Hopke, 2014; Chung et al., 2005; Kothai et al., 2008; Lestari & Dwi Mauliadi, 2009; Lim, Lee, Moon, Chung, & Kim, 2010; Rahman et al., 2015; Santoso, Hopke, Hidayat, & Dwiana, 2008; Siddique,

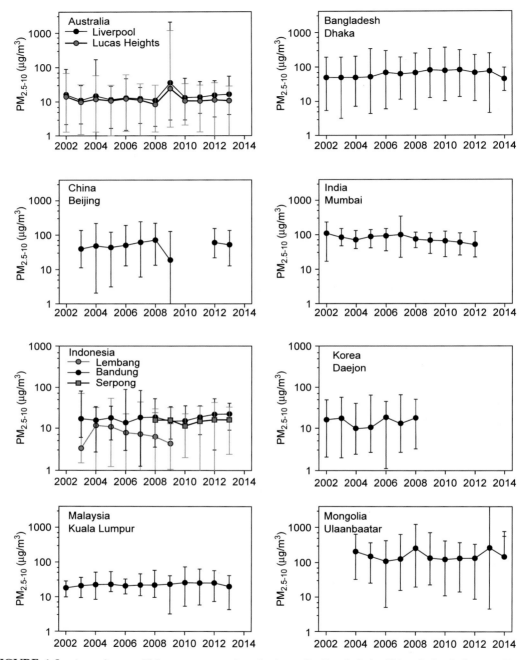

FIGURE 1.8 Annual mean $PM_{2.5-10}$ concentrations in Australia, Bangladesh, China, India, Indonesia, Korea, Malaysia, and Mongolia measured using samples collected with the Gent sampler.

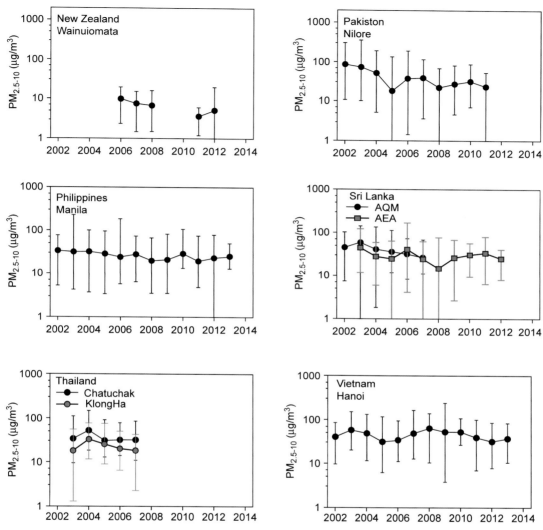

FIGURE 1.9 Annual mean PM$_{2.5-10}$ concentrations in New Zealand, Pakistan, Philippines, Thailand, Sri Lanka, and Vietnam measured using samples collected with the Gent sampler.

Waheed, Daud, Markwitz, & Hopke, 2012; Waheed, Jaafar, Siffique, Markwitz, & Brereton, 2012; Wimolwattanapun, Hopke, & Pongkiatkul, 2011). They have reported that the major coarse particle sources were road dust and soil. Thus, traffic-generated, non-exhaust emissions from motor vehicles are an important source of coarse particles in many urban areas in developing countries.

References

AIRUSE-LIFE+. (2015). *LIFE + project on testing and development of air quality mitigation measures in southern Europe*. PM speciation and source apportionment. Report B2-04, 141 pp. http://airuse.eu/wp-content/uploads/2013/11/04_B2_PM-Speciation-and-source-apportionment.pdf

Amato, F., Alastuey, A., Karanasiou, A., Lucarelli, F., Nava, S., Calzolai, G., et al. (2016). AIRUSE-LIFE+: A harmonized PM speciation and source apportionment in five southern European cities. *Atmospheric Chemistry and Physics, 16*, 3289—3309.

Barmpadimos, I., Keller, J., Oderbolz, D., Hueglin, C., & Prevot, A. S. H. (2012). One decade of parallel fine ($PM_{2.5}$) and coarse ($PM_{10}-PM_{2.5}$) particulate matter measurements in Europe: Trends and variability. *Atmospheric Chemistry and Physics, 12*, 3189—3203.

Begum, B. A., Biswas, S. K., Markwitz, A., & Hopke, P. K. (2010). Identification of sources of fine and coarse particulate matter in Dhaka, Bangladesh. *Aerosol and Air Quality Research, 10*, 345—353.

Begum, B. A., Hopke, P. K., & Markwitz, A. (2013). Air pollution by fine particulate matter in Bangladesh. *Atmospheric Pollution Research, 4*, 75—86.

Begum, B. A., Nasiruddin, M. D., Randal, S., Sivertsen, B., & Hopke, P. K. (2014). Identification and apportionment of sources from air particulate matter at urban environments in Bangladesh. *British Journal of Applied Science & Technology, 4*, 3930—3955.

Chung, Y. S., Kim, S. H., Moon, J. H., Kim, Y. J., Lim, J. M., & Lee, J. H. (2005). Source identification and long—term monitoring of airborne particulate matter ($PM_{2.5}/PM_{10}$) in an urban region of Korea. *Journal of Radioanalytical and Nuclear Chemistry, 267*, 35—48.

EEA. (2016). *Air quality in Europe 2016*. European Environmental Agency, EEA Report 28/2016, ISBN 978-92-9213-824-0. https://doi.org/10.2800/413142, 83 pp. http://www.eea.europa.eu/publications/air-quality-in-europe-2016.

EAA, AirBase-EEA. (2017). *AirBase — the European air quality database*. European Environment Agency. https://www.eea.europa.eu/data-and-maps/data/airbase-the-european-air-quality-database-8.

EMEP. (2016). *Air pollution trends in the EMEP region between 1990 and 2012*. EMEP/CCC-Report 1/2016, O-7726, ISBN : 978-82-425-2833-9 (printed), ISBN : 978-82-425-2834-6 (electronic), 105 pp.

Gilbert, R. O. (1987). *Statistical methods for environmental pollution monitoring*. New York: Van Nostrand Reinhold.

Guerreiro, C. B. B., Foltescu, V., & de Leeuw, F. (2014). Air quality status and trends in Europe. *Atmospheric Environment, 98*, 376—384.

Hopke, P. K., Cohen, D. D., Begum, B. A., Biswas, S. K., Ni, B., Pandit, G. G., et al. (2008). Urban air quality in the Asian region. *The Science of the Total Environment, 404*, 103—112.

Hopke, P. K., Xie, Y., Raunemaa, T., Bieglski, S., Landsberger, S., Maenhaut, W., et al. (1997). Characterization of gent stacked filter unit PM_{10} sampler. *Aerosol Science and Technology, 27*, 726—735.

Kothai, P., Sardhi, I. V., Prathibha, P., Hopke, P. K., Pandit, G. G., & Puranik, V. E. (2008). Source apportionment of coarse and fine particulate matter at Navi Mumbai, India. *Aerosol and Air Quality Research, 8*, 423—436.

Lestari, P., & Dwi Mauliadi, Y. (2009). Source apportionment of particulate matter at urban mixed site in Indonesia using PMF. *Atmospheric Environment, 43*, 1760—1770.

Lim, J. M., Lee, J. H., Moon, J. H., Chung, Y. S., & Kim, K. H. (2010). Source apportionment of PM_{10} at a small industrial area using positive matrix factorization. *Atmospheric Research, 95*, 88—100.

Loomis, D., Grosse, Y., Lauby-Secretan, B., El Ghissassi, F., Bouvard, V., Benbrahim-Tallaa, L., et al.(On behalf of the International Agency for research on cancer monograph). (2013). The carcinogenicity of outdoor air pollution. *The Lancet Oncology, 14*, 1262—1263.

Querol, X., Alastuey, A., Pandolfi, M., Reche, C., Pérez, N., Minguillón, M. C., et al. (2014). 2001—2012 trends on air quality in Spain. *The Science of the Total Environment, 490*, 957—969.

Rahman, S. A., Hamzah, M. S., Elias, Md S., Salim, N. A. A., Hashim, A., Shukor, S., et al. (2015). A long term study on characterization and source apportionment of particulate pollution in Klang valley, Kuala Lumpur. *Aerosol and Air Quality Research, 15*, 2291—2304.

Salako, G. O., & Hopke, P. K. (2012). Impact of percentile computation method on PM 24-h air quality standard. *Journal of Environmental Management, 107*, 110—113.

Santoso, M., Hopke, P. K., Hidayat, A., & Dwiana, L. D. (2008). Sources identification of the atmospheric aerosol at urban and suburban sites in Indonesia by positive matrix factorization. *Science of the Total Environment, 397*, 229—237.

Siddique, N., Waheed, S., Daud, M., Markwitz, A., & Hopke, P. K. (2012). Air quality study of Islamabad: Preliminary results. *Journal of Radioanalytical and Nuclear Chemistry, 293*, 351–358.

Waheed, S., Jaafar, M. Z., Siffique, N., Markwitz, A., & Brereton, R. G. (2012). PIXE analysis of $PM_{2.5}$ and $PM_{2.5-10}$ for air quality assessment of Islamabad, Pakistan: Application of chemometrics for source identification. *Journal of Environmental Science and Health, Part A, 47*, 2016–2027.

WHO. (2000). *Air quality guidelines for Europe* (2nd ed.) Copenhagen.

WHO. (2006). *Air quality guidelines: Global update 2005.* Copenhagen.

Wimolwattanapun, W., Hopke, P. K., & Pongkiatkul, P. (2011). Source apportionment and potential source locations of $PM_{2.5}$ and $PM_{2.5-10}$ at residential sites in metropolitan Bangkok. *Atmospheric Pollution Research, 2*, 172–181.

CHAPTER

2

Vehicle Non-Exhaust Emissions: Impact on Air Quality

Elio Padoan[1,2], Fulvio Amato[1]

[1]Institute of Environmental Assessment and Water Research (IDÆA), Spanish National Research Council (CSIC), Barcelona, Spain; [2]Università degli Studi di Torino, DiSAFA - Chimica agraria, Grugliasco (TO), Italy

OUTLINE

1. Introduction	21	3.2 Spatial Variability	51
		3.3 Other Metrics	54
2. Review Method and Limitations	24		
		4. Conclusions	55
3. Results	24		
3.1 Impact on Ambient Air Particulate	24	References	57

1. INTRODUCTION

The evidence for adverse health effects of ambient air pollution has grown dramatically in the past 20 years. The Global Burden of Disease study ranked exposure to ambient fine particulate matter ($PM_{2.5}$) as the seventh most important risk factor, contributing to total mortality with 2.9 million premature deaths in 2013, globally, and as fifth largest contributor in East Asia (GBD 2013 Risk Factors Collaborators et al., 2015). Although the overall toxicity of bulk PM is known, researchers still face a long way toward the understanding of the most harmful species (or mixtures) and sources, which would help dramatically the prioritization of control actions at international and local levels (West et al., 2016). Road traffic has been often pointed as one of the most harmful source category (Bell et al., 2010; Ostro et al., 2011) in multipollutants epidemiological studies. However, the road traffic sector comprises a number of different emission processes (combustion, wear, suspension), which need different control actions. Among these processes, it is necessary to identify those of higher impact, both on terms of air quality (contribution to the particle mass or number) and health outcomes (relative toxicity, Amato, Cassee et al., 2014; Denier van der Gon et al., 2013).

In comparison with vehicle exhaust, there are a few in vivo toxicity studies focused on non-exhaust sources; however, the most recent data indicate that non-exhaust particles can be as hazardous as exhaust PM depending on the nature of the health effect studied. Particle mass, size, and surface chemistry affect PM toxicity. Oxidative stress is one of the main biological mechanisms causing toxicity and is often related to transition metals and/or redox active organics such as quinone (Borm, Kelly, Kunzli, Schins, & Donaldson, 2007; Kelly, 2003). Brake and tire wear particles (TWPs) have higher oxidative potential than other traffic-related sources and their effect is very local (50−100 m from the source) (Yanosky, Tonne, Beevers, Wilkinson, & Kelly, 2012), yielding more oxidant PM (per $\mu g/m^3$) at road sites rather than at urban background sites. TWPs have been shown to induce reactive oxygen species formation and inflammatory reaction in human alveolar cells (Gualtieri, Mantecca, Cetta, & Camatini, 2008; Gualtieri, Rigamonti, Galeotti, & Camatini, 2005) as well as inflammatory response in mouse lungs (Mantecca et al., 2009, 2010). Other important factors to be investigated are PM size and size distribution, particle number (PN), composition (including coating and surface modifications), shape, surface area/specific surface area, surface chemistry, and charge and solubility/dispersibility.

Happo et al. (2010) found significant inflammatory response in rats' lungs exposed to coarse PM in Helsinki and correlated this with Fe and Cu content. A recent assessment using ascorbic acid depletion (marker for presence of redox active metals), electron spin resonance (marker for OH• radical), and dithiothreitol consumption (marker for redox active organics) showed a clearly higher oxidation potential of brake pad particles compared with diesel engine exhaust and tire or road dust (Amato, Cassee et al., 2014). Gustafsson et al. (2008) showed at least as high inflammatory potential from road wear PM_{10} compared with diesel engine exhaust particles. Brake wear particles damages have been linked to oxidative stress and inflammatory responses in the lung using incubations of lung cells with brake wear particles.

An established terminology and hierarchy of non-exhaust emissions does not exist, hampering the comparison of literature studies. In this book we therefore aim to promote the use of the following terminology:

Vehicle non-exhaust emissions include the following:

1. Direct brake wear: the fraction of pad, disc, and clutch wear particles that are directly airborne
2. Direct tire wear: the fraction of TWPs that are directly airborne
3. Direct road wear: the fraction of road wear particles (RWPs) that are directly airborne
4. Road dust suspension: any particle on (paved) road surface that is suspended to air by vehicles or wind, including deposited brake/tire/road wear particles and other deposited particles from various origin (salt, sand, exhaust, secondary, other mineral dust, etc.)

Another part of traffic emissions arises from unpaved road dust suspension, but because of the different emission dynamics and possible remediation, unpaved roads remain out of the scopes of this book, even if their contribution can be important in cities of several developing countries.

The abovementioned distinction is however only hypothetical, as separating the contribution of the single sources has been a difficult task for researchers over the past 20 years. In spite of the quite large number of studies on aerosol source apportionment, a clear separation has been rarely found. This chapter aims at reviewing the body of literature on non-exhaust emissions contribution to air pollution concentrations. In particular we aim to shed light on the following:

- Impact on WHO standard parameters
- Impact on different types of environment (industrial, urban, traffic, etc.)
- Impact of climate on road dust emissions

Non-exhaust sources differ widely in terms of chemical composition and particle size distribution (Amato et al., 2011; Duong & Lee, 2011; Grigoratos & Martini, 2015; Pant & Harrison, 2013). For a comprehensive discussion on their physicochemical characteristics, please see Chapters 8—11.

Briefly, brake pads and discs are composed of many ingredients, used as binders, fillers, lubricants, fibers, and frictional additives (Grigoratos & Martini, 2015). Thus, chemical profiles vary largely although Fe, Ba, Sb, Sn, and Cu are commonly reported as reliable tracers in brake linings (Amato et al., 2009; Gietl, Lawrence, Thorpe, & Harrison, 2010; Hulskotte, Roskam, & Denier van der Gon, 2014; Kukutschová et al., 2011; Thorpe & Harrison, 2008). As different materials crumble in different periods, wear debris composition can be significantly different from brake lining material (Thorpe & Harrison, 2008). Von Uexküll et al. (2005), for example, found that Ni and Cu were in much greater concentrations in the emitted dust than in the bulk brake material. Iijima et al. (2008) reported, for nonasbestos organic pads, similar composition for bulk pad and dusts, with a slight increase for Fe and Ti.

TWPs composition is mainly carbonaceous, but aluminum, silicon, zinc, and sulfur have been found (Harrison, Jones, Gietl, Yin, & Green, 2012; Rogge, Hildemann, Mazurek, Cass, & Simoneit, 1993; Thorpe & Harrison, 2008). Zinc, in particular, is used in rubber as ZnS and considered as tracer for tire wear, along with organic compounds (e.g., styrene), as it is also emitted from other sources (Kreider, Panko, McAtee, Sweet, & Finley, 2010; Pant & Harrison, 2013).

Road surface wear is the most difficult source to trace, as the composition is primarily mineral aggregates bond by bitumen. The higher variability of geologic materials used, as its usual similarity with soil dust (Wahlin, Berkowicz, & Palmgren, 2006), prevents the identification of unique elemental tracers, while organic tracers as asphaltenes and maltenes have been suggested by Fauser (1999).

Road dust particles can arise not only from all abovementioned wear sources but also from exhaust emissions, road sanding and salting, and geogenic material carried onto the road. Chemical composition has found to be dominated from elements associated to crustal material, as soil and road wear (Gu et al., 2011; Kupiainen et al., 2005; Thorpe & Harrison, 2008), and therefore is likely to be site dependent. Where road salting procedures and studded tire use is common, winter seasonality can be evident (Kupiainen et al., 2016).

2. REVIEW METHOD AND LIMITATIONS

In Scopus (www.scopus.com), we searched for publications on vehicle non-exhaust emission impact, using the following search terms: PM_{10}, $PM_{2.5}$, coarse PM, fine PM, PM, air quality, wear, resuspension, suspension, non-exhaust, tire, brake, road dust, road wear, traffic, crustal, suspension, resuspension, airborne, PN, particle volume, emission model, emission inventory, source apportionment, receptor model, and binary and ternary combinations of these. The main tool used in the reviewed studies was a receptor modeling technique based on measured PM concentrations at some receptor site, coupled with chemical characterization. However, we also found studies based on emission inventories, emission models, and chemical transport models. The contributions to PN and volume concentrations were also reviewed, although their number is much smaller than those with PM mass. However, based on this methodology, some relevant information might be still missing. Our conclusions might still be affected by some limitations:

- As previously mentioned, an established terminology does not exist; therefore source/factors labeling by individual authors may not coincide.
- An established hierarchy does not exist yet; therefore the non-exhaust label does not contain implicitly contributions from all brake, tire wear, and road dust resuspension.
- Source identification: This chapter does not evaluate the goodness of the source apportionment study procedure in the published studies.

3. RESULTS

3.1 Impact on Ambient Air Particulate

In this review, a total of 146 articles worldwide were included, where at least one air pollution source/factor related to vehicle non-exhaust emissions was identified. Among them 47 could not well disentangle a non-exhaust source (namely road dust/crustal, road dust/construction, road dust/harbor, road dust/tram, road dust and salt, traffic/wear, traffic/road dust, exhaust/wear, among others) and were therefore discarded. As one article often included more than one metric and/or monitoring station, we collected a total number of 256 contributions estimates. The methods used for estimating non-exhaust contributions were grouped as follows:

- Receptor modeling based on PM chemical characterization:
  - Positive matrix factorization (PMF or ME-2, 55 articles)
  - Chemical mass balance (19 articles)
  - Principal component analysis (8 articles)
  - COPREM (1 article)
  - Unmix (1 article)
- Other PM observations:
  - Metal size distribution (1 article)
  - Scanning electron microscope (SEM) (1 article)
  - Aerosol Time-Of-Flight Mass Spectrometry (ToF-AMS) (1 article)
  - Rubber tracer concentration (1 article)

- Dispersion modeling/emission inventory:
  - CAMx (1 article)
  - URBIS (1 article)
  - KCLurban (1 article)
  - LOTOS-EUROS (1 article)
  - CALIOPE (1 article)
  - ATMOs (1 article)
  - UCD/CiT (1 article)
  - NORTRIP (1 article)
  - VAPI (1 article)
  - USEPA ISCST3 (1 article)

The 256 estimates of non-exhaust contributions (Table 2.1) span different climatic conditions (using the Köppen climate classification; Peel, Finlayson, & McMahon, 2007): Arid (3 estimates), continental (51 estimates), Mediterranean (70 estimates), oceanic (60 estimates), tropical monsoon (28 estimates), and temperate-subtropical (43 estimates). The studies are mostly settled in Europe (150, Fig. 2.1), Asia (51), and North America (48), whereas much fewer studies are available in Oceania (5) and Africa (1). The first studies were performed in the 1980s in the United States, followed by the first few European studies in the 1990s (Fig. 2.1). The number of studies increased sensibly in the new century, including the rest of continents. Different types of environment were investigated: rural/regional (16), suburban (6), urban (undefined [113], commercial [3], residential [11], background [44], and roadside [41]), industrial (11), freeway (2), and tunnels (6).

Most of the estimates concerned the road dust emissions (71%), while only 9% and 8% refer to brake and tire wear, respectively. The rest of cases refer to a generic non-exhaust source (12%). Other studies, identifying miscellaneous factors (e.g., road dust/traffic, road dust/crustal), were discarded. The direct road wear has been rarely found, due to the high chemical similarity to road dust, and was thus incorporated into the "road dust" category. We have grouped labels according to Table 2.2.

There are several variables influencing the impact of non-exhaust sources: activity (vehicle flow), climate (mostly for road dust and direct road wear), type of station (proximity to roadways), sampled PM size, and time of the year. However, quantifying the influence of all these factors in terms of source contributions is hampered by the dearth of a sufficient number of studies covering such variability. As an example, we found only three studies estimating tire wear contribution to $PM_{2.5}$, which does not allow investigating the variability with climate, season, or type of site. At a first stage, we have grouped all studies, regardless of climate, type of station, and period of the year to evaluate the overall impact of each source.

The duration of the study is also crucial, mostly for road dust emissions, which can vary largely with meteorology. In the case of road dust, around three-fourths of the studies were performed at least during 1 year.

The particle sizes under study were not always corresponding to the regulated metrics (i.e., PM_{10} and $PM_{2.5}$) but also include PM_1 and PM coarse. Moreover, additional cutoff sizes were found and, for sake of simplicity, grouped as follows:

- PM_1 (<1 μm and 0.3–1 μm)
- $PM_{2.5}$ (<2 μm; <2.1 μm; <2.5 μm; <3 μm; and 1–2.5 μm)

TABLE 2.1 Results for Exhaust and Non-Exhaust Contributions to Size-Fractionated PM as Reported in the Literature

Location	Climate	Year	Type of Station	PM Metric	Study Duration	Method	Source	Share (%)	Share (μg/m³)	Exhaust (%)	Exhaust (μg/m³)	References
Australia, Brisbane	Temperate	1993–94	Urban	PM 2.5	Yearly	RM, CMB	Road dust	2	0.14			Chan et al. (1999)
Australia, Brisbane	Temperate	1993–94	Urban	PM 2.5 –10	Yearly	RM, CMB	Road dust	15	1.5			Chan et al. (1999)
Australia, Brisbane	Temperate	2003–04	Urban	PM 10	Yearly	RM, PMF	Road dust	28	3.6	40	5.1	Chan et al. (2008)
Australia, Melbourne	Oceanic	2003–04	Urban	PM 10	Yearly	RM, PMF	Road dust	21	3.4	37	6	Chan et al. (2008)
Bangladesh, Chittagong	Tropical monsoon	2007	Urban background	PM 2.5	Yearly	RM, PMF	Road dust	19	8.9	9.8	4.5	Begum, Biswas, and Nasiruddin (2009)
Belgium, Antwerp	Oceanic	2013–14	Urban background	PM 10	Yearly	RM, PMF	Brake wear	4.9	1.5	3.9	1.2	Mooibroek, Staelens, Cordell, and Panteliadis (2016)
Belgium, Antwerp	Oceanic	2011–12	Urban background	PM 10	Yearly	RM, PMF	Non-exhaust	1	0.22	13	2.9	Maenhaut, Vermeylen, Claeys, Vercauteren, and Roekens (2016)
Belgium, Bruges	Oceanic	2011–12	Urban background	PM 10	Yearly	RM, PMF	Non-exhaust	1	0.23	9	2.1	Maenhaut et al. (2016)
Belgium, Ghent	Oceanic	2011–12	Urban background	PM 10	Yearly	RM, PMF	Non-exhaust	1	0.25	11	2.8	Maenhaut et al. (2016)
Belgium, Ostend	Oceanic	2011–12	Urban background	PM 10	Yearly	RM, PMF	Non-exhaust	1	0.24	8	1.9	Maenhaut et al. (2016)
Canada, Edmonton	Continental	2004–08	Urban	PM 2.5	Yearly	RM, PMF	Road dust	7	0.6	12	1	Jeong, McGuire, Herod, and Dann (2011)

Location	Climate	Period	Setting	Size	Term	Method	Source					Reference
Canada, Montreal	Continental	2004–08	Urban	PM 2.5	Yearly	RM, PMF	Road dust	4	0.4	14	1.4	Jeong et al. (2011)
Canada, Toronto	Continental	2004–08	Urban	PM 2.5	Yearly	RM, PMF	Road dust	3	0.4	10	1.3	Jeong et al. (2011)
China, Beijing	Continental	2013	Urban	Particle number	Short term	RM, PMF	Road dust	4		53.8		Liu et al. (2016)
China, Beijing	Continental	2013	Urban	PM 10 volume	Short term	RM, PMF	Road dust	11		13.4		Liu et al. (2016)
China, Beijing	Continental	2000	Urban	PM 2.5	Short term	RM, UNMIX	Road dust	9	8.4	6	5.7	Song et al. (2006)
China, Beijing	Continental	2013–14	Urban	PM 2.1	Yearly	RM, PMF	Road dust	8	5.6	19.6	13.1	Tian, Pan, and Wang (2016)
China, Beijing	Continental	2013–14	Urban	PM 2.1–9	Yearly	RM, PMF	Road dust	11	6.8			Tian et al. (2016)
China, Beijing	Temperate	2010	Urban	PM 2.5	Yearly	RM, PMF	Road dust	13	6.6	17.1	8.9	Yu et al. (2013)
China, Handan	Temperate	2012–13	Urban	PM 2.5	Yearly	RM, PMF	Road dust	11	17.45	7.7	11.5	Wei, Wang, Chen, and Zheng (2014)
China, Harbin	Continental	2006–07	Urban	PM 10	Yearly	RM, CMB	Road dust	16	27.2	61.7	103.6	Huang, Wang, Yuan, and Wang (2010)
China, Jinan	Temperate	2002–04	Urban	PM 10	Yearly	RM, CMB	Road dust	30	42.6	10	14.2	Zhao, Feng, Zhu, and Wu (2006)
China, Nanjing	Temperate	2013	Urban	PM 2.5	Yearly	RM, PMF	Road dust	9	10.9			Li et al. (2016)
China, Panzhihua	Temperate	2007	Urban	PM 10	Short term	RM, CMB	Road dust	7	9.2	13.7	18.9	Xue et al. (2010)
China, Shijiazhuang	Temperate	2002–04	Urban	PM 10	Yearly	RM, CMB	Road dust	43	68.5	10	15.9	Zhao et al. (2006)

(Continued)

TABLE 2.1 Results for Exhaust and Non-Exhaust Contributions to Size-Fractionated PM as Reported in the Literature—cont'd

Location	Climate	Year	Type of Station	PM Metric	Study Duration	Method	Source	Share (%)	Share (µg/m³)	Exhaust (%)	Exhaust (µg/m³)	References
China, Taiyuan	Temperate	2002–04	Urban	PM 10	Yearly	RM, CMB	Road dust	32	54.6	15	25.6	Zhao et al. (2006)
China, Taiyuan	Temperate	2001	Urban	PM 10	Yearly	RM, PCA/MLR-CMB	Road dust	26	79.3	13	39.7	Zeng et al. (2010)
China, Tianjin	Temperate	2002–04	Urban	PM 10	Yearly	RM, CMB	Road dust	34	43	13	16.4	Zhao et al. (2006)
China, Urumqi	Temperate	2002–04	Urban	PM 10	Yearly	RM, CMB	Road dust	30	40.4	9	12.1	Zhao et al. (2006)
China, Yinchuan	Temperate	2002–04	Urban	PM 10	Yearly	RM, CMB	Road dust	59	65.9	7	7.8	Zhao et al. (2006)
Croatia, Rijeka	Oceanic	2013–15	Urban	PM 2.5	Yearly	RM, PMF	Road dust	6	1.2	33	6.6	Ivošević, Stelcer, Orlić, BogdanovićRadović, and Cohen (2016)
Cyprus, Larnaca	Mediterranean	2012	Urban	PM 2.5	Yearly	RM, PMF	Road dust	2	0.3	17	2.3	Achilleos et al. (2016)
Cyprus, Larnaca	Mediterranean	2012	Urban	PM 2.5–10	Yearly	RM, PMF	Road dust	33	5.1			Achilleos et al. (2016)
Cyprus, Limassol	Mediterranean	2012	Urban	PM 2.5	Yearly	RM, PMF	Road dust	4	0.5	32	4.1	Achilleos et al. (2016)
Cyprus, Limassol	Mediterranean	2012	Urban	PM 2.5–10	Yearly	RM, PMF	Road dust	47	7.8			Achilleos et al. (2016)
Cyprus, Nicosia	Mediterranean	2012	Urban	PM 2.5	Yearly	RM, PMF	Road dust	3	0.4	21	3.1	Achilleos et al. (2016)
Cyprus, Nicosia	Mediterranean	2012	Urban	PM 2.5–10	Yearly	RM, PMF	Road dust	45	5.5			Achilleos et al. (2016)
Cyprus, Paphos	Mediterranean	2012	Urban	PM 2.5	Yearly	RM, PMF	Road dust	2	0.3	14	1.7	Achilleos et al. (2016)

Location	Climate	Year	Site type	PM size	Term	Method	Source					Reference
Cyprus, Paphos	Mediterranean	2012	Urban	PM 2.5–10	Yearly	RM, PMF	Road dust	29	3.6			Achilleos et al. (2016)
Czech Republic	Continental	2008–10	Rural	PM 1–10	Short term	RM, PMF	Road dust/brake wear	3.5	0.13			Pokorna et al. (2013)
Czech Republic, Ostrava	Continental	2012	Urban residential	PM 1.5–10	Short term	RM, PMF	Road dust	62	11.6			Pokorna et al. (2015)
Denmark, Copenhagen	Oceanic	1999	Urban	PM 10	Short term		Tire wear	5				Fauser, 1999
Denmark, Copenhagen	Continental	2003	Urban	PM 2.5	Short term	RM, COPREM	Brake wear	9.3	0.72	55.2	4.3	Wahlin et al. (2006)
Denmark, Copenhagen	Continental	2003	Urban	PM 2.5–10	Short term	RM, COPREM	Brake wear	6.7	0.8	13.3	1.6	Wahlin et al. (2006)
Denmark, Copenhagen	Continental	2003	Urban	PM 2.5	Short term	RM, COPREM	Road dust	28	2.2	55.2	4.3	Wahlin et al. (2006)
Denmark, Copenhagen	Continental	2003	Urban	PM 2.5–10	Short term	RM, COPREM	Road dust	48	5.8	13.3	1.6	Wahlin et al. (2006)
Denmark, Copenhagen	Oceanic	2006–08	Urban roadside	PM 10		EM, NORTRIP	Tire wear	5.8	1.0			Denby et al. (2013)
Denmark, Copenhagen	Continental	2006–08	Urban roadside	PM 10		EM, NORTRIP	Brake wear	6.4	1.1			Denby et al. (2013)
Denmark, Copenhagen	Continental	2006–08	Urban roadside	PM 10		EM, NORTRIP	Road wear	57.2	9.6			Denby et al. (2013)
Finland, Helsinki	Continental	2012	Suburban	PM 10	Short term	RM,CMB	Road wear	43				Kupiainen et al. (2016)
France, Lille	Oceanic	2013–14	Urban background	PM 10	Yearly	RM, PMF	Brake wear	5	1.5	5.3	1.6	Mooibroek et al. (2016)
France, Dunkirk	Oceanic	2012	Industrial	PM 2.5	Short term	RM, PCA-MLRA	Road dust	24	3.8	2	0.32	Mbengue, Alleman, and Flament (2017)

(Continued)

TABLE 2.1 Results for Exhaust and Non-Exhaust Contributions to Size-Fractionated PM as Reported in the Literature—cont'd

Location	Climate	Year	Type of Station	PM Metric	Study Duration	Method	Source	Share (%)	Share (µg/m³)	Exhaust (%)	Exhaust (µg/m³)	References
France, Nice	Mediterranean	2006	Tunnel	PM 2.5	Short term	RM, PMF	Road dust	43		21		Fabretti, Sauret, Gal, Maria, and Schärer (2009)
France, Paris	Oceanic	2012–13	Urban	PM 10	Yearly	RM, PMF	Road dust	13	6.3			Amato, Favez, Pandolfi, and Alastuey (2016c)
Germany, Augsburg	Continental	2006–07	Urban background	Particle number	Short term	RM, PMF	Road dust	3		65.2		Gu et al. (2011)
Germany, Augsburg	Continental	2006–07	Urban background	PM 10 volume	Short term	RM, PMF	Road dust	9		10.6		Gu et al. (2011)
Germany, Augsburg	Continental	2006–07	Urban	PM 10	Short term	RM, PMF	Road dust	20	6.4	16.6	5.3	Gu et al. (2011)
Germany, Essen	Oceanic	2009	Urban background	PM 1	Yearly	SEM	Road dust	5		6		Weinbruch et al. (2014)
Germany, Essen	Oceanic	2009	Urban roadside	PM 1	Yearly	SEM	Road dust	21		15		Weinbruch et al. (2014)
Germany, Essen	Oceanic	2009	Urban background	PM 10	Yearly	SEM	Road dust	15		6		Weinbruch et al. (2014)
Germany, Essen	Oceanic	2009	Urban roadside	PM 10	Yearly	SEM	Road dust	28		13		Weinbruch et al. (2014)
Germany, Essen	Oceanic	2009	Urban background	PM 1–10	Yearly	SEM	Road dust	31		1		Weinbruch et al. (2014)
Germany, Essen	Oceanic	2009	Urban roadside	PM 1–10	Yearly	SEM	Road dust	53		6		Weinbruch et al. (2014)
Germany, Essen	Oceanic	2009	Urban background	PM 1	Yearly	SEM	Brake wear/ tire wear	4		6		Weinbruch et al. (2014)
Germany, Essen	Oceanic	2009	Urban roadside	PM 1	Yearly	SEM	Brake wear/ tire wear	3		15		Weinbruch et al. (2014)

Location	Climate	Year	Environment	PM	Period	Method	Source					Reference
Germany, Essen	Oceanic	2009	Urban background	PM 10	Yearly	SEM	Brake wear/tire wear	6		6		Weinbruch et al. (2014)
Germany, Essen	Oceanic	2009	Urban roadside	PM 10	Yearly	SEM	Brake wear/tire wear	7		13		Weinbruch et al. (2014)
Germany, Essen	Oceanic	2009	Urban background	PM 1–10	Yearly	SEM	Brake wear/tire wear	4		1		Weinbruch et al. (2014)
Germany, Essen	Oceanic	2009	Urban roadside	PM 1–10	Yearly	SEM	Brake wear/tire wear	15		6		Weinbruch et al. (2014)
Germany, North Rhine-Westphalia	Oceanic	2008–09	Rural background	PM 10	Yearly	RM, PMF	Road dust	2	0.3			Beuck, Quass, Klemm, and Kuhlbusch (2011)
Germany, North Rhine-Westphalia	Oceanic	2008–09	Urban background	PM 10	Yearly	RM, PMF	Road dust	8	2.4			Beuck et al. (2011)
Greece, Athens	Mediterranean	2002	Urban	PM 10	Yearly	RM, PMF	Road dust	34	3.62	19	1.84	Karanasiou, Siskos, and Eleftheriadis (2009)
Greece, Athens	Mediterranean	2002	Urban	PM 2	Yearly	RM, PMF	Road dust	27	1.5	27	1.54	Karanasiou et al. (2009)
Greece, Athens	Mediterranean	2002	Urban	PM 2–10	Yearly	RM, PMF	Road dust	53	2.09	8	0.3	Karanasiou et al. (2009)
Greece, Athens	Mediterranean	2013	Urban background	PM 10	Yearly	RM, PMF	Non-exhaust	8	1.8	10	2.1	Amato, Alastuey, Karanasiou, and Lucarelli (2016a)
Greece, Athens	Mediterranean	2013	Urban background	PM 2.5	Yearly	RM, PMF	Non-exhaust	5	0.6	15	1.7	Amato et al. (2016a)
Greece, Attica	Mediterranean	2006		PM 10		DM, CAMx	Road dust	23	12.5			Athanasopoulou et al. (2010)

(Continued)

TABLE 2.1 Results for Exhaust and Non-Exhaust Contributions to Size-Fractionated PM as Reported in the Literature—cont'd

Location	Climate	Year	Type of Station	PM Metric	Study Duration	Method	Source	Share (%)	Share (µg/m³)	Exhaust (%)	Exhaust (µg/m³)	References
Greece, Megalopolis	Mediterranean	2009–11	Urban	PM 10	Yearly	RM, PMF	Road dust	15	3.6	12	2.9	Manousakas, Diapouli, Papaefthymiou, Migliori, and Karydas (2015)
Greece, Thessaloniki	Mediterranean	1994	Urban commercial	PM 3	Yearly	RM, PCA-APCS	Road dust	28	27.2	38	36.9	Manoli, Voutsa, and Samara (2002)
Greece, Thessaloniki	Mediterranean	1994	Urban commercial	PM 3–10	Yearly	RM, PCA-APCS	Road dust	57	17.1	9	2.7	Manoli et al. (2002)
Greece, Thessaloniki	Mediterranean	1997–98	Industrial	PM 10	Yearly	RM, CMB	Road dust	19	13.7	46.7	34.6	Samara, Kouimtzis, Tsitouridou, Kanias, and Simeonov (2003)
Greece, Thessaloniki	Mediterranean	1997–98	Urban	PM 10	Yearly	RM, CMB	Road dust	21	17.4	62.8	63.6	Samara et al. (2003)
Greece, Thessaloniki	Mediterranean	1997–98	Urban residential	PM 10	Yearly	RM, CMB	Road dust	22	19.7	64.3	57.3	Samara et al. (2003)
Hong Kong	Temperate	2005	Urban	PM 2.5	Yearly	RM, PMF	Road dust	1	0.55	29	16.1	Cheng et al. (2015)
Hong Kong	Temperate	2005	Urban	PM 2.5–10	Yearly	RM, PMF	Road dust	17	4.4	11	2.8	Cheng et al. (2015)
India, Bangalore	Tropical monsoon	2013	Urban	PM 10	Yearly	RM, CMB	Road dust	51	47	18	16.8	Sharma and Patil (2016)
India, Chennai	Tropical monsoon	2009	Urban	PM 10	Yearly	RM, CMB	Road dust	1	0.7	58.5	31.9	Srimuruganandam and Shiva Nagendra (2012b)
India, Chennai	Tropical monsoon	2009	Urban	PM 2.5	Yearly	RM, CMB	Road dust	1	0.59	57.7	23.5	Srimuruganandam and Shiva Nagendra (2012b)

Location	Climate	Year	Setting	PM	Term	Method	Source						Reference
India, Chennai	Tropical monsoon	2009	Urban	PM 10	Yearly	RM, CMB	Brake wear	0.14	0.14	0.12	58.5	31.9	Srimuruganandam and Shiva Nagendra (2012b)
India, Chennai	Tropical monsoon	2009	Urban	PM 2.5	Yearly	RM, CMB	Brake wear		0.14	0.09	57.7	23.5	Srimuruganandam and Shiva Nagendra (2012b)
India, Chennai	Tropical monsoon	2009	Urban	PM 10	Short term	RM, PMF	Brake wear/tire wear		4.1	3.4	15.8	13.1	Srimuruganandam and Shiva Nagendra (2012a)
India, Chennai	Tropical monsoon	2009	Urban	PM 2.5	Short term	RM, PMF	Brake wear/tire wear		5.4	3.4	6	3.8	Srimuruganandam and Shiva Nagendra (2012a)
India, Chennai	Tropical monsoon	2009	Urban	PM 10	Short term	RM, PMF	Tire wear/brake wear		4.1	3.4	15.8	13.1	Srimuruganandam and Shiva Nagendra (2012a)
India, Chennai	Tropical monsoon	2009	Urban	PM 2.5	Short term	RM, PMF	Tire wear/brake wear		5.4	3.4	6	3.8	Srimuruganandam and Shiva Nagendra (2012a)
India, Delhi	Tropical monsoon	2001	Urban	PM 1.6–10.9	Short term	RM, CMB	Road dust		26				Srivastava and Jain (2007)
India, Delhi	Tropical monsoon	2010		PM 2.5		EI + DM, ATMOs	Road dust		8	6.5			Guttikunda and Calori (2013)
India, Hyderabad	Tropical monsoon	2004–05	Urban	PM 10	Yearly	RM, CMB	Road dust		40	48.5	22	26.7	Gummeneni, Yusup, Chavali, and Samadi (2011)
India, Hyderabad	Tropical monsoon	2004–05	Urban	PM 2.5	Yearly	RM, CMB	Road dust		26	16.2	31	19.3	Gummeneni et al. (2011)
India, Hyderabad	Tropical monsoon	2005–06	Urban background	PM 10	Yearly	RM, CMB	Road dust		36	26.9	41.6	31.6	Guttikunda, Kopakka, Dasari, and Gertler, (2013)
India, Hyderabad	Tropical monsoon	2005–06	Urban residential	PM 10	Yearly	RM, CMB	Road dust		31	39.6	38.4	48.6	Guttikunda et al. (2013)

(Continued)

TABLE 2.1 Results for Exhaust and Non-Exhaust Contributions to Size-Fractionated PM as Reported in the Literature—cont'd

Location	Climate	Year	Type of Station	PM Metric	Study Duration	Method	Source	Share (%)	Share (µg/m³)	Exhaust (%)	Exhaust (µg/m³)	References
India, Hyderabad	Tropical monsoon	2005–06	Urban residential	PM 10	Yearly	RM, CMB	Road dust	36	40.3	41.5	46.8	Guttikunda et al. (2013)
India, Hyderabad	Tropical monsoon	2005–06	Urban background	PM 2.5	Yearly	RM, CMB	Road dust	20	7.5	38.1	14.3	Guttikunda et al. (2013)
India, Hyderabad	Tropical monsoon	2005–06	Urban residential	PM 2.5	Yearly	RM, CMB	Road dust	13	8.7	34	22.6	Guttikunda et al. (2013)
India, Hyderabad	Tropical monsoon	2005–06	Urban residential	PM 2.5	Yearly	RM, CMB	Road dust	17	8.9	38.1	20.2	Guttikunda et al. (2013)
India, Kanpur	Tropical monsoon	2007		PM 10		EI + DM, USEPA ISCST3	Road dust	14				Behera, Sharma, Dikshit, and Shkula (2011)
India, Kolkata	Tropical monsoon	2003–04	Urban residential	PM 10	Yearly	RM, CMB	Road dust	21	36.6			Gupta, Karar, and Srivastava (2007)
India, Kolkata	Tropical monsoon		Industrial	PM 10	Short term	RM, PCA-MLRA	Tire wear	8	15.8	37	73.1	Karar and Gupta (2007)
India, Mumbai	Tropical monsoon	2007–08	Urban residential	PM 10	Yearly	RM, PMF	Road dust	18	33.3	23	42.6	Gupta, Salunkhe, and Kumar (2012)
India, Nagpur	Tropical monsoon	2009–10	Urban residential	PM 2.5	Short term	RM, CMB	Road dust	6	4	57	38.2	Pipalatkar, Khaparde, Gajghate, and Bawase (2014)
India, Nagpur	Tropical monsoon	2009–10	Urban roadside	PM 2.5	Short term	RM, CMB	Road dust	10	9.6	62	59.8	Pipalatkar et al. (2014)
India, Nagpur	Tropical monsoon	2009–10	Industrial	PM 2.5	Short term	RM, CMB	Road dust	9	7.7	65	55.4	Pipalatkar et al. (2014)
Indonesia, Serpong	Tropical monsoon		Urban	PM 2.5		RM, PMF	Road dust	17	3.5	30	6.2	Santoso et al. (2011)
Iran, Ahvaz	Arid	2010–11	Urban	PM 10	Yearly	RM, PMF	Road dust	6	17.6	11.5	36.8	Sowlat et al. (2013)

Location	Climate	Year	Site type	PM	Method	Method 2	Source					Reference
Italy, Bologna	Temperate	2006	Urban background	PM 10	Yearly	RM, PMF	Road dust	11	4.9	35	15.6	Tositti et al. (2014)
Italy, Civitavecchia	Mediterranean	2010–14	Rural	PM 10	Yearly	RM, PMF	Road dust	6	1	7.2	1.2	Cesari, Donateo, Conte, and Contini (2016)
Italy, Civitavecchia	Mediterranean	2010–14	Urban	PM 10	Yearly	RM, PMF	Road dust	13	2.8	16.9	3.7	Cesari et al. (2016)
Italy, Civitavecchia	Mediterranean	2010–14	Urban background	PM 10	Yearly	RM, PMF	Road dust	7	1.1	9	1.5	Cesari et al. (2016)
Italy, Florence	Mediterranean	2009	Urban	PM 2.5	Yearly	RM, ME-2	Non-exhaust	7	1.4	37	8	Crespi et al. (2016)
Italy, Florence	Mediterranean	2013	Urban background	PM 10	Yearly	RM, PMF	Non-exhaust	9	1.8	13	2.5	Amato et al. (2016a)
Italy, Florence	Mediterranean	2013	Urban background	PM 2.5	Yearly	RM, PMF	Non-exhaust	2	0.3	18	2.5	Amato et al. (2016a)
Italy, Milan	Oceanic	2006–09	Urban	PM 10	Yearly	RM, CMB + Calculations	Road dust	6	2.6	22	11.1	Perrone et al. (2012)
Italy, Milan	Oceanic	2006–09	Urban	PM 2.5	Yearly	RM, CMB + Calculations	Road dust	3	0.65	20.5	4.5	Perrone et al. (2012)
Italy, Milan	Oceanic	2006–09	Rural	PM 2.5	Yearly	RM, CMB + Calculations	Road dust	1	0.15	7.7	1.7	Perrone et al. (2012)
Italy, Milan	Oceanic	2013	Urban background	PM 10	Yearly	RM, PMF	Non-exhaust	14	5.8	7	2.8	Amato et al. (2016a)
Italy, Milan	Oceanic	2013	Urban background	PM 2.5	Yearly	RM, PMF	Non-exhaust	8	2.5	6	1.8	Amato et al. (2016a)
Italy, Po Valley	Temperate		Regional background	PM 10	ToF-AMS		Tire wear	0.5				Dall'Osto et al. (2014)
Korea, Daejeon	Temperate	2000–02	Industrial	PM 10	Yearly	RM, PMF	Road dust	12	10.3	9	7.7	Lim, Lee, Moon, Chung, and Kim (2010)

(Continued)

TABLE 2.1 Results for Exhaust and Non-Exhaust Contributions to Size-Fractionated PM as Reported in the Literature—cont'd

Location	Climate	Year	Type of Station	PM Metric	Study Duration	Method	Source	Share (%)	Share (µg/m³)	Exhaust (%)	Exhaust (µg/m³)	References
Korea, Seoul	Temperate	2006–07	Urban	PM 10	Yearly	RM, PMF	Road dust	18	10.2	16.6	9.5	Yi and Hwang (2014)
New Zealand, Auckland	Oceanic	2003	Urban background	PM 2.5 –10	Yearly	RM, PMF	Road dust	19	1.67			Wang and Shooter, (2005)
Pakistan, Karachi	Arid	2006–07	Urban	PM 2.5	Yearly	RM, PMF	Road dust	16	12.8	18.5	14.7	Mansha, Ghauri, Rahman, and Amman (2012)
Poland, Krakow	Continental	2014–15	Urban	PM 2.5	Yearly	RM, PMF	Road dust	3	0.96	43	14.2	Samek, Stegowski, Furman, and Fiedor (2017)
Portugal, Lisbon	Mediterranean	2001	Urban	PM 2.5	Yearly	RM, PCA-MLRA	Road dust	14	3.4	22	5.3	Almeida, Pio, Freitas, Reis, and Trancoso (2005)
Portugal, Lisbon	Mediterranean	2001	Urban	PM 2.5 –10	Yearly	RM, PCA-MLRA	Road dust	13	2.1			Almeida et al. (2005)
Portugal, Porto	Mediterranean	2013	Urban roadside	PM 10	Yearly	RM, PMF	Non-exhaust	8	2.9	23	7.9	Amato et al. (2016a)
Portugal, Porto	Mediterranean	2013	Urban roadside	PM 2.5	Yearly	RM, PMF	Non-exhaust	5	1.3	32	8.1	Amato et al. (2016a)
Spain	Mediterranean	2004		PM 10		EI + DM, CALIOPE, HERMES	Road dust	27	3.5			Pay, Jimenez-Guerrero, and Baldasano (2011)
Spain, Barcelona	Mediterranean	2003–07	Urban background	PM 1	Yearly	RM, ME-2	Road dust	2	0.3	36	6.2	Amato et al. (2009)
Spain, Barcelona	Mediterranean	2003–07	Urban background	PM 10	Yearly	RM, ME-2	Road dust	17	6.9	21	8.5	Amato et al. (2009)
Spain, Barcelona	Mediterranean	2003–07	Urban background	PM 2.5	Yearly	RM, ME-2	Road dust	8	2.2	32	8.8	Amato et al. (2009)

Location	Region	Year	Site type	PM	Term	Method	Source					Reference
Spain, Barcelona	Mediterranean	2010	Urban background	PM 10	Short term	RM, PMF	Road dust	12	3.3	18	5	Brines et al. (2016)
Spain, Barcelona	Mediterranean	2010	Urban roadside	PM 10	Short term	RM, PMF	Road dust	12	3.8	27	8.7	Brines et al. (2016)
Spain, Barcelona	Mediterranean	2010	Urban	PM 10	Short term	RM, PMF	Road dust	9	2.3	11	2.9	Brines et al. (2016)
Spain, Barcelona	Mediterranean	2010	Urban background	PM 10	Short term	RM, PMF	Road dust	8	1.6	10	1.9	Brines et al. (2016)
Spain, Barcelona	Mediterranean	2009	Urban background	PM 10		DM, URBIS	Road dust	12	10			Amato et al. (2016b)
Spain, Barcelona	Mediterranean	2009	Urban traffic	PM 10		DM, URBIS	Road dust	33.5	10			Amato et al. (2016b)
Spain, Barcelona	Mediterranean	2013	Urban background	PM 10	Yearly	RM, PMF	Non-exhaust	12	2.6	14	3.2	Amato et al. (2016a)
Spain, Barcelona	Mediterranean	2013	Urban background	PM 2.5	Yearly	RM, PMF	Non-exhaust	1	0.2	19	2.9	Amato et al. (2016a)
Spain, Barcelona	Mediterranean		Urban roadside	PM 10	ToF-AMS		Tire wear	2				Dall'Osto et al. (2014)
Spain, Càdiz	Mediterranean	2003–10	Industrial	PM 10	Yearly	RM, PMF	Road dust	19	6.7	12.2	4.4	Amato et al. (2014)
Spain, Càdiz	Mediterranean	2003–10	Industrial	PM 2.5	Yearly	RM, PMF	Road dust	10	2.3	15.7	3.6	Amato et al. (2014)
Spain, Cordoba	Mediterranean	2003–10	Urban	PM 10	Yearly	RM, PMF	Road dust	29	10.4	16	5.6	Amato et al. (2014)
Spain, Cordoba	Mediterranean	2003–10	Urban	PM 2.5	Yearly	RM, PMF	Road dust	11	1.9	22	3.8	Amato et al. (2014)
Spain, Granada	Mediterranean	2003–10	Urban roadside	PM 10	Yearly	RM, PMF	Road dust	24	10.4	20	8.8	Amato et al. (2014)
Spain, Granada	Mediterranean	2003–10	Urban roadside	PM 2.5	Yearly	RM, PMF	Road dust	22	8.2	18	6.5	Amato et al. (2014)

(Continued)

TABLE 2.1 Results for Exhaust and Non-Exhaust Contributions to Size-Fractionated PM as Reported in the Literature—cont'd

Location	Climate	Year	Type of Station	PM Metric	Study Duration	Method	Source	Share (%)	Share ($\mu g/m^3$)	Exhaust (%)	Exhaust ($\mu g/m^3$)	References
Spain, Granada	Mediterranean	2003–10	Urban roadside	PM 10	Yearly	RM, PMF	Tire wear	8	3.4	20	8.8	Amato et al. (2014)
Spain, Granada	Mediterranean	2003–10	Urban roadside	PM 2.5	Yearly	RM, PMF	Tire wear	18	6.6	18	6.5	Amato et al. (2014)
Spain, Madrid	Continental	2009	Urban	PM 10	Short term	RM, PMF	Road dust	29	14.6	31	15.6	Karanasiou, Moreno, Amato, Lumbreras, and Narros (2011)
Spain, Málaga	Mediterranean	2003–10	Rural	PM 10	Yearly	RM, PMF	Road dust	9	1.3	23	3.5	Amato et al. (2014)
Spain, Málaga	Mediterranean	2003–10	Urban roadside	PM 10	Yearly	RM, PMF	Road dust	21	9	19	8	Amato et al. (2014)
Spain, Málaga	Mediterranean	2003–10	Rural	PM 2.5	Yearly	RM, PMF	Road dust	7	1.4	20	3.9	Amato et al. (2014)
Spain, Málaga	Mediterranean	2003–10	Urban roadside	PM 2.5	Yearly	RM, PMF	Road dust	21	4.8	12	2.8	Amato et al. (2014)
Spain, Seville	Mediterranean	2003–10	Urban	PM 10	Yearly	RM, PMF	Road dust	34	13.9	12	4.7	Amato et al. (2014)
Spain, Seville	Mediterranean	2003–10	Urban roadside	PM 10	Yearly	RM, PMF	Road dust	35	14.4	20	8.1	Amato et al. (2014)
Spain, Seville	Mediterranean	2003–10	Urban	PM 2.5	Yearly	RM, PMF	Road dust	31	9.5	10	3.2	Amato et al. (2014)
Spain, Seville	Mediterranean	2003–10	Urban roadside	PM 2.5	Yearly	RM, PMF	Road dust	31	9.5	19	5.8	Amato et al. (2014)
Sri Lanka, Colombo	Tropical monsoon	2000–05	Urban	PM 2.5	Yearly	RM, PMF	Road dust	27	7.8	48	13.9	Seneviratne et al. (2011)
Sri Lanka, Colombo	Tropical monsoon	2003–08	Urban residential	PM 2.5	Yearly	RM, PMF	Road dust	9	2.1	17	4	Seneviratne et al. (2011)
Sweden, Lycksele	Continental	2002	Rural	PM 2.5	Short term	RM, PMF	Brake wear	13.9	1			Hedberg, Johansson, Johansson, Swietlicki, and Brorström-Lundén (2012)

Location	Climate	Year	Site type	PM size	Period	Method	Source category					Reference
Sweden, Stockholm	Continental	2003–04	Freeway	PM 10	Short term	RM, PMF	Road dust	11	2.1	13	2.4	Furusjö, Sternbeck, and Cousins (2007)
Sweden, Stockholm	Continental	2003–04	Urban roadside	PM 10	Short term	RM, PMF	Road dust	13	4.6	36	13	Furusjö et al. (2007)
Sweden, Stockholm	Continental	2006–07	Urban roadside	PM 10		EM, NORTRIP	Brake wear	5.5	1.7			Denby et al. (2013)
Sweden, Stockholm	Continental	2006–07	Urban roadside	PM 10		EM, NORTRIP	Road wear	76	23.3			Denby et al. (2013)
Switzerland, Basel	Continental	1999	Suburban	PM 10	Yearly	RM	Road dust	7	1.8	18.2	4.9	Gehrig, Huglin, and Hofer (2001)
Switzerland, Bern	Continental	1999	Urban roadside	PM 10	Yearly	RM	Road dust	16	6.4	18.4	7.6	Gehrig et al. (2001)
Switzerland, Bern	Continental	1999	Urban roadside	PM 10	Yearly	RM	Tire wear	7.5		18.4	7.6	Gehrig et al. (2001)
Switzerland, Erstfeld	Continental	2008	Rural	PM 10	Yearly	RM, PMF	Brake wear/tire wear	4	0.56	16	2.2	Ducret-Stich et al. (2013)
Switzerland, Erstfeld	Continental	2007–09	Rural	PM 10	Yearly	RM, PMF	Road dust	8	1.1	16	2.2	Ducret-Stich et al. (2013)
Switzerland, Zurich	Continental	1999	Urban background	PM 10	Yearly	RM	Road dust	10	2.5	13.2	3.4	Gehrig et al. (2001)
Switzerland, Zurich	Continental	1999	Urban background	PM 10	Yearly	RM	Tire wear	1.9		13.2	3.4	Gehrig et al. (2001)
The Netherlands, Amsterdam	Oceanic	2013–14	Urban background	PM 10	Yearly	RM, PMF	Brake wear	2.8	0.7	2	0.5	Mooibroek et al. (2016)
The Netherlands, Wijk aan Zee	Oceanic	2013–14	Industrial	PM 10	Yearly	RM, PMF	Brake wear	0.67	0.2	1.8	0.5	Mooibroek et al. (2016)
Turkey, Aliaga	Mediterranean	2009	Urban	PM 10	Yearly	RM, PMF	Road dust	23	11.8			Kara et al. (2015)

(Continued)

TABLE 2.1 Results for Exhaust and Non-Exhaust Contributions to Size-Fractionated PM as Reported in the Literature—cont'd

Location	Climate	Year	Type of Station	PM Metric	Study Duration	Method	Source	Share (%)	Share (µg/m³)	Exhaust (%)	Exhaust (µg/m³)	References
UK, Birmingham	Oceanic	1992	Urban	PM 2.1	Yearly	RM, PCA-MLRA	Road dust	32		25		Harrison, Smith, Piou, and Castro (1997)
UK, Hatfield	Oceanic	2006	Tunnel	PM 10	Short term	RM, PCA-MLRA	Brake wear	11		12		Lawrence et al. (2013)
UK, Hatfield	Oceanic	2006	Tunnel	PM 10	Short term	RM, PCA-MLRA	Road dust	27		12		Lawrence et al. (2013)
UK, Leicester	Oceanic	2013–14	Urban background	PM 10	Yearly	RM, PMF	Brake wear	3.8	0.8	4.3	0.9	Mooibroek et al. (2016)
UK, London	Oceanic	2007	Urban	Particle number	Short term	RM, PMF	Brake wear	1.7		65.4		Harrison, Beddows, and Dall'Osto (2011)
UK, London	Oceanic	2007	Urban	PM 20 volume	Short term	RM, PMF	Brake wear	13.7		22.4		Harrison et al. (2011)
UK, London	Oceanic	2007	Urban	Particle Number	Short term	RM, PMF	Road dust	5		65.4		Harrison et al. (2011)
UK, London	Oceanic	2007–11	Urban roadside	PM 0.9–11.5	Short term	Extrapolation	Brake wear	13.5	3.3			Harrison, Jones, Gietl, Yin, and Green (2012)
UK, London	Oceanic	2007–11	Urban roadside	PM 0.9–11.5	Short term	Extrapolations	Tire wear	2.7	0.65			Harrison et al. (2012)
UK, London	Oceanic	2007–11	Urban roadside	PM 0.9–11.5	Short term	Extrapolations	Road dust	9	2.3			Harrison et al. (2012)
UK, London	Oceanic	2008	Urban roadside	PM 10		EI, LAEI + DM, KCLurban	Tire wear	5	2			Beevers et al. (2013)
UK, London	Oceanic	2008	Urban roadside	PM 10		EI, LAEI + DM, KCLurban	Brake wear	20	8			Beevers et al. (2013)
UK, London	Oceanic	2008	Urban roadside	PM 10		EI, LAEI + DM, KCLurban	Non-exhaust	42.5	17			Beevers et al. (2013)

UK, London	Oceanic	2008	Suburban	PM 10		EI, LAEI + DM, KCLurban	Non-exhaust	9.8	2		2	Beevers et al. (2013)
UK, London	Oceanic	2008	Urban roadside	PM 2.5		EI, LAEI + DM, KCLurban	Non-exhaust	44.6	9		9	Beevers et al. (2013)
UK, London	Oceanic	2008	Suburban	PM 2.5		EI, LAEI + DM, KCLurban	Non-exhaust	11.4	1.5			Beevers et al. (2013)
UK, London	Oceanic	2012	Urban roadside	PM 2.5–10	Short term	RM, PMF	Road dust	31		13.4		Crilley et al. (2017)
USA, Albuquerque	Arid	2007–08	Urban	PM 2.5	Yearly	RM, PMF	Road dust	11	0.59	5.9	0.33	Kavouras, DuBois, Nikolich, and Etyemezian (2015)
USA, Atlanta	Temperate	2001–05	Urban	PM 2.5	Yearly	RM, CMB	Road dust	2	0.34	15.9	2.7	Chen et al. (2012)
USA, Atlanta	Temperate	2001	Urban	PM 2.5	Short term	RM, PMF	Road dust	7	1.1	17	2.84	Ke et al. (2008)
USA, Atlanta	Temperate	2008–10	Urban	PM 2.5	Yearly	RM, PMF-CMB	Road dust	4	0.5	9	0.92	Watson et al. (2015)
USA, Atlanta	Temperate	2001–05	Rural	PM 2.5	Yearly	RM, CMB	Road dust	2	0.35	12.7	1.85	Chen et al. (2012)
USA, Birmingham	Temperate	2001–05	Urban	PM 2.5	Yearly	RM, CMB	Road dust	3	0.57	15.7	2.88	Chen et al. (2012)
USA, Bimingham	Temperate	2002–04	Urban	PM 2.5	Yearly	RM, PMF	Road dust	6	1.2	25	5.1	Baumann, Jayanty, and Flanagan (2008)
USA, Bimingham	Temperate	2008–10	Urban	PM 2.5	Yearly	RM, PMF-CMB	Road dust	6	0.8	18.7	2	Watson et al. (2015)
USA, California	Temperate	1993	Industrial	PM 2.5	Short term	RM, CMB	Road dust	6	2.5	30.1	11.9	Schauer, Fraser, Cass, and Simoneit (2002)
USA, California	Temperate	1993	Urban roadside	PM 2.5	Short term	RM, CMB	Road dust	14	9.26	39.5	25.9	Schauer et al. (2002)
USA, California	Temperate	1993	Urban background	PM 2.5	Short term	RM, CMB	Road dust	7	4.1	37.4	22.2	Schauer et al. (2002)
USA, California	Temperate	1993	Urban background	PM 2.5	Short term	RM, CMB	Road dust	20	8.7	26.5	13.7	Schauer et al. (2002)

(Continued)

TABLE 2.1 Results for Exhaust and Non-Exhaust Contributions to Size-Fractionated PM as Reported in the Literature—cont'd

Location	Climate	Year	Type of Station	PM Metric	Study Duration	Method	Source	Share (%)	Share (µg/m³)	Exhaust (%)	Exhaust (µg/m³)	References
USA, California	Temperate	1993	Industrial	PM 2.5	Short term	RM, CMB	Tire wear	1	0.4	30.1	11.9	Schauer et al. (2002)
USA, California	Temperate	1993	Urban roadside	PM 2.5	Short term	RM, CMB	Tire wear	1.1	0.73	39.5	25.9	Schauer et al. (2002)
USA, California	Temperate	1993	Urban background	PM 2.5	Short term	RM, CMB	Tire wear	3.3	2	37.4	22.2	Schauer et al. (2002)
USA, California	Mediterranean	1987	Urban	PM 10	Yearly	RM, CMB	Road dust	24	15.6			Watson et al. (1994)
USA, California	Mediterranean	2000–04	Rural background	PM 2.5	Yearly	RM, PMF	Road dust	9	0.27	10.7	0.34	Green, Antony Chen, DuBois, and Molenar (2012)
USA, California	Mediterranean	2005–09	Rural background	PM 2.5	Yearly	RM, PMF	Road dust	5	0.14	8.5	0.26	Green et al. (2012)
USA, California	Mediterranean	2002–2012	Urban commercial	PM 2.5	Yearly, 10	RM, PMF	Road dust	5	0.58	24.2	2.74	Wang and Hopke (2013)
USA, Centreville	Temperate	2001–05	Rural	PM 2.5	Yearly	RM, CMB	Road dust	2	0.31	9.6	1.3	Chen et al. (2012)
USA, Chicago	Continental	2009	Urban	PM 2.5–10	Short term	RM, PMF	Brake wear	23.2	1.29			Sturtz, Adar, Gould, and Larson (2014)
USA, Chicago	Continental	2009	Urban	PM 2.5–10	Short term	RM, PMF	Tire wear	7	0.39			Sturtz et al. (2014)
USA, Connecticut	Continental	2000–04	Urban	PM 2.5	Yearly	RM, PMF	Road dust	9	1.2	31.3	4.2	Lee, Gent, Leadere, and Koutrakis (2011)
USA, Connecticut	Continental	2000–04	Urban	PM 2.5	Yearly	RM, PMF	Road dust	15	2	25	3.3	Lee et al. (2011)

USA, Connecticut	Continental	2000–04	Urban	PM 2.5	Yearly	RM, PMF	Road dust	7	0.8	26	3.1	Lee et al. (2011)
USA, Connecticut	Continental	2000–04	Urban	PM 2.5	Yearly	RM, PMF	Road dust	17	2.9	29.4	5	Lee et al. (2011)
USA, Detroit	Continental	2005	Urban	PM 2.5	Yearly	RM, CMB	Road dust	4	0.7	30	5.3	Duvall et al. (2012)
USA, Detroit	Continental	2005	Urban background	PM 2.5	Yearly	RM, CMB	Road dust	2	0.34	40	6.1	Duvall et al. (2012)
USA, Gulfport	Temperate	2001–05	Urban	PM 2.5	Yearly	RM, CMB	Road dust	4	0.48	7.5	0.9	Chen et al. (2012)
UK, Hatfield	Oceanic	2006	Tunnel	PM 10	Short term	RM, PCA-MLRA	Road wear	11		12		Lawrence et al. (2013)
USA, Houston	Temperate	2000	Urban	PM 2.5	Yearly	RM, PMF	Road dust	17	0.8			Buzcu, Fraser, Kulkarni, and Chellam (2003)
USA, Little Rock	Temperate	2002–10	Urban	PM 2.5	Yearly	RM, PMF	Road dust	8	1	11.7	1.5	Chalbot, McElroy, and Kavouras (2013)
USA, Los Angeles	Mediterranean	1982	Urban	PM 2	Yearly	RM, CMB	Road dust	12	3.9	22.7	20.2	Schauer et al. (1996)
USA, Los Angeles	Mediterranean	1982	Urban	PM 2	Yearly	RM, CMB	Tire wear	0.9	0.25	22.7	20.2	Schauer et al. (1996)
USA, Los Angeles	Mediterranean	1993		PM 2.5		EI + DM, UCD/CIT	Road dust	13	7.7			Held, Ying, Kleeman, Schauer, and Fraser (2005)
USA, Massachussets	Continental	2000–04	Urban	PM 2.5	Yearly	RM, PMF	Road dust	11	1.4	27.7	3.6	Lee et al. (2011)
USA, New York	Continental	2011	Urban	PM 2.5	Short term	RM, PMF	Road dust	25	4.13	3.2	0.52	Li et al. (2004)
USA, Oak Grove	Temperate	2001–05	Rural	PM 2.5	Yearly	RM, CMB	Road dust	5	0.6	13.2	1.65	Chen et al. (2012)

(Continued)

TABLE 2.1 Results for Exhaust and Non-Exhaust Contributions to Size-Fractionated PM as Reported in the Literature—cont'd

Location	Climate	Year	Type of Station	PM Metric	Study Duration	Method	Source	Share (%)	Share (µg/m³)	Exhaust (%)	Exhaust (µg/m³)	References
USA, Pensacola	Temperate	2001–05	Suburban	PM 2.5	Yearly	RM, CMB	Road dust	3	0.41	11.3	1.36	Chen et al. (2012)
USA, Pensacola	Temperate	2001–05	Urban	PM 2.5	Yearly	RM, CMB	Road dust	4	0.53	13.4	1.82	Chen et al. (2012)
USA, Sacramento	Mediterranean	2010	Urban	PM 0.1	Yearly	RM, PMF	Brake wear	4	0.007	2	0.003	Kuwayama, Ruehl, and Kleeman (2013)
USA, San Joaquin Valley	Mediterranean	1996		PM 2.5		EI + DM, UCD/CIT	Road dust	1	1.05			Held et al. (2005)
USA, St. Paul	Continental	2009	Urban	PM 2.5 –10	Short term	RM, PMF	Brake wear	5.6	0.27			Sturtz et al. (2014)
USA, St. Paul	Continental	2009	Urban	PM 2.5 –10	Short term	RM, PMF	Tire wear	1.9	0.09			Sturtz et al. (2014)
USA, Winston –Salem	Temperate	2009	Urban	PM 2.5 –10	Short term	RM, PMF	Brake wear	18.4	0.61			Sturtz et al. (2014)
USA, Winston –Salem	Temperate	2009	Urban	PM 2.5 –10	Short term	RM, PMF	Tire wear	1.6	0.05			Sturtz et al. (2014)

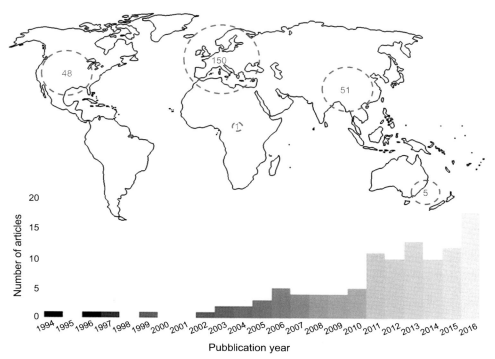

FIGURE 2.1 The number of study estimates for non-exhaust contributions per continent and the number of articles per year.

TABLE 2.2 Grouping of Literature Studies According to Sources Labeling

Source Category	Non-Exhaust—Miscellaneous	Brake Wear	Tire Wear	Road Dust
Percentage of total number of non-exhaust estimates	12	9	8	71
	Non-exhaust	Brake wear	Tire wear	Resuspension
	Non-exhaust (sum)			Road dust
	Non-exhaust (wear + resuspension)			Road wear
	Brake/tire wear			Resuspended dust
	Tire and brake wear			
	Brake and tire wear			
	Road dust/brake wear			

- PM coarse (2.5−10 μm; 1.5−10 μm; 1.6−10.9 μm; 2.1−9 μm; 2−10 μm; and 3−10 μm)
- PM_{10} (<10 μm; 1−10 μm and 0.9−11.5 μm)

Averaging all studies on PM_{10}, road dust resulted to have a far higher impact than other reviewed sources. Mean contribution to PM_{10} (74 estimates in total) was 22%, ranging from 1% to 76% among the different studies, versus a range of 1%−64% (21% as mean) from vehicle exhaust emissions (Fig. 2.2). In absolute concentrations, the mean (among different studies) road dust contribution was 16.4 μg/m^3 (ranging from 0.3 to 79.3 μg/m^3), whereas it was 17.3 μg/m^3 for exhaust (ranging from 1.2 to 103.6 μg/m^3). For $PM_{2.5}$ (81 estimates in total), road dust contributions were significantly lower, with a mean of 11%, ranging from 1% to 43% depending on the study, versus a mean of 24% (range of 2%−65%) due to vehicle exhaust (Fig. 2.2). In absolute terms the mean (among different studies) road dust contribution to $PM_{2.5}$ was 3.8 μg/m^3 (8.8 μg/m^3 for exhaust), ranging from 0.1 to 27.2 μg/m^3 (0.3−59.8 μg/m^3 for exhaust). For PM coarse (15 estimates in total), the relative contribution to PM mass was the highest, 34% (11%−62%). In absolute concentrations, the average contribution was 5.8 μg/m^3, ranging from 1.5 to 17.1 μg/m^3. Only few studies allowed direct comparison with exhaust for PM coarse (Fig. 2.2), revealing average contribution of 44% (7.3 μg/m^3) for road dust versus 10% (1.9 μg/m^3) for exhaust. For PM_1, we only found two studies with non-exhaust contributions of 2% and 5%−21%, although the latter is more uncertain as it was estimated by means of SEM.

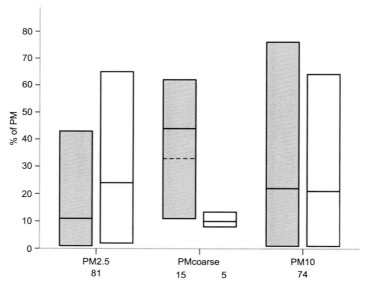

FIGURE 2.2   Range and mean road dust (in gray) and vehicle exhaust (in white) contribution (%) to different particulate matter (PM) size fractions. *Solid lines* represent the average of the studies where both exhaust and non-exhaust sources were found, whereas *dotted lines* represent the mean of all studies. The number of studies available is shown below each PM size fraction.

This comparison allows concluding that globally, regardless of the environment studied, non-exhaust emissions are already at least as important as exhaust ones for PM_{10}, considering also that the aforementioned contributions are derived mostly from receptor modeling, thus including the contribution of secondary PM for vehicle exhaust (while they are discarded in emission inventories). For $PM_{2.5}$ they represent at least a third of total traffic contribution while they dominate the PM coarse ($>1.5\ \mu m$) mode ($>80\%$).

Direct brake wear contributions to PM_{10} (11 estimates) were found to vary from 0.1% to 20% (mean 7% of all studies) and, in absolute terms, from 0.1 to $8.0\ \mu g/m^3$ (mean of $1.9\ \mu g/m^3$). Only two studies allowed comparing with vehicle exhaust providing the same (5%) contribution (ranges of 1%–11% for non-exhaust and 2%–12% for exhaust, Fig. 2.3).

For $PM_{2.5}$ we only found two studies with brake wear contribution of 9% ($0.7\ \mu g/m^3$) and 14% surprisingly with the latter at a rural site. In any case, only the former, performed at an urban site, allows comparing with exhaust emissions that contributed to 55% ($4.3\ \mu g/m^3$, Fig. 2.3).

For coarse particles, four estimates were found providing a contribution range of 7%–23% (average 13%) and $0.7\ \mu g/m^3$ varying from 0.3 to $1.3\ \mu g/m^3$ (Fig. 2.3). Only one study compared exhaust versus non-exhaust, providing a surprising ratio of 2 (1.6 vs. $0.8\ \mu g/m^3$, respectively), but as mentioned by authors, results were high uncertain due to the high correlation between the contributions of two factors (Wahlin et al., 2006).

Several studies identified a contribution of disc wear to total brake wear emissions. Iijima et al. (2008) reported that approximately 30% of dust originated from disc in their

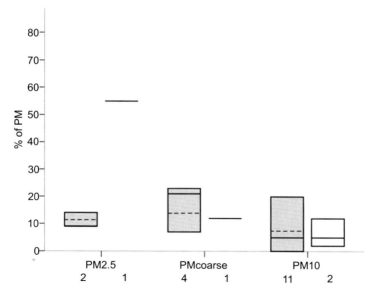

FIGURE 2.3 Range and mean brake wear (in gray) and vehicle exhaust (in white) contribution (%) to different particulate matter (PM) size fractions. *Solid lines* represent the average of the studies where both exhaust and non-exhaust sources were found, whereas *dotted lines* represent the mean of all studies. The number of studies available is shown below each PM size fraction.

experiments, while Hulskotte et al. (2014) calculated that 70% of brake wear mass originate from disc wear. This wide variability is also reported for Fe, C, and Cu from studies analyzing brake pads and discs composition. However, average Fe concentration is about 40%, whereas it is >90% in gray cast iron disc (Amato et al., 2009; Hulskotte et al., 2014; Kukutschová et al., 2011; Thorpe & Harrison, 2008). In Fig. 2.2, we compare the average composition for brake pads (from Amato et al., 2009; Figi et al., 2010; Hjortenkrans, Bergback, & Haggerud, 2007; Hulskotte et al., 2014; Iijima et al., 2008; Kukutschová et al., 2011; Von Uexküll et al., 2005) and brake discs (Hulskotte et al., 2014) with that of PMF brake wear factor profiles (Hedberg, Johansson, Johansson, Swietlicki, & Brorström-Lundén, 2012; Mooibroek et al, 2016; Sturtz, Adar, Gould, & Larson, 2014; Visser et al., 2015; Wahlin et al., 2006).

Brake wear factor profiles are generally similar to those of brake pads, with Fe as main components in mass. The typical tracers Ba, Cu, Sb, and Sn have similar concentrations in brake pads and in PMF profiles. However, iron content in PMF profile results to be intermediate between disc and pads composition, confirming the importance of disc wear (Fig. 2.4). On the contrary, sulfur content is slightly higher in PMF profiles rather than in both pads and discs (Fig. 2.4). This can be interpreted as poor representativeness of pads experimental profile or to a mix of PMF brake factors with other traffic-related sources (exhaust for example).

Direct tire wear contribution (in %) was found to be lower than brake wear for all PM size fractions. In PM_{10} (11 estimates) an average value of 4% (0.5%−8%) and 1.8 $\mu g/m^3$ (0.6−3.4 $\mu g/m^3$) was found (Fig. 2.5), except an outlier data with a mean contribution of 15.4 $\mu g/m^3$ in India. Only four studies allowed the intercomparison with vehicle exhaust contribution, showing 22% (13%−37%) from exhaust and 6% (2%−8%) from tire wear. In $PM_{2.5}$ we only found three studies; one in Spain that found, surprisingly, a 1:1 ratio with exhaust (18%), probably due to a colinearity among the two PMF factors (Fig. 2.5). The other two studies in the United States found a contribution ranging from 0.9% to 3.3% (mean of 1.6%), with a mean contribution of 0.8 $\mu g/m^3$ (0.3−2.0 $\mu g/m^3$) (Fig. 2.5). In the same studies, exhaust emissions contributed 32% (23%−40%) and 20 $\mu g/m^3$ (12−26 $\mu g/m^3$). For coarse PM

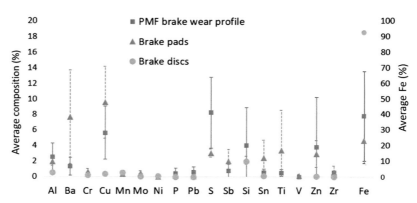

FIGURE 2.4 Average compositions of brake pads (Amato et al., 2009; Figi et al., 2010; Hjortenkrans et al., 2007; Hulskotte et al., 2014; Iijima et al., 2008; Kukutschová et al., 2011; Von Uexküll et al., 2005), brake discs (Hulskotte et al., 2014), and positive matrix factorization (PMF) brake wear profile as found from source apportionment studies (Hedberg et al., 2012; Mooibroek et al., 2016; Sturtz et al., 2014; Visser et al., 2015; Wahlin et al., 2006).

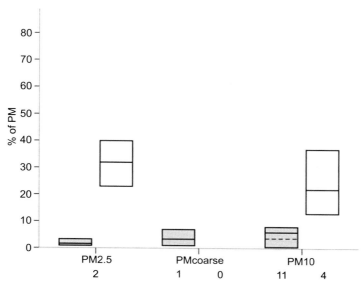

FIGURE 2.5 Range and mean tire wear (in gray) and vehicle exhaust (in white, not available for particulate matter [PM] coarse) contribution (%) to different PM size fractions. *Solid lines* represent the average of the studies where both exhaust and non-exhaust sources were found, whereas *dotted lines* represent the mean of all studies. The number of studies available is shown below each PM size fraction.

only one study was found (Fig. 2.5), estimating contribution at three cities in the United States: The mean contribution was 3.4%, varying from 1% to 7%, while in absolute concentrations, 0.27 $\mu g/m^3$ (from 0.05 to 0.39 $\mu g/m^3$).

Another category of studies included 12 articles identifying a generic "non-exhaust" source or similar (Table 2.2, Fig. 2.6). The mean contribution of these studies to PM_{10} (18 estimates) was 8% (range: 1%−14%, plus a 42% value found in a roadside station UK), being the absolute contribution estimated only by a part of them in 3 $\mu g/m^3$ (range: 0.2−17 $\mu g/m^3$). Selecting those studies that also identified the exhaust contribution, the comparison shows a mean 6% (1%−15%) contribution of non-exhaust versus a mean 11% (1%−23%) from exhaust (Fig. 2.6). However, we have to remind that, given the lack of definition on the term "non-exhaust," this category cannot be intended a priori as the "total" non-exhaust contribution (i.e., road dust contribution might be hidden in "mineral" or "crustal" factors).

In $PM_{2.5}$, a mean contribution of 6% was found (range: 1%−11%, Fig. 2.6) except one study with 47% at a roadside site in the UK being the absolute contribution estimated only by a part of them in 1.4 $\mu g/m^3$ (range: 0.2−3.4 $\mu g/m^3$), whereas the exhaust contribution was found to be in average 19% (6%−37%) and to vary within 1.7−8.1 $\mu g/m^3$ (4.2 $\mu g/m^3$ as average). In PM_1 only a German study was found with a mean contribution of 3% versus 10% from exhaust (Fig. 2.6).

Non-exhaust emissions have been predicted to increase in relative importance to urban PM mass during the last two decades due to the reduction of exhaust emissions (Keuken, Roemer, Zandveld, Verbeek, & Velders, 2012; Kousoulidou, Ntziachristos, Mellios, &

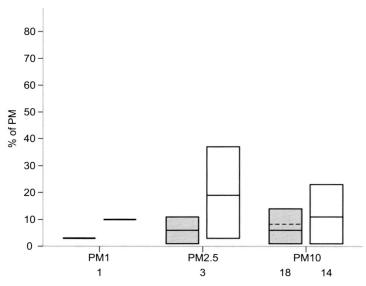

FIGURE 2.6 Range and mean non-exhaust (in gray) and vehicle exhaust (in white) contribution (%) to different particulate matter (PM) size fractions. *Solid lines* represent the average of the studies where both exhaust and non-exhaust sources were found, whereas *dotted lines* represent the mean of all studies. The number of studies available is shown below each PM size fraction.

Samaras, 2008; Rexeis & Hausberger, 2009). Recently, analysis of trends in atmospheric emissions reported first evidences of this increase of the share of coarse particles in PM_{10}, as $PM_{2.5}$ decreases more rapidly than PM_{10} (Masri, Kang, & Koutrakis, 2015). Similarly, at various European sites, $PM_{2.5}$ trends over last decade declined more rapidly than coarse PM (Barmpadimos, Keller, Oderbolz, Hueglin, & Prévôt, 2012). Font and Fuller (2016) also reported a general downward trend in $PM_{2.5}$ in London, proving the usefulness of exhaust abatement policies, whereas for PM_{10} such trend was not noted. In southern Spain, for example, from 2004 to 2011 road dust contributions to PM_{10} levels measured at a number of sites did not decrease, whereas vehicle exhaust contributions decreased ($P < .001$) by 0.4 (0.57−0.24) $\mu g/m^3$ year (Amato, Alastuey et al., 2014). Further increase in Europe in the coming years may also be boosted by the economic crisis, the consequent poor maintenance of vehicles and roads, and the possible increase of low-cost materials and technologies used, with worse quality and faster degradation/erosion.

The aforementioned figures on air quality (PM_{10} and $PM_{2.5}$) impact are due to the coarser size distribution of non-exhaust particle volume/mass compared with exhaust PM.

Size distribution of brake wear particles was recently reviewed by Grigoratos and Martini (2015), who classified three types of methodologies used to assessing it: (1) emission studies in laboratory (e.g., brake dynamometer, or pin-on-disc; Garg, Cadle, Mulawa, & Groblicki, 2000; Gasser et al., 2009; Iijima et al., 2007, 2008; Kukutschová et al., 2011; Mosleh, Blau, & Dumitrescu, 2004; Sanders, Xu, Dalka, & Maricq, 2003; Wahlström, Söderberg, Olander, Jansson, & Olofsson, 2010); (2) receptor modeling studies (Bukowiecki, Gehrig et al., 2009; Bukowiecki, Lienemann et al., 2009; Dongarra, Manno, & Varrica, 2009; Fabretti, Sauret,

Gal, Maria, & Schärer, 2009; Gietl et al., 2010; Harrison et al., 2012; Hjortenkrans et al., 2007; Wahlin et al., 2006); and (3) "real-world" emission studies (Mathissen, Scheer, Vogt, & Benter, 2011). Most of the studies found unimodal brake wear mass distributions with maxima ranging between 1.0 and 6.0 μm, confirming the high impact of brake wear emissions also in $PM_{2.5}$. For instance, Kukutschová et al. (2011) found unimodal distribution with maxima at 2—4 μm using low-metallic pads, whereas Von Uexküll et al. (2005) conducted tests on front and rear truck brakes and found unimodal mass distributions with maxima at 2—3 μm. Similar conclusions were reached by Harrison et al. (2012), who collected size-fractionated samples of airborne PM and found a unimodal PM_{10} mass distribution with a peak at 2—3 μm.

Regarding tire wear mass size distribution, only three studies were found: Kwak, Kim, Lee, and Lee (2013) investigated the mass size distributions of RWPs and TWPs during constant speed conditions (50, 80, 110, and 140 km/h). The mode diameters of TWPs were between 2 and 3 μm. Harrison et al. (2012) also estimated a bimodal distribution with maxima at 2 and 8 μm, respectively, using zinc as tracer of mass size distribution. Differently, Gustafsson et al. (2008) indicated tire wear as the source of particles <1 μm on a road simulator.

Concerning road wear, Kwak et al. (2013) found that peak values of RWPs generated at vehicle speeds of 50, 80, 110, and 140 km/h ranged between 2 and 3 μm, whereas for speeds greater than 80 km/h, RWPs showed a bimodal distribution with peaks at 2—3 μm and >10 μm. Aatmeeyata et al. (2009) found that about 32% of particle mass emitted from road—tire friction of PM_{10} was present below 1 μm. The number, as well as mass size distribution, for PM_{10} was observed to be bimodal with peaks at 0.3 μm and 4—5 μm. Kupiainen et al. (2005) found however particle mass size distribution with maxima well above 2.5 μm, both with friction tires and studded tires.

Concerning strictly road dust, most of the studies identified only impact on $PM_{2.5}$ or PM_{10} mass fraction, without further subtle fractionation. As it is generally identified as a mixture of the previously mentioned sources plus suspended soil, it is obviously enriched mostly in the 2.5—10 μm fraction. Most of the studies, based on various methods, found a unimodal mass size distribution above, at least, 3—4 μm (Harrison et al., 2012; McKenzie, Wong, Green, Kayhanian, & Young, 2008; Tian, Pan, & Wang, 2016; Zhu, Gilles, Etyemezian, Nikolich, & Shaw, 2015). For example, Harrison et al. (2012) collected size-fractionated samples of airborne PM and found a unimodal PM_{10} mass distribution of road dust particles peaking around 4 μm. Tian et al. (2016), using receptor modeling, report that road dust profile has the major relative contribution to PM (>10%) above 3.3 μm. The particle mass size distribution of roadway particles (which are assumable to be road dust) was also studied by Lee, Kwak, Kim, and Lee (2013), who found a bimodal distribution at lower sizes, being the maxima at 0.7 and 2 μm.

3.2 Spatial Variability

The microscale (rural to traffic locations) spatial variability of source contributions according to receptor sites shows the following:

- **Road dust suspension**: At rural locations, PM_{10} mean contribution (4 estimates) was 8% (6%—9%) versus a 15% (7%—23%) from vehicle exhaust, 1.1 μg/m³ versus 2.3 μg/m³ in

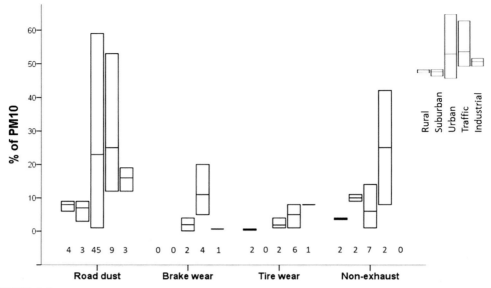

FIGURE 2.7   Range of contributions (%) to PM$_{10}$ of different non-exhaust sources at different receptor sites.

absolute contributions, respectively (Fig. 2.7). This is likely due to the lower traveling distances of coarse particles. In PM$_{2.5}$, the proportion of road dust particles is about three times lower than exhaust: 4% (1%−9%) versus 12% (8%−20%), 0.5 μg/m^3 versus 1.6 μg/m^3 because the coarse fraction of road dust is not captured. At suburban, although only a very limited number of studies were found (three estimates), PM$_{10}$ contribution increased to 7% (Fig. 2.7) (except one study in Finland finding 43%, Kupiainen et al., 2016) versus 18% of exhaust emissions, corresponding to 1.8 μg/m^3 versus 4.9 μg/m^3. In PM$_{2.5}$ the same pattern of rural background is observed: 3% versus 11% and 0.4 μg/m^3 versus 1.4 μg/m^3. The largest body of literature was found for urban locations (128 estimates): PM$_{10}$ contributions averaged 23% (range of 1%−59%) versus 22% of exhaust (range of 1%−64%), 22.1 μg/m^3 versus 20.7 μg/m^3, respectively (Fig. 2.7). In PM$_{2.5}$ mean contributions of 11% (range of 1%−32%) versus 24% (3%−58%) and 3.8 μg/m^3 versus 8.2 μg/m^3 were found, whereas in PM coarse (1.5−10 μm approximately): 44% (17%−47%) versus 10% (8%−14%) from exhaust, 5.0 μg/m^3 (7.3 vs. 1.8 μg/m^3).

A lower number of studies were found at roadside locations (19 estimates). PM$_{10}$ contributions were on average 25% (12%−53%), with 20% for exhaust (6%−36%), but reaching 57%−76% in Sweden and Denmark (Fig. 2.7); in absolute concentrations, mean values were 8.1 μg/m^3 for road dust versus 9.0 μg/m^3 for exhaust (higher due to the lack of absolute contribution in one study with SEM). Only one study quantified road dust contribution in freeway (11% vs. 13%, 2.1 vs. 2.4 μg/m^3) and other two in tunnels (19% vs. 12%) (Fig. 2.7). For PM$_{2.5}$, mean contribution at roadside locations was 20% (10%−31%) versus 30% for exhaust (12%−62%), corresponding to 8.3 μg/m^3 versus

$20.0 \mu g/m^3$, while the only study in tunnel found 43% versus 21%. For coarse PM only one study compared road dust contribution share with exhaust (31% vs. 13% for exhaust).

Other seven estimates were found in industrial settings: In PM_{10} 16% (12%−19%) was the average road dust contribution versus 22% (9%−47%) of exhaust (Fig. 2.7), corresponding to $10 \mu g/m^3$ versus $15 \mu g/m^3$, whereas in $PM_{2.5}$ mean contribution of road dust was 12% (6%−24%) versus 28% (2%−65%), corresponding to $4.0 \mu g/m^3$ versus $17.8 \mu g/m^3$ respectively.

- **Brake wear**: At rural locations, only one estimate was found (14% of $PM_{2.5}$, $1 \mu g$ $PM_{2.5}/m^3$). The largest number of studies was found for urban locations (16 estimates). For PM_{10} only two studies were found (Fig. 2.7): one study in northern Europe, where contributions averaged 4% (3%−5%) versus 4% (2%−5%) of exhaust, and one study in India (58% of PM_{10} from exhaust vs. 0.1% from brake wear), who also found the same values for $PM_{2.5}$. Another study in Denmark (Wahlin et al., 2006) found a mean contribution of brake wear of 9% versus 55% of exhaust. For urban PM coarse the average brake wear contribution was found to be 13% (6%−23%) corresponding to $0.7 \mu g/m^3$.

  A lower number of studies were found at roadside locations (four estimates). PM_{10} contributions were on average 11% (5%−20%), $3.6 \mu g/m^3$ for brake wear (Fig. 2.7). The comparison with exhaust was made only in one tunnel study (11% vs. 12%). Another study in an industrial setting found in PM_{10} 0.7% as the brake wear contribution versus 1.8% of exhaust, whereas in $PM_{0.3-3}$ a brake wear share of 0.4% was found.

- **Tire wear**: Very few studies were found for each location category; on average, it contributes 0.5% to rural PM_{10} and 1.9% to PM_{10} in urban sites, versus 13% of vehicle exhaust (Fig. 2.7). In urban $PM_{2.5}$ its share is 2% (1%−3%) versus 30% (23%−37%) from exhaust. In urban PM coarse, it increases slightly to 3% (2%−7%), corresponding to $0.3 \mu g/m^3$ on average. More studies are located roadside, providing a mean contribution of 5% of PM_{10} (Fig. 2.7), although it increases to 8% (7%−8%) in those studies that also estimated contribution of exhaust, at 19% (18%−20%), $2.1 \mu g/m^3$ on average. One roadside $PM_{2.5}$ study showed a mean 1.1% contribution versus 39% contribution from exhaust, although another study found a 18% contribution, but likely due to a mix of sources. Only one estimate in industrial sites was found, displaying a value of 8% of PM_{10}.

- **Non-exhaust**: 29 estimates for the generic non-exhaust source were found (only 24 times together with exhaust). At rural sites, contribution to PM_{10} was found on average 3.75% (3.5%−4%) versus 16% of exhaust (Fig. 2.7). At suburban locations, only two studies revealed about 10% contributions for both PM_{10} and $PM_{2.5}$ (approximately $1.8 \mu g/m^3$). At urban sites, 6% contribution (1%−14%) versus 10% contribution (1%−16%) was found in PM_{10} and 5% (1%−8%) versus 17% (6%−37%) in $PM_{2.5}$. At roadside, only two studies were found, calculating a contribution of 42% and 8% (vs. 23% from exhaust) in PM_{10} (Fig. 2.7) and 45% and 5% (vs. 32%) in $PM_{2.5}$.

The macroscale variability of non-exhaust emissions is mostly driven by climate. Road dust suspension was the only source that could be evaluated, being the number of studies was large enough to span well over different climatic conditions. To better visualize the climate dependence, we only compared PM_{10} among urban and roadside sites (Table 2.3).

TABLE 2.3   Average Share of PM$_{10}$ Explained by Road Dust and Exhaust Factor in Urban and Roadside Sites According to Climate

	Urban		Roadside	
	Road Dust	**Vehicle Exhaust**	**Road Dust**	**Vehicle Exhaust**
Continental (8)	16% (9%−29%)	24% (11%−62%)	14% (13%−16%)	27% (18%−36%)
Oceanic (7)	13% (6%−21%)	22% (6%−37%)	19% (9%−28%)	Only 1 study (13%)
Mediterranean (18)	19% (7%−34%)	22% (9%−64%)	23% (11%−35%)	21% (19%−27%)
Temperate (11, mostly in China)	29% (7%−59%)	16% (7%−40%)	na	na
Tropical monsoon (8, in India)	29% (1%−51%)	34% (18%−58%)	na	na

The lowest contributions of road dust relatively to exhaust emissions were found in oceanic urban sites, where five studies found a mean road dust contribution of 13% versus 22% from exhaust (ratio, 0.60). Slightly higher ratios (0.66) were found in urban continental sites with 16% versus 24% mean contributions. The importance of road dust increases considerably in Mediterranean and tropical monsoon climates where ratios increase to 0.85 and even above 1 at traffic sites. The worst scenario for road dust emissions was found in China where road dust contributions exceed exhaust ones with a ratio of 1.8.

3.3 Other Metrics

Although regulated air quality metrics only refer to particle mass, there is an increasing concern on PN concentration (PNC), due to the capability of finer particles to penetrate into the alveolar system and into the blood system. For brake wear particles, Garg et al. (2000) conducted dynamometer tests and found the highest number of emitted particles to lie into diameters smaller than 30 nm. This is in agreement with Mathissen et al. (2011), who exhibited a bimodal PN distribution with a nucleation mode at 10 nm and a second mode between 30 and 50 nm. Other studies found that, despite the negligible generation of small wear particles (<500 nm) at low rotor temperatures, the concentration of nanoparticles smaller than 100 nm significantly increases with the increase of the cast iron disc temperature (up to 340°C) (Kukutschová et al., 2011; Nosko & Olofsson, 2017). They proposed that submicron particles are rather formed by the evaporation/condensation process, with subsequent aggregation of primary nanoparticles, than by an abrasive type of wear. Riediker et al. (2008) tested pad materials of six different passenger cars under controlled environmental conditions and found a bimodal PN distribution with peaks at 80 nm (depending on the tested car and braking behavior) and at 200−400 nm. They found that, compared with normal deceleration, full stops result in higher nanoparticle production. More details can be found in Chapter 8.

Three studies were found investigating the size distribution of tire wear PN. Dall'Osto et al. (2014) found a bimodal particle distribution below 100 nm with modes at about 35 nm and 85 nm. Unimodal (70−90 nm) and bimodal (<10 and 30−60 nm) number size distributions in the nanosize range have been reported for tire particles under low- and high-speed conditions, respectively (Mathissen et al., 2011). Dahl et al. (2006) found an enrichment of tire particles between 15 and 50 nm using a road simulator. They identified two types of particle: one comprising mineral oils from the softening filler, and the other of sootlike agglomerates from the carbon-reinforcing filler material. Kreider et al. (2010) characterized the physical and chemical properties of particles generated from the interaction of tires and road surfaces. However, these particles were distributed (in number) in the very large coarse mode, spanning from 4−6 μm to 265−280 μm, with the mode centered at approximately 50−75 μm. Kreider et al. (2010) found, interestingly, that the mode of PN size distribution of RWPs was at 11−12 microns. Dahl et al. (2006) investigated PN size distributions of the aerosol generated at different speeds with studded and nonstudded tires on quartzite and granite pavements, finding in all cases maxima concentration below 50 nm. More details are available in Chapters 8−11.

Concerning the contribution of non-exhaust emissions to PN concentrations, four studies were found investigating non-exhaust contribution to PN concentration (Gu et al., 2011; Harrison, Beddows, & Dall'Osto, 2011; Liu et al., 2016; Sowlat, Hasheminassab, & Sioutas, 2016). All four sites were urban. The particle size range analyzed was 3 nm−10 μm, 15 nm−20 μm, 15 nm−20 μm, and 14 nm−10 μm, respectively. As expected, the non-exhaust contribution is negligible compared with that from exhaust. Brake wear and road dust were found to contribute within 1%−5% of PNC, while exhaust contributed 54%−67%.

Given the high metal content on brake and tire materials, and the strict integrated pollution prevention and control regulation on industrial emissions, non-exhaust sources are becoming one of the largest source of some metals and metalloids in urban air. Using tunnel measurements could be useful to compare only traffic-related emission; Fabretti et al. (2009), for example, calculated that resuspension and vehicular abrasion contributed 43% and 36% to the global emission of metals in $PM_{2.5}$ in a tunnel, while combustion, only 21%. Also Lawrence et al. (2013) assessed source contribution through metal concentrations and found resuspension to be the major contributor, followed by the sum of exhaust, brake, and tire wear and road wear. Visser et al. (2015) identified three factors, namely resuspended dust, brake wear, and traffic related (composed almost exclusively from Fe) that account for the majority of the total trace elements mass, together with sea/road salt.

4. CONCLUSIONS

Our literature review on the impact of non-exhaust emissions on air quality resulted in 99 peer-reviewed articles who estimated contributions of at least one of the non-exhaust sources, providing a total of 256 estimates due to the fact that several studies analyzed >1 PM fraction or >1 location. Most of the studies were carried out with receptor modeling, and more than half by means of PMF and performed in the last 15 years, with a significant increasing trend. Geographically, most of the studies are carried out in Europe, East Asia, and the United States. There is a dearth of studies in central and South America, Africa, Middle East, and Oceania.

We observed the lack of a common terminology and hierarchy for non-exhaust sources. Therefore, based on the majority of available studies, we proposed the following: vehicle non-exhaust emissions include "direct brake wear," "direct tire wear," "direct road wear," and "road dust suspension." However, "direct road wear" falls often within the "road

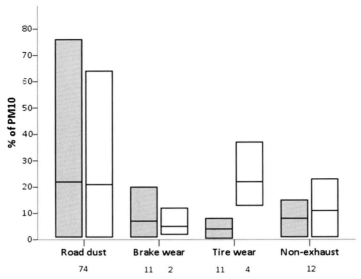

FIGURE 2.8   Range of contributions (%) to PM_{10} of different non-exhaust sources.

dust" category due to their similar chemical composition. The road dust category was the most identified source (Fig. 2.8), with slightly higher contributions range (22% as mean) than vehicle exhaust (21%) in PM_{10}, and sensibly higher than brake wear (7%) and tire wear (4%). In $PM_{2.5}$ the road dust contribution is still the highest (mean of 11% of $PM_{2.5}$) among non-exhaust ones (9% from brake wear and 2% from tire wear), but sensibly lower than vehicle exhaust (24%). This comparison allows concluding that globally, regardless of the environment studied, non-exhaust emissions are already at least as important as exhaust ones for PM_{10}, considering also that the aforementioned contributions are derived mostly from receptor modeling, thus including the contribution of secondary PM for vehicle exhaust (while they are discarded in emission inventories). For $PM_{2.5}$ they represent at least a third of total traffic contribution while they dominate the PM coarse (>1.5 µm) mode ($>80\%$).

However, there is still limited information on brake (17 estimates) and tire wear (15 estimates) contributions. Based on these few studies, there is evidence that brake wear particles include an important portion (about half of the mass) of disc wear. Backward trend analysis of source contributions and PM concentrations indicate a relative increment of the share of non-exhaust particles versus the exhaust ones.

In spite of the limited number of observations for each location category, it was possible to observe an increasing gradient of contributions (% of PM_{10}) from rural to traffic locations for all non-exhaust sources. For road dust the largest increase was found from suburban to urban location (due to the relatively coarser size distribution), indicating the need of local measures at the municipal level.

Climate dependence was also found for road dust contribution: The lowest contributions of road dust relatively to exhaust emissions were found in oceanic urban sites, where a mean road dust contribution of 13% versus 22% from exhaust was found (ratio 0.60). Slightly higher ratios (0.66) were found in urban continental sites with 16% versus 24% mean contributions. The importance of road dust increases considerably in Mediterranean and tropical monsoon climates where ratios increase to 0.85 and even above 1 at traffic sites. The worst scenario for road dust emissions was found in China where road dust contributions exceed exhaust ones with a ratio of 1.8.

Further research is necessary to better separate individual contributions from road dust resuspension, brake, and tire and road wear, given that the relative toxicity and mitigation measures are different. In this sense, valuable information can be offered by size and time-resolved PM chemical characterization and particle size distribution measurement, as well as improved source apportionment tools.

References

Aatmeeyata, Kaul, D. S., & Sharma, M. (2009). Traffic generated non-exhaust particulate emissions from concrete pavement: A mass and particle size study for two-wheelers and small cars. *Atmospheric Environment, 43*(35), 5691−5697.

Achilleos, S., Wolfson, J. M., Ferguson, S. T., Kang, C.-M., Hadjimitsis, D. G., Hadjicharalambous, M., et al. (2016). Spatial variability of fine and coarse particle composition and sources in Cyprus. *Atmospheric Research, 169*, 255−270. http://dx.doi.org/10.1016/j.atmosres.2015.10.005.

Almeida, S. M., Pio, C. A., Freitas, M. C., Reis, M. A., & Trancoso, M. A. (2005). Source apportionment of fine and coarse particulate matter in a sub-urban area at the Western European Coast. *Atmospheric Environment, 39*, 3127−3138.

Amato, F., Alastuey, A., Karanasiou, A., Lucarelli, F., Nava, S., Calzolai, G., et al. (2016). AIRUSE-LIFEC: A harmonized PM speciation and source apportionment in five southern European cities. *Atmospheric Chemistry and Physics, 16*, 3289−3309. http://dx.doi.org/10.5194/acp-16-3289-2016.

Amato, F., Alastuey, A., de la Rosa, J., Gonzalez-Castanedo, Y., Sánchez de la Campa, A. M., Pandolfi, M., et al. (2014). Trends of road dust emissions contributions on ambient air particulate levels at rural, urban and industrial sites in southern Spain. *Atmospheric Chemistry and Physics, 14*, 3533−3544.

Amato, F., Cassee, F. R., Denier van der Gon, H. A. C., Gehrig, R., Gustafsson, M., Hafner, W., et al. (2014). Urban air quality: The challenge of traffic non-exhaust emissions. *Journal of Hazardous Materials, 275*, 31−36. http://dx.doi.org/10.1016/j.jhazmat.2014.04.053.

Amato, F., Favez, O., Pandolfi, M., Alastuey, A., Querola, X., Moukhtarc, S., et al. (2016). Traffic induced particle resuspension in Paris: Emission factors and source contributions. *Atmospheric Environment, 129*, 114−124. http://dx.doi.org/10.1016/j.atmosenv.2016.01.022.

Amato, F., Pandolfi, M., Moreno, T., Furger, M., Pey, J., Alastuey, A., et al. (2011). Sources and variability of inhalable road dust particles in three European cities. *Atmospheric Environment, 45*(37), 6777−6787. http://dx.doi.org/10.1016/j.atmosenv.2011.06.003.

Amato, F., Pandolfi, M., Viana, M., Querol, X., Alastuey, A., & Moreno, T. (2009). Spatial and chemical patterns of PM_{10} in road dust deposited in urban environment. *Atmospheric Environment, 43*(9), 1650−1659. http://dx.doi.org/10.1016/j.atmosenv.2008.12.009.

Amato, F., Zandveld, P., Keuken, M., Jonkers, S., Querol, X., Rche, C., et al. (2016). Improving the modeling of road dust levels for Barcelona at urban scale and street level. *Atmospheric Environment, 125*(A), 231−242. http://dx.doi.org/10.1016/j.atmosenv.2015.10.078.

Athanasopoulou, E., Tombrou, M., Russell, A. G., Karanasiou, A., Eleftheriadis, K., & Dandou, A. (2010). Implementation of road and soil dust emission parameterizations in the aerosol model CAMx: Applications over the greater Athens urban area affected by natural sources. *Journal of Geophysical Research, 115*(D17), 301. http://dx.doi.org/10.1029/2009JD013207.

Barmpadimos, I., Keller, J., Oderbolz, D., Hueglin, C., & Prévôt, A. S. H. (2012). One decade of parallel fine ($PM_{2.5}$) and coarse ($PM_{10}−PM_{2.5}$) particulate matter measurements in Europe: Trends and variability. *Atmospheric Chemistry and Physics, 12*(7), 3189−3203. http://dx.doi.org/10.5194/acp-12-3189-2012.

Baumann, K., Jayanty, R. K. M., & Flanagan, J. B. (2008). Fine particulate matter source apportionment for the chemical speciation trends network site at Birmingham, Alabama, using positive matrix factorization. *Journal of the Air & Waste Management Association, 58*(1), 27–44. http://dx.doi.org/10.3155/1047-3289.58.1.27.

Beevers, S. D., Kitwiroon, N., Williams, M. L., Kelly, F. J., Anderson, H. R., & Carslaw, D. C. (2013). Air pollution dispersion models for human exposure predictions in London. *Journal of Exposure Science & Environmental Epidemiology, 23*, 647–653.

Begum, B. A., Kim, E., Biswas, S. K., & Hopke, P. K. (2004). Investigation of sources of atmospheric aerosol at urban and semi-urban areas in Bangladesh. *Atmospheric Environment*.

Behera, S. N., Sharma, M., Dikshit, O., & Shkula, S. P. (2011). Gis-based emission inventory, dispersion modeling, and assessment for source contributions of particulate matter in an urban environment. *Water, Air and Soil Pollution, 218*, 423–436.

Bell, M. L., Belanger, K., Ebisu, K., Gent, J. F., Lee, H. J., Koutrakis, P., et al. (2010). Prenatal exposure to fine particulate matter and birth weight: Variations by particulate constituents and sources. *Epidemiology, 21*(6), 884–891.

Beuck, H., Quass, U., Klemm, O., & Kuhlbusch, T. A. J. (2011). Assessment of sea salt and mineral dust contributions to PM_{10} in NW Germany using tracer models and positive matrix factorization. *Atmospheric Environment, 45*, 5813–5821. http://dx.doi.org/10.1016/j.atmosenv.2011.07.010.

Borm, P. J. A., Kelly, F., Kunzli, N., Schins, R. P. F., & Donaldson, K. (2007). Oxidant generation by particulate matter: From biologically effective dose to a promising, novel metric. *Occupational and Environmental Medicine, 64*(2), 73–74.

Brines, M., Dall'Osto, M., Amato, F., Minguillón, M. C., Karanasiou, A., Alastuey, A., et al. (2016). Vertical and horizontal variability of PM_{10} source contributions in Barcelona during SAPUSS. *Atmospheric Chemistry and Physics, 16*, 6785–6804. http://dx.doi.org/10.5194/acp-16-6785-2016.

Bukowiecki, N., Gehrig, R., Lienemann, P., Hill, M., Figi, R., Buchmann, B., et al. (2009). *PM_{10} emission factors of abrasion particles from road traffic (APART)*. Swiss Association of Road and Transportation Experts (VSS).

Bukowiecki, N., Lienemann, P., Hill, M., Figi, R., Richard, A., Furger, M., et al. (2009b). Real-world emission factors for antimony and other brake wear related trace elements: Size segregated values for light and heavy duty vehicles. *Environmental Science & Technology, 43*, 8072–8078.

Buzcu, B., Fraser, M. P., Kulkarni, P., & Chellam, S. (2003). Source identification and apportionment of fine particulate matter in Houston, TX, using positive matrix factorization. *Environmental Engineering Science, 20*(6), 533–545.

Cesari, D., Donateo, A., Conte, M., & Contini, D. (2016). Inter-comparison of source apportionment of PM_{10} using PMF and CMB in three sites nearby an industrial area in central Italy. *Atmospheric Research, 182*, 282–293.

Chalbot, M. C., McElroy, B., & Kavouras, I. G. (2013). Sources, trends and regional impacts of fine particulate matter in southern Mississippi valley: Significance of emissions from sources in the gulf of Mexico coast. *Atmospheric Chemistry and Physics, 13*, 3721–3732. http://dx.doi.org/10.5194/acp-13-3721-2013.

Chan, Y. C., Cohen, D. D., Hawas, O., Stelcer, E., Simpson, R., Denison, L., et al. (2008). Apportionment of sources of fine and coarse particles in four major Australian cities by positive matrix factorisation. *Atmospheric Environment, 42*(2), 374–389. http://dx.doi.org/10.1016/j.atmosenv.2007.09.030.

Chan, Y. C., Simpson, R. W., Mctainsh, G. H., Vowles, P. D., Cohen, D. D., & Bailey, G. M. (1999). Source apportionment of $PM_{2.5}$ and PM_{10} aerosols in Brisbane (Australia) by receptor modelling. *Atmospheric Environment, 33*, 3251–3268.

Chen, Y., Zheng, M., Edgerton, E. S., Ke, L., Sheng, G., & Fu, J. (2012). $PM_{2.5}$ source apportionment in the southeastern U.S.: Spatial and seasonal variations during 2001–2005. *Journal of Geophysical Research, 117*, D08304. http://dx.doi.org/10.1029/2011JD016572.

Cheng, Y., Lee, S. C., Gu, Z. L., Ho, K. F., Zhang, Y. W., Huang, Y., et al. (2015). $PM_{2.5}$ and $PM_{10-2.5}$ chemical composition and source apportionment near a Hong Kong roadway. *Particuology, 18*, 96–104. http://dx.doi.org/10.1016/j.partic.2013.10.003.

Crespi, A., Bernardoni, V., Calzolai, G., Lucarelli, F., Nava, S., Valli, G., et al. (2016). Implementing constrained multi-time approach with bootstrap analysis in ME-2: An application to $PM_{2.5}$ data from Florence (Italy). *Science of the Total Environment, 541*, 502–511. http://dx.doi.org/10.1016/j.scitotenv.2015.08.159.

Crilley, L. R., Lucarelli, F., Bloss, W. J., Harrison, R. M., Beddows, D. C., Calzolai, G., et al. (2017). Source apportionment of fine and coarse particles at a roadside and urban background site in London during the 2012 summer ClearfLo campaign. *Environmental Pollution, 220*, 766–778. http://dx.doi.org/10.1016/j.envpol.2016.06.002.

Dahl, A., Gharibi, A., Swietlicki, E., Gudmundsson, A., Bohgard, M., Ljungman, A., et al. (2006). Traffic-generated emissions of ultrafine particles from pavement-tire interface. *Atmospheric Environment, 40*(7), 1314–1323. http://dx.doi.org/10.1016/j.atmosenv.2005.10.029.

Dall'Osto, M., Beddows, D. C. S., Gietl, J. K., Olatunbosun, O. A., Yang, X., & Harrison, R. M. (2014). Characteristics of tyre dust in polluted air: Studies by single particle mass spectrometry (ATOFMS). *Atmospheric Environment, 94*, 224–230. http://dx.doi.org/10.1016/j.atmosenv.2014.05.026.

Denby, B. R., Sundvor, I., Johansson, C., Pirjola, L., Ketzel, M., Norman, M., et al. (2013). A coupled road dust and surface moisture model to predict non-exhaust road traffic induced particle emissions (NORTRIP). Part 1: Road dust loading and suspension modelling. *Atmospheric Environment, 77*, 283–300. http://dx.doi.org/10.1016/j.atmosenv.2013.04.069.

Denier van der Gon, H. A. C., Gerlofs-Nijland, M. E., Gehrig, R., Gustafsson, M., Janssen, N., Harrison, R. M., et al. (2013). The policy relevance of wear emissions from road transport, now and in the future—an international workshop report and consensus statement. *Journal of the Air & Waste Management Association, 63*(2), 136–149. http://dx.doi.org/10.1080/10962247.2012.741055.

Dongarra, G., Manno, E., & Varrica, D. (2009). Possible markers of traffic related emissions. *Environmental Monitoring and Assessment, 154*, 117–125.

Ducret-Stich, R. E., Tsai, M. Y., Thimmaiah, D., Künzli, N., Hopke, P. K., & Phuleria, H. C. (2013). PM_{10} source apportionment in a Swiss Alpine valley impacted by highway traffic. *Environmental Science and Pollution Research, 20*, 6496–6508. http://dx.doi.org/10.1007/s11356-013-1682-1.

Duong, T. T. T., & Lee, B.-K. (2011). Determining contamination level of heavy metals in road dust from busy traffic areas with different characteristics. *Journal of Environmental Management, 92*(3), 554–562. http://dx.doi.org/10.1016/j.jenvman.2010.09.010.

Duvall, R. M., Norris, G. A., Burke, J. M., Olson, D. A., Vedantham, R., & Williams, R. (2012). Determining spatial variability in $PM_{2.5}$ source impacts across Detroit, MI. *Atmospheric Environment, 47*, 491–498. http://dx.doi.org/10.1016/j.atmosenv.2011.09.071.

Fabretti, J. F., Sauret, N., Gal, J. F., Maria, P. C., & Schärer, U. (2009). Elemental characterization and source identification of $PM_{2.5}$ using positive matrix factorization: The malraux road tunnel, Nice, France. *Atmospheric Research, 94*(2), 320–329. http://dx.doi.org/10.1016/j.atmosres.2009.06.010.

Fauser, P. (1999). *Particulate air pollution with emphasis on traffic generated aerosols.* Riso, 1999: ALL.

Figi, R., Nagel, O., Tuchschmid, M., Lienemann, P., Gfeller, U., & Bukowiecki, N. (2010). Quantitative analysis of heavy metals in automotive brake linings: A comparison between wet-chemistry based analysis and insitu screening with a handheld X-ray fluorescence spectrometer. *Analytica Chimica Acta, 676*(1–2), 46–52.

Font, A., & Fuller, G. W. (2016). Did policies to abate atmospheric emissions from traffic have a positive effect in London? *Environmental Pollution, 218*, 463–474. http://dx.doi.org/10.1016/j.envpol.2016.07.026.

Furusjö, E., Sternbeck, J., & Cousins, A. P. (2007). PM_{10} source characterization at urban and highway roadside locations. *Science of the Total Environment, 387*, 206–219.

Garg, B. D., Cadle, S. H., Mulawa, P. A., & Groblicki, P. J. (2000). Brake wear particulate matter emissions. *Environmental Science & Technology, 34*, 4463–4469.

Gasser, M., Riediker, M., Mueller, L., Perrenoud, A., Blank, F., Gehr, P., et al. (2009). Toxic effects of brake wear particles on epithelial lung cells in vitro. *Particle and Fibre Toxicology, 6*(30).

GBD 2013 Risk Factors Collaborators, Forouzanfar, M. H., Alexander, L., Anderson, H. R., Bachman, V. F., Biryukov, S., et al. (2015). Global, regional, and national comparative risk assessment of 79 behavioural, environmental and occupational, and metabolic risks or clusters of risks in 188 countries, 1990–2013: A systematic analysis for the global burden of disease study 2013. *Lancet (London, England), 386*(10010), 2287–2323. http://dx.doi.org/10.1016/S0140-6736(15)00128-2.

Gehrig, R., Huglin, C., & Hofer, P. (2001). Contributions of road traffic to ambient PM_{10} and $PM_{2.5}$ concentrations. In *1st Swiss transport research conference, Monte Verita/Ascona.*

Gietl, J. K., Lawrence, R., Thorpe, A. J., & Harrison, R. M. (2010). Identification of brake wear particles and derivation of a quantitative tracer for brake dust at a major road. *Atmospheric Environment, 44*(2), 141–146. http://dx.doi.org/10.1016/j.atmosenv.2009.10.016.

Green, M. C., Antony Chen, L. W., DuBois, D. W., & Molenar, J. V. (2012). Fine particulate matter and visibility in the Lake Tahoe Basin: Chemical characterization, trends, and source apportionment. *Journal of the Air & Waste Management Association, 62*(8), 953–965. http://dx.doi.org/10.1080/10962247.2012.690362.

Grigoratos, T., & Martini, G. (2015). Brake wear particle emissions: A review. *Environmental Science and Pollution Research, 22*(4), 2491−2504. http://dx.doi.org/10.1007/s11356-014-3696-8.

Gualtieri, M., Mantecca, P., Cetta, F., & Camatini, M. (2008). Organic compounds in tire particle induce reactive oxygen species and heat-shock proteins in the humanalveolar cell line A549. *Environment International, 34*(4), 437−442.

Gualtieri, M., Rigamonti, L., Galeotti, V., & Camatini, M. (2005). Toxicity of tire debris extracts on human lung cell line A549. *Toxicology In Vitro, 19*(7), 1001−1008.

Gu, J., Pitz, M., Schnelle-Kreis, J., Diemer, J., Reller, A., Zimmermann, R., et al. (2011). Source apportionment of ambient particles: Comparison of positive matrix factorization analysis applied to particle size distribution and chemical composition data. *Atmospheric Environment, 45*(10), 1849−1857. http://dx.doi.org/10.1016/j.atmosenv.2011.01.009.

Gummeneni, S., Yusup, Y. B., Chavali, M., & Samadi, S. Z. (2011). Source apportionment of particulate matter in the ambient air of Hyderabad city, India. *Atmospheric Environment, 101*(3), 752−764. http://dx.doi.org/10.1016/j.atmosres.2011.05.002.

Gupta, A. K., Karar, K., & Srivastava, A. (2007). Chemical mass balance source apportionment of PM$_{10}$ and TSP in residential and industrial sites of an urban region of Kolkata, India. *Journal of Hazardous Materials, 142*, 279−287.

Gupta, I., Salunkhe, A., & Kumar, R. (2012). Source apportionment of PM$_{10}$ by positive matrix factorization in urban area of Mumbai, India. *The Scientific World Journal, 2012*, 585791. http://dx.doi.org/10.1100/2012/585791.

Gustafsson, M., Blomqvist, G., Gudmundsson, A., Dahl, A., Swietlicki, E., Bohgard, M., et al. (2008). Properties and toxicological effects of particles from the interaction between tyres, road pavement and winter traction material. *Science of the Total Environment, 393*(2−3), 226−240.

Guttikunda, S. K., & Calori, G. (2013). A GIS based emissions inventory at 1 km × 1 km spatial resolution for air pollution analysis in Delhi, India. *Atmospheric Environment, 67*, 101−111. http://dx.doi.org/10.1016/j.atmosenv.2012.10.040.

Guttikunda, S. K., Kopakka, R. V., Dasari, P., & Gertler, A. W. (2013). Receptor model-based source apportionment of particulate pollution in Hyderabad, India. *Environmental Monitoring and Assessment, 185*, 5585−5593. http://dx.doi.org/10.1007/s10661-012-2969-2.

Happo, M. S., Salonen, R. O., Hlinen, A. I., Jalava, P. I., Pennanen, A. S., Dor-mans, J. A. M. A., et al. (2010). Inflammation and tissue damage in mouse lung by single and repeated dosing of urban air coarse and fine particles collected from six European cities. *Inhalation Toxicology, 22*(5), 402−416.

Harrison, R. M., Beddows, D. C. S., & Dall'Osto, M. (2011). PMF analysis of wide-range particle size spectra collected on a major highway. *Environmental Science & Technology, 45*, 5522−5528. http://dx.doi.org/10.1021/es2006622.

Harrison, R. M., Jones, A. M., Gietl, J., Yin, J., & Green, D. C. (2012). Estimation of the contributions of brake dust, tire wear, and resuspension to nonexhaust traffic particles derived from atmospheric measurements. *Environmental Science & Technology, 46*(12), 6523−6529. http://dx.doi.org/10.1021/es300894r.

Harrison, R. M., Smith, D. J. T., Piou, C. A., & Castro, L. M. (1997). Comparative receptor modelling study of airborne particulate pollutants in Birmingham (United Kingdom), Coimbra (Portugal) and Lahore (Pakistan). *Atmospheric Environment, 31*(20), 3309−3321. http://dx.doi.org/10.1016/S1352-2310(97)00152-0.

Hedberg, E., Johansson, C., Johansson, L., Swietlicki, E., & Brorström-Lundén, E. (2012). Is levoglucosan a suitable quantitative tracer for wood burning? Comparison with receptor modeling on trace elements in Lycksele, Sweden. *Journal of the Air & Waste Management Association, 56*(12), 1669−1678. http://dx.doi.org/10.1080/10473289.2006.10464572.

Held, T., Ying, Q., Kleeman, M. J., Schauer, J. J., & Fraser, M. P. (2005). A comparison of the UCD/CIT air quality model and the CMB source−receptor model for primary airborne particulate matter. *Atmospheric Environment, 39*(12), 2281−2297. http://dx.doi.org/10.1016/j.atmosenv.2004.12.034.

Hjortenkrans, D. S. T., Bergback, B. G., & Haggerud, A. V. (2007). Metal emissions from brake linings and tires: Case studies of Stockholm, Sweden 1995/1998 and 2005. *Environmental Science & Technology, 41*, 5224−5230.

Huang, L., Wang, K., Yuan, C. S., & Wang, G. (2010). Study on the seasonal variation and source apportionment of PM$_{10}$ in Harbin, China. *Aerosol and Air Quality Research, 10*, 86−93. http://dx.doi.org/10.4209/aaqr.2009.04.0025.

Hulskotte, J. H. J., Roskam, G. D., & Denier van der Gon, H. A. C. (2014). Elemental composition of current automotive braking materials and derived air emission factors. *Atmospheric Environment, 99*, 436−445. http://dx.doi.org/10.1016/j.atmosenv.2014.10.007.

Iijima, A., Sato, K., Yano, K., Kato, M., Kozawa, K., & Furuta, N. (2008). Emission factor for antimony in brake abrasion dusts as one of the major atmospheric antimony sources. *Environmental Science & Technology, 42*(8), 2937−2942. http://dx.doi.org/10.1021/es702137g.

Iijima, A., Sato, K., Yano, K., Kato, M., Tago, H., Kato, M., et al. (2007). Particle size and composition distribution analysis of automotive brake abrasion dusts for the evaluation of antimony sources of airborne particulate matter. *Atmospheric Environment, 41*, 4908−4919.

Ivošević, T., Stelcer, E., Orlić, I., Bogdanović Radović, I., & Cohen, D. (2016). Characterization and source apportionment of fine particulate sources at Rijeka, Croatia from 2013 to 2015. *Nuclear Instruments and Methods in Physics Research B, 371*, 376−380. http://dx.doi.org/10.1016/j.nimb.2015.10.023.

Jeong, C. H., McGuire, M. L., Herod, D., Dann, T., Dabek−Zlotorzynska, E., Wang, D., et al. (2011). Receptor model based identification of $PM_{2.5}$ sources in Canadian cities. *Atmospheric Pollution Research, 2*(2), 158−171. http://dx.doi.org/10.5094/APR.2011.021.

Kara, M., Hopke, P. K., Dumanoglu, Y., Altiok, H., Elbir, T., Odabasi, M., et al. (2015). Characterization of PM using multiple site data in a heavily industrialized region of Turkey. *Aerosol and Air Quality Research, 15*, 11−27. http://dx.doi.org/10.4209/aaqr.2014.02.0039.

Karanasiou, A., Moreno, T., Amato, F., Lumbreras, J., Narros, A., Borge, R., et al. (2011). Road dust contribution to PM levels − evaluation of the effectiveness of street washing activities by means of Positive Matrix Factorization. *Atmospheric Environment, 45*(13), 2193−2201. http://dx.doi.org/10.1016/j.atmosenv.2011.01.067.

Karanasiou, A. A., Siskos, P. A., & Eleftheriadis, K. (2009). Assessment of source apportionment by Positive Matrix Factorization analysis on fine and coarse urban aerosol size fractions. *Atmospheric Environment, 43*(21), 3385−3395. http://dx.doi.org/10.1016/j.atmosenv.2009.03.051.

Karar, K., & Gupta, A. K. (2007). Source apportionment of PM_{10} at residential and industrial sites of an urban region of Kolkata, India. *Atmospheric Research, 84*, 30−41.

Kavouras, I. G., DuBois, D. W., Nikolich, G., & Etyemezian, V. (2015). Monitoring, source identification and health risks of air toxics in Albuquerque, New Mexico, U.S.A. *Aerosol and Air Quality Research, 15*, 556−571. http://dx.doi.org/10.4209/aaqr.2014.04.0075.

Ke, L., Liu, W., Wang, Y., Russell, A. G., Edgerton, E. S., & Zheng, M. (2008). Comparison of $PM_{2.5}$ source apportionment using positive matrix factorization and molecular marker-based chemical mass balance. *Science of the Total Environment, 394*, 290−302. http://dx.doi.org/10.1016/j.scitotenv.2008.01.030.

Kelly, F. J. (2003). Oxidative stress: Its role in air pollution and adverse health effects. *Occupational and Environmental Medicine, 60*(8), 612−616.

Keuken, M. P., Roemer, M. G. M., Zandveld, P., Verbeek, R. P., & Velders, G. J. M. (2012). Trends in primary NO_2 and exhaust PM emissions from road traffic for the period 2000−2020 and implications for air quality and health in the Netherlands. *Atmospheric Environment, 54*, 313−319. http://dx.doi.org/10.1016/j.atmosenv.2012.02.009.

Kousoulidou, M., Ntziachristos, L., Mellios, G., & Samaras, Z. (2008). Road-transport emission projections to 2020 in European urban environments. *Atmospheric Environment, 42*(32), 7465−7475. http://dx.doi.org/10.1016/j.atmosenv.2008.06.002.

Kreider, M. L., Panko, J. M., McAtee, B. L., Sweet, L. I., & Finley, B. L. (2010). Physical and chemical characterization of tire-related particles: Comparison of particles generated using different methodologies. *Science of the Total Environment, 408*(3), 652−659. http://dx.doi.org/10.1016/j.scitotenv.2009.10.016.

Kukutschová, J., Moravec, P., Tomášek, V., Matějka, V., Smolík, J., Schwarz, J., et al. (2011). On airborne nano/microsized wear particles released from low-metallic automotive brakes. *Environmental Pollution, 159*(4), 998−1006. http://dx.doi.org/10.1016/j.envpol.2010.11.036.

Kupiainen, K., Ritola, R., Stojiljkovic, A., Pirjola, L., Malinen, A., & Niemi, J. (2016). Contribution of mineral dust sources to street side ambient and suspension PM_{10} samples. *Atmospheric Environment, 147*, 178−189. http://dx.doi.org/10.1016/j.atmosenv.2016.09.059.

Kupiainen, K. J., Tervahattu, H., Räisänen, M., Mäkelä, T., Aurela, M., & Hillamo, R. (2005). Size and composition of airborne particles from pavement wear, tires, and traction sanding. *Environmental Science & Technology, 39*(3), 699−706.

Kuwayama, t., Ruehl, C. R., & Kleeman, M. J. (2013). Daily trends and source apportionment of ultrafine particulate mass ($PM_{0.1}$) over an annual cycle in a typical California city. *Environmental Science & Technology, 47*, 13957−13966. http://dx.doi.org/10.1021/es403235c.

Kwak, J. H., Kim, H., Lee, J., & Lee, S. (2013). Characterization of non-exhaust coarse and fine particles from on-road driving and laboratory measurements. *Science of the Total Environment, 458–460*, 273–282.

Lawrence, S., Sokhi, R., Ravindra, K., Mao, H., Prain, H. D., & Bull, I. D. (2013). Source apportionment of traffic emissions of particulate matter using tunnel measurements. *Atmospheric Environment, 77*, 548–557. http://dx.doi.org/10.1016/j.atmosenv.2013.03.040.

Lee, H. J., Gent, J. F., Leadere, B. P., & Koutrakis, P. (2011). Spatial and temporal variability of fine particle composition and source types in five cities of Connecticut and Massachusetts. *Science of the Total Environment, 409*, 2133–2142. http://dx.doi.org/10.1016/j.scitotenv.2011.02.025.

Lee, S., Kwak, J., Kim, H., & Lee, J. (2013). Properties of roadway particles from interaction between the tire and road pavement. *International Journal of Automotive Technology, 14*(1), 163–173.

Li, Z., Hopke, P. K., Husain, L., Qureshi, S., Dutkiewicz, V. A., Schwab, J. J., et al. (2004). Sources of fine particle composition in New York city. *Atmospheric Environment, 38*, 6521–6529. http://dx.doi.org/10.1016/j.atmosenv.2004.08.040.

Li, H., Wang, Q., Yang, M., Li, F., Wang, J., Sun, Y., et al. (2016). Chemical characterization and source apportionment of $PM_{2.5}$ aerosols in a megacity of Southeast China. *Atmospheric Research, 181*, 288–299. http://dx.doi.org/10.1016/j.atmosres.2016.07.005.

Lim, J. M., Lee, J. H., Moon, J. H., Chung, Y. S., & Kim, K. H. (2010). Source apportionment of PM_{10} at a small industrial area using Positive Matrix Factorization. *Atmospheric Research, 95*, 88–100. http://dx.doi.org/10.1016/j.atmosres.2009.08.009.

Liu, Z., Wang, Y., Hu, B., Ji, D., Zhang, J., Wu, F., et al. (2016). Source appointment of fine particle number and volume concentration during severe haze pollution in Beijing in January 2013. *Environmental Science and Pollution Research, 23*(7), 6845–6860. http://dx.doi.org/10.1007/s11356-015-5868-6.

Maenhaut, W., Vermeylen, R., Claeys, M., Vercauteren, J., & Roekens, E. (2016). Sources of the PM_{10} aerosol in Flanders, Belgium, and re-assessment of the contribution from wood burning. *Science of the Total Environment, 562*, 550–560. http://dx.doi.org/10.1016/j.scitotenv.2016.04.074.

Manoli, E., Voutsa, D., & Samara, C. (2002). Chemical characterization and source identification/apportionment of fine and coarse air particles in Thessaloniki, Greece. *Atmospheric Environment, 36*(6), 949–961. http://dx.doi.org/10.1016/S1352-2310(01)00486-1.

Manousakas, M., Diapouli, E., Papaefthymiou, H., Migliori, A., Karydas, A. G., Padilla-Alvarez, R., et al. (2015). Source apportionment by PMF on elemental concentrations obtained by PIXE analysis of PM_{10} samples collected at the vicinity of lignite power plants and mines in Megalopolis, Greece. *Nuclear Instruments and Methods in Physics Research Section B: Beam Interactions with Materials and Atoms, 349*, 114–124. http://dx.doi.org/10.1016/j.nimb.2015.02.037.

Mansha, M., Ghauri, B., Rahman, S., & Amman, A. (2012). Characterization and source apportionment of ambient air particulate matter ($PM_{2.5}$) in Karachi. *Science of the Total Environment, 425*, 176–183. http://dx.doi.org/10.1016/j.scitotenv.2011.10.056.

Mantecca, P., Farina, F., Moschini, E., Gallinotti, D., Gualtieri, M., Rohr, A., et al. (2010). Comparative acute lung inflammation induced by atmospheric PM and size-fractionated tire particles. *Toxicology Letters, 198*(2), 244–254.

Mantecca, P., Sancini, G., Moschini, E., Farina, F., Gualtieri, M., Rohr, A., et al. (2009). Lung toxicity induced by intra-tracheal instillation of size-fractionated tire particles. *Toxicology Letters, 189*(3), 206–214.

Masri, S., Kang, C. M., & Koutrakis, P. (2015). Composition and sources of fine and coarse particles collected during 2002–2010 in Boston, MA. *Journal of the Air & Waste Management Association, 65*(3), 287–297. http://dx.doi.org/10.1080/10962247.2014.982307.

Mathissen, M., Scheer, V., Vogt, R., & Benter, T. (2011). Investigation on the potential generation of ultrafine particles from the tire–road interface. *Atmospheric Environment, 45*, 6172–6179.

Mbengue, S., Alleman, L. Y., & Flament, P. (2017). Metal-bearing fine particle sources in a coastal industrialized environment. *Atmospheric Research, 183*, 202–211. http://dx.doi.org/10.1016/j.atmosres.2016.08.014.

McKenzie, E. R., Wong, C. M., Green, P. G., Kayhanian, M., & Young, T. M. (2008). Size dependent elemental composition of road-associated particles. *Science of the Total Environment, 398*(1–3), 145–153.

Mooibroek, D., Staelens, J., Cordell, R., Panteliadis, P., Delaunay, T., Weijers, E., et al. (2016). PM_{10} source apportionment in five North Western European cities—outcome of the Joaquin Project. In R. E. Hester, R. M. Harrison, & X. Querol (Eds.), *Airborne particulate Matter: Sources, atmospheric processes and health* (pp. 264–292). Royal Society of Chemistry. http://dx.doi.org/10.1039/9781782626589-00264.

Mosleh, M., Blau, P. J., & Dumitrescu, D. (2004). Characteristics and morphology of wear particles from laboratory testing of disc brake materials. *Wear, 256*, 1128–1134.

Nosko, O., & Olofsson, U. (2017). Quantification of ultrafine airborne particulate matter generated by the wear of car brake materials. *Wear, 374–375*, 92–96. http://dx.doi.org/10.1016/j.wear.2017.01.003.

Ostro, B., Tobias, A., Querol, X., Alastuey, A., Amato, F., Pey, J., et al. (2011). The effects of particulate matter sources on daily mortality: A case-crossover study of Barcelona, Spain. *Environmental Health Perspectives, 119*(12), 1781–1787.

Pant, P., & Harrison, R. M. (2013). Estimation of the contribution of road traffic emissions to particulate matter concentrations from field measurements: A review. *Atmospheric Environment, 77*, 78–97. http://dx.doi.org/10.1016/j.atmosenv.2013.04.028.

Pay, M. T., Jimenez-Guerrero, P., & Baldasano, J. M. (2011). Implementation of resuspension from paved roads for the improvement of CALIOPE air quality system in Spain. *Atmospheric Environment, 45*(3), 802–807. http://dx.doi.org/10.1016/j.atmosenv.2010.10.032.

Peel, M. C., Finlayson, B. L., & McMahon, T. A. (2007). Updated world map of the Köppen-Geiger climate classification. *Hydrology and Earth System Sciences, 11*, 1633–1644.

Perrone, M. G., Larsen, B. R., Ferrero, L., Sangiorgi, G., De Gennaro, G., Udisti, R., et al. (2012). Sources of high $PM_{2.5}$ concentrations in Milan, Northern Italy: Molecular marker data and CMB modelling. *Science of the Total Environment, 414*, 343–355. http://dx.doi.org/10.1016/j.scitotenv.2011.11.026.

Pipalatkar, P., Khaparde, V. V., Gajghate, D. G., & Bawase, M. A. (2014). Source apportionment of $PM_{2.5}$ using a CMB model for a centrally located indian city. *Aerosol and Air Quality Research, 14*, 1089–1099. http://dx.doi.org/10.4209/aaqr.2013.04.0130.

Pokorná, P., Hovorka, J., Klan, M., & Hopke, P. K. (2015). Source apportionment of size resolved particulate matter at a European air pollution hot spot. *Science of the Total Environment, 502*, 172–183. http://dx.doi.org/10.1016/j.scitotenv.2014.09.021.

Pokorná, P., Hovorka, J., Kroužek, J., & Hopke, P. K. (2013). Particulate matter source apportionment in a village situated in industrial region of Central Europe. *Journal of the Air & Waste Management Association, 63*(12), 1412–1421. http://dx.doi.org/10.1080/10962247.2013.825215.

Rexeis, M., & Hausberger, S. (2009). Trend of vehicle emission levels until 2020-Prognosis based on current vehicle measurements and future emission legislation. *Atmospheric Environment, 43*(31), 4689–4698. http://dx.doi.org/10.1016/j.atmosenv.2008.09.034.

Riediker, M., Gasser, M., Perrenoud, A., Gehr, P., & Rothen-Rutishauser, B. (June 2008). A system to test the toxicity of brake wear particles. *12th International ETH-Conference on Combustion Generated Nanoparticles*, 23–25. Zurich, Switzerland.

Rogge, W. F., Hildemann, L. M., Mazurek, M. A., Cass, G. R., & Simoneit, B. R. T. (1993). Sources of fine organic aerosol. 3. Road dust, tire debris, and organometallic brake lining dust: Roads as sources and sinks. *Environmental Science & Technology, 27*(9), 1892–1904.

Samara, C., Kouimtzis, T., Tsitouridou, R., Kanias, G., & Simeonov, V. (2003). Chemical mass balance source apportionment of PM_{10} in an industrialized urban area of Northern Greece. *Atmospheric Environment, 37*, 41–54.

Samek, L., Stegowski, Z., Furman, L., & Fiedor, J. (2017). Chemical content and estimated sources of fine fraction of particulate matter collected in Krakow. *Air Quality, Atmosphere & Health, 10*(1), 47–52. http://dx.doi.org/10.1007/s11869-016-0407-2.

Sanders, P. G., Xu, N., Dalka, T. M., & Maricq, M. M. (2003). Airborne brake wear debris: Size distributions, composition, and a comparison of dynamometer and vehicle tests. *Environmental Science & Technology, 37*, 4060–4069.

Santoso, M., Lestiani, D. D., Mukhtar, R., Hamonangan, E., Syafrul, H., Markwitz, A., et al. (2011). Preliminary study of the sources of ambient air pollution in Serpong, Indonesia. *Atmospheric Pollution Research, 2*, 190–196. http://dx.doi.org/10.5094/APR.2011.024.

Schauer, J. J., Fraser, M. P., Cass, G. R., & Simoneit, B. R. T. (2002). Source reconciliation of atmospheric gas-phase and particle-phase pollutants during a severe photochemical smog episode. *Environmental Science & Technology, 36*, 3806–3814. http://dx.doi.org/10.1021/es011458j.

Schauer, J. J., Rogge, W. F., Hildemann, L. M., Mazurek, M. A., Cass, G. R., & Simoneit, B. R. T. (1996). Source apportionment of airborne particulate matter using organic compounds as tracers. *Atmospheric Environment, 30*(22), 3837–3855.

Seneviratne, M. C. S., Waduge, V. A., Hadagiripathira, L., Sanjeewani, S., Attanayake, T., Jayaratne, N., et al. (2011). Characterization and source apportionment of particulate pollution in Colombo, Sri Lanka. *Atmospheric Pollution Research, 2*(2), 207−212. http://dx.doi.org/10.5094/APR.2011.026.

Sharma, S., & Patil, K. V. (2016). Emission scenarios and health impacts of air pollutants in Goa. *Aerosol and Air Quality Research, 16,* 2474−2487. http://dx.doi.org/10.4209/aaqr.2015.12.0664.

Sharma, S., Panwar, T. S., & Hooda, R. K. (2013). Quantifying sources of particulate matter pollution at different categories of landuse in an urban setting using receptor modelling. *Sustainable Environment Research, 23*(6), 393−402.

Song, Y., Zhang, Y., Xie, S., Zeng, L., Zheng, M., Salmon, L. G., et al. (2006). Source apportionment of $PM_{2.5}$ in Beijing by positive matrix factorization. *Atmospheric Environment, 40*(8), 1526−1537. http://dx.doi.org/10.1016/j.atmosenv.2005.10.039.

Sowlat, M. H., Hasheminassab, S., & Sioutas, C. (2016). Source apportionment of ambient particle number concentrations in central Los Angeles using positive matrix factorization (PMF). *Atmospheric Chemistry and Physics, 16*(8), 4849−4866. http://dx.doi.org/10.5194/acp-16-4849-2016.

Sowlat, M. H., Naddafi, K., Yuneslan, M., Jackson, P. L., Lotfi, S., & Shahsavani, A. (2013). PM_{10} source apportionment in Ahvaz, Iran, using positive matrix factorization. *Clean − Soil, Air, Water, 41*(12), 1143−1151.

Srimuruganandam, B., & Shiva Nagendra, S. M. (2012a). Application of positive matrix factorization in characterization of PM_{10} and $PM_{2.5}$ emission sources at urban roadside. *Chemosphere, 88,* 120−130. http://dx.doi.org/10.1016/j.chemosphere.2012.02.083.

Srimuruganandam, B., & Shiva Nagendra, S. M. (2012b). Source characterization of PM_{10} and $PM_{2.5}$ mass using a chemical mass balance model at urban roadside. *Science of the Total Environment, 433,* 8−19. http://dx.doi.org/10.1016/j.scitotenv.2012.05.082.

Srivastava, A., & Jain, V. K. (2007). Seasonal trends in coarse and fine particle sources in Delhi by the chemical mass balance receptor model. *Journal of Hazardous Materials, 144,* 283−291.

Sturtz, T. M., Adar, S. D., Gould, T., & Larson, T. V. (2014). Constrained source apportionment of coarse particulate matter and selected trace elements in three cities from the multi-ethnic study of atherosclerosis. *Atmospheric Environment, 84,* 65−77. http://dx.doi.org/10.1016/j.atmosenv.2013.11.031.

Thorpe, A. J., & Harrison, R. M. (2008). Sources and properties of non-exhaust particulate matter from road traffic: A review. *Science of the Total Environment, 400*(1−3), 270−282. http://dx.doi.org/10.1016/j.scitotenv.2008.06.007.

Tian, S. L., Pan, Y. P., & Wang, Y. S. (2016). Size-resolved source apportionment of particulate matter in urban Beijing during haze and non-haze episodes. *Atmospheric Chemistry and Physics, 16*(1). http://dx.doi.org/10.5194/acp-16-1-2016.

Tositti, L., Brattich, E., Masiol, M., Baldacci, D., Ceccato, D., Parmeggiani, S., et al. (2014). Source apportionment of particulate matter in a large city of southeastern Po Valley (Bologna, Italy). *Environmental Science and Pollution Research, 21,* 872−890.

Visser, S., Slowik, J. G., Furger, M., Zotter, P., Bukowiecki, N., Canonaco, F., et al. (2015). Advanced source apportionment of size-resolved trace elements at multiple sites in London during winter. *Atmospheric Chemistry and Physics, 15,* 11291−11309. http://dx.doi.org/10.5194/acp-15-11291-2015.

Von Uexküll, O., Skerfving, S., Doyle, R., & Braungart, M. (2005). Antimony in brake pads-a carcinogenic component? *Journal of Cleaner Production, 13*(1), 19−31. http://dx.doi.org/10.1016/j.jclepro.2003.10.008.

Wahlin, P., Berkowicz, R., & Palmgren, F. (2006). Characterisation of traffic-generated particulate matter in Copenhagen. *Atmospheric Environment, 40,* 2151−2159.

Wahlström, J., Söderberg, A., Olander, L., Jansson, A., & Olofsson, U. (2010). A pin-on-disc simulation of airborne wear particles from disc brakes. *Wear, 268,* 763−769.

Wang, Y., & Hopke, P. K. (2013). A ten−year source apportionment study of ambient fine particulate matter in San Jose, California. *Atmospheric Pollution Research, 4,* 398−404.

Wang, H., & Shooter, D. (2005). Source apportionment of fine and coarse atmospheric particles in Auckland, New Zealand. *Science of the Total Environment, 340,* 189−198.

Watson, J. G., Chow, J. C., Lowenthal, D. H., Antony Chen, L. W., Shaw, S., Edgerton, E. S., et al. (2015). $PM_{2.5}$ source apportionment with organic markers in the Southeastern Aerosol Research and Characterization (SEARCH) study. *Journal of the Air & Waste Management Association, 65*(9), 1104−1118. http://dx.doi.org/10.1080/10962247.2015.1063551.

Watson, J. G., Chow, J. C., Lu, Z., Fujita, E. M., Lowenthal, D. H., Lawson, D. R., et al. (1994). Chemical mass balance source apportionment of PM_{10} during the southern California air quality study. *Aerosol Science and Technology, 21*(1), 1−36. http://dx.doi.org/10.1080/02786829408959693.

Wei, Z., Wang, L. T., Chen, M. Z., & Zheng, Y. (2014). The 2013 severe haze over the southern Hebei, China: $PM_{2.5}$ composition and source apportionment. *Atmospheric Pollution Research, 5*, 759−768. http://dx.doi.org/10.5094/APR.2014.085.

Weinbruch, S., Worringen, A., Ebert, M., Scheuvens, D., Kandler, K., Pfeffer, U., et al. (2014). A quantitative estimation of the exhaust, abrasion and resuspension components of particulate traffic emissions using electron microscopy. *Atmospheric Environment, 99*, 175−182. http://dx.doi.org/10.1016/j.atmosenv.2014.09.075.

West, J. J., Cohen, A., Dentener, F., Brunekreef, B., Zhu, T., Armstrong, B., et al. (2016). What we breathe impacts uur health: Improving understanding of the link between air pollution and health. *Environmental Science & Technology, 50*(10), 4895−4904.

Xue, Y. H., Wu, J. H., Feng, Y. C., Dai, L., Bi, X. H., Li, X., et al. (2010). Source characterization and apportionment of PM_{10} in Panzhihua, China. *Aerosol and Air Quality Research, 10*, 367−377. http://dx.doi.org/10.4209/aaqr.2010.01.0002.

Yanosky, J. D., Tonne, C. C., Beevers, S. D., Wilkinson, P., & Kelly, F. J. (2012). Modeling exposures to the oxidative potential of PM_{10}. *Environmental Science & Technology, 46*(14), 7612−7620.

Yi, S. M., & Hwang, I. (2014). Source identification and estimation of source apportionment for ambient PM_{10} in Seoul, Korea. *Asian Journal of Atmospheric Environment, 8*(3), 115−125. http://dx.doi.org/10.5572/ajae.2014.8.3.115.

Yu, L., Wang, G., Zhang, R., Zhang, L., Song, Y., Wu, B., et al. (2013). Characterization and source apportionment of $PM_{2.5}$ in an urban environment in Beijing. *Aerosol and Air Quality Research, 13*, 574−583. http://dx.doi.org/10.4209/aaqr.2012.07.0192.

Zeng, F., Shi, G. L., Li, X., Feng, Y. C., Bi, X. H., Wu, J. H., et al. (2010). Application of a combined model to study the source apportionment of PM_{10} in Taiyuan, China. *Aerosol and Air Quality Research, 10*, 177−184. http://dx.doi.org/10.4209/aaqr.2009.09.0058.

Zhao, P., Feng, Y., Zhu, T., & Wu, J. (2006). Characterizations of resuspended dust in six cities of North China. *Atmospheric Environment, 40*, 5807−5814. http://dx.doi.org/10.1016/j.atmosenv.2006.05.026.

Zhu, D., Gilles, J. A., Etyemezian, V., Nikolich, G., & Shaw, W. J. (2015). Evaluation of the surface roughness effect on suspended particle deposition near unpaved roads. *Atmospheric Environment, 122*, 541−551. http://dx.doi.org/10.1016/j.atmosenv.2015.10.009.

CHAPTER

3

Impact on Public Health—Epidemiological Studies
A Review of Epidemiological Studies on Non-Exhaust Particles: Identification of Gaps and Future Needs

Massimo Stafoggia, Annunziata Faustini

Department of Epidemiology, Lazio Region Health Service/ASL Roma 1, Rome, Italy

OUTLINE

1. Introduction   67

2. Non-Exhaust Particles: A Brief Overview   68
   2.1 Main Sources and Components   68
   2.2 Exposure Assessment of Non-Exhaust Particulate Matter   69
     2.2.1 Short-Term Exposure   70
     2.2.2 Long-Term Exposure   70

3. Health Effects of Non-Exhaust Particles: Epidemiological Evidence   71

   3.1 Short-Term Effects   73
   3.2 Long-Term Effects   77
   3.3 Main Challenges and Novel Approaches for Estimating Effects of Particulate Matter Components   77
   3.4 Health Effects of Non-Exhaust Particles: A Summary   80

Epidemiological Study Designs for the Analysis of Air Pollution and Health   81

References   83

1. INTRODUCTION

During the last decades, exposure to air pollution, especially particulate matter (PM), has been causally linked to acute and chronic health effects (US EPA, 2009). A recent update of the Global Burden of Disease project compared 79 behavioral, environmental, and

Non-Exhaust Emissions
https://doi.org/10.1016/B978-0-12-811770-5.00003-0

67

occupational and metabolic risk factors in terms of attributable deaths and disability-adjusted life-years and ranked ambient PM air pollution exposure as the sixth leading risk factor worldwide, stationary from 1990 to 2015 (GBD 2015 Risk Factors Collaborators, 2016). The International Agency for Research on Cancer has recently classified outdoor air pollution in general, and PM in particular, as carcinogenic to humans, as there is sufficient evidence for air pollution to cause lung cancer (IARC, 2013).

Most of the studies relating air pollution to health endpoints have been conducted in urban areas, and traffic-generated pollutants have been suggested as especially relevant to human health, because of their inherent toxicity and the smaller size fraction, which facilitates the deposition of the finer particles on the respiratory tract and their translocation into the circulatory system. However, it is still unclear what is the relative contribution of exhaust particles (those directly emitted by vehicles as by-products of incomplete fuel combustion) compared with non-exhaust ones (those generated by mechanical processes or resuspended from the road surface) on human health.

In the last 10 years there has been a generalized decrease in air pollution emissions from the transport sector in most of the cities in Europe and the United States, also as a result of more stringent control measures on exhaust emissions (EEA, 2016; US EPA, 2017a,b). In contrast, no standardized actions have been adopted to reduce non-exhaust emission sources, which are projected to exceed the exhaust emissions on urban PM concentrations in the forthcoming years (Denier Van der Gon et al., 2012).

The aim of this chapter is to provide an updated overview of the epidemiologic evidence linking exposure to non-exhaust particle emissions and health effects in the general population. A short overview of the main sources and chemical components of non-exhaust PM in urban cities will be provided, and major challenges in the estimation of population exposure will be discussed. Then, epidemiological results on the health effects of non-exhaust PM will be presented, distinguishing between those originating from short-term (e.g., daily) exposures and those related to long-term (e.g., annual) exposures. The plausible biological mechanisms responsible for the health effects will be described, and suggestions for the design of future epidemiological studies will be advanced.

2. NON-EXHAUST PARTICLES: A BRIEF OVERVIEW

2.1 Main Sources and Components

Non-exhaust traffic-related particles include those generated from non-exhaust traffic sources (wear particles from brakes, tires, and road surface) and those already existing in the environment from natural sources (crustal, desert, and sea) or from humans activities (construction sites) and resuspended from traffic induced turbulence (Grigoratos & Martini, 2014). The estimation of the contribution of non-exhaust emissions to traffic-related PM concentrations in urban settings is very difficult, because studies have adopted different approaches and obtained heterogeneous results. However, it has been suggested that non-exhaust and exhaust emissions might contribute similarly to traffic-generated PM concentrations (Amato et al., 2011; Bukowiecki et al., 2009). In addition, while tailpipe emissions are decreasing because of stringent regulations and technological upgrades, there are no regulated strategies for the abatement of non-exhaust PM emission, increasing their relative contribution in the forthcoming years (Denier Van der Gon et al., 2012). It has been estimated that nearly 90% of total PM emissions from road traffic will come from non-exhaust sources by the end of the decade (EEA, 2016).

Among the non-exhaust sources, road dust resuspension provides the highest contribution to traffic non-exhaust PM_{10} emissions (between 28% and 59% by mass), whereas brake wear contribution is estimated to range between 16% and 55%, and tire wear between 5% and 30% (Grigoratos & Martini, 2014). The corresponding values with regard to PM_{10} concentrations are in the order of $1-10 \, \mu g/m^3$, lower than the $10 \, \mu g/m^3$ increase usually analyzed for health effects (Grigoratos & Martini, 2014). Although there is consensus that tailpipe emissions produce particles mainly in the fine fraction (Harrison, Jones, Gietl, Yin, & Green, 2012), several studies have shown that non-exhaust emissions contribute to both the fine and coarse modes of PM_{10}, with the coarse fraction being predominant. This is extremely relevant from the public health perspective, because recent evidence has linked exposure to the coarse fraction of PM with adverse mortality and morbidity effects (Brunekreef & Forsberg, 2005; Stafoggia et al., 2013), and natural sources of PM, such as desert dust, are partially responsible for these associations (Stafoggia et al., 2016).

Recent research on the health effects of PM has focused on its chemical composition, with the aim of identifying the constituents responsible for the highest effects and identifying their sources. It has been shown that components associated with non-exhaust sources of traffic PM might be as toxic as those associated with exhaust sources, making an accurate characterization of the chemical profile of non-exhaust PM of primary importance. Trace metals have been commonly associated to non-exhaust PM emissions. These include Zn for tire wear and Cu, Fe, Sb, Ba, Zr, Sn, and Zn for brake wear (Pant & Harrison, 2013). A few metals have been identified as tracers of sources: Zn has been identified as the inorganic marker of tires, whereas Cu, and recently the Cu:Sb ratio, has been used as a diagnostic tool to identify brake wear source of non-exhaust PM (Pant & Harrison, 2013). The chemical characterization of resuspended dust is more complicated, as it depends on many factors highly heterogeneous across locations and periods, such as characteristics of the road surface, presence of natural dust, and meteorological conditions. Several studies associated the presence of Al, Fe, Ca, and Si to road dust, but this list is far from being exhaustive (Pant & Harrison, 2013). In addition, several organic compounds, including polycyclic aromatic hydrocarbons (PAHs), have been detected in tire, brake, and road particles (Grigoratos & Martini, 2014). Despite PAHs marginal contribution to the overall PM concentrations, they are highly toxic as they are involved in carcinogenic processes. In addition, their contribution to total emissions from the transport sector has steadily increased in Europe over the last 15 years, whereas most of the trace metals and gases routinely monitored in the EU have been stable or have decreased substantially (EEA, 2016).

2.2 Exposure Assessment of Non-Exhaust Particulate Matter

Several methods have been used to estimate non-exhaust emissions from road traffic. These include measurements in tunnels/highways (Gertler et al., 2002; Wang et al., 2010) or near roadside (Westerdahl, Wang, Pan, & Zhang, 2009), twin-site studies comparing traffic and background site concentrations (Harrison, 2009; Wang et al., 2010), and receptor models of traffic PM, which indirectly "interpret measurements of physical and chemical properties taken at different times and places to infer the possible sources of excessive concentrations and to quantify the contributions from those sources" (Watson & Chow, 2007). Receptor models are based on factor analysis tools, such as chemical mass balance, principal components analysis, positive matrix factorization (PMF), multilinear engine, and UNMIX. These methods vary in the way they group PM components into different sources but are all based on the common principle that a chemically speciated sample is composed of a linear combination of contributions from a limited number of unique factors (Stanek, Sacks, Dutton, & Dubois, 2011). All these methods are highly dependent on the conditions under which measurements are taken, or on the assumptions made to derive sources from species; therefore caution is needed in generalizing the results. As a consequence, these estimates might be of limited value when interest lies in estimating the health effects of an "exposed" population. To understand the implications for epidemiological evaluations, it is necessary to distinguish between study designs focusing on short-term exposures to non-exhaust PM, which might trigger acute effects, versus long-term exposures causing chronic effects.

2.2.1 Short-Term Exposure

The hypothesis is that short-term (e.g., daily, hourly) exposure to some toxicant might trigger an acute response. These studies rely on the assumption that the exposure under investigation (e.g., concentrations of trace metals from non-exhaust PM emissions) is accurate enough to capture the temporal variability within a specific study population. As long as such variability is similar across locations, large spatial contrasts in average exposure are not of concern for the estimate of short-term effects. However, if such an exposure displays different temporal patterns across space (more variability close to roads, less in residential areas), it might introduce bias, often toward the null value, in the effect estimates. The methods previously described for the estimate of non-exhaust PM emissions are prone to exposure misclassification for short-term studies, as they rely on measurements taken over few days on specific seasons, limiting the power of the study to evaluate short-term temporal variability and its generalizability on longer periods of observations. In addition, measurements are often taken in hot spots such as tunnels, highways, and roadside, where larger short-term fluctuations are expected, compared with residential areas, where most of the population lives. Twin-site and source apportionment methods provide indirect estimates of non-exhaust PM emissions, and it is not clear to what extent such estimates reflect the temporal variability of the underlying sources of interest. Despite these limitations, source apportionment methods represent a promising tool for the design of future epidemiological analyses of short-term effects of non-exhaust PM emissions because they are able to separate out the fraction of each PM species attributable to different emission sources, allowing to estimate independent effects of each source, while accounting for the others (multiexposure models).

2.2.2 Long-Term Exposure

The hypothesis is that long-term (e.g., annual) exposure to some toxicant might contribute to health deterioration through different chronic mechanisms ultimately leading to observable adverse endpoints, such as cause-specific mortality. From the exposure assessment point of view, these studies assume that the exposure under investigation is accurate enough to capture the spatial variability within the study population. Otherwise, bias, toward or away from the null value, can be introduced in the health effect estimates because of exposure misclassification. Instead, opposite to the short-term case, differential temporal variability in exposure across locations is not an issue because exposure is averaged over longer periods (e.g., annual average). In addition, in the case of studies related to long-term effects, most of the methods applied for measuring or estimating non-exhaust PM emissions might produce errors when their results are applied as proxy for population exposure: measurements taken in hot spots do not adequately represent the annual average exposure of the population either because they are overly influenced by local sources or because they are taken on time windows (specific hours of the day, days of the week or seasons) not representative of annual averages. Similar limitations apply to source apportionment methods, when they are based on PM species measured at locations not representative of the average population exposure.

3. HEALTH EFFECTS OF NON-EXHAUST PARTICLES: EPIDEMIOLOGICAL EVIDENCE

The health effects of air pollution are well known to the scientific community since the early 20th century, when serious epidemics of mortality and respiratory diseases were linked to concurrent extraordinary peaks of air pollution (Ebelt et al., 2001; Firket, 1936; Logan, 1953; Pope, 1989; Wichmann et al., 2000). Toward the end of the last century, researchers started to focus their attention on the possible adverse health effects induced by "usual" levels of outdoor air pollutants (Brunekreef & Holgate, 2002; Dockery et al., 1993; Pope & Dockery, 1999, 2006), while clinicians acknowledged the causal role of air pollution on human health since 2000, when the American Thoracic Society (ATS) first defined the concept of an "adverse respiratory effect of air pollution" (ATS, 2000) as "… medically significant physiologic or pathologic changes (such as) (1) interference with the normal activity, (2) episodic respiratory illness, (3) incapacitating illness, (4) permanent respiratory injury and/or progressive respiratory dysfunction." The respiratory effects were firstly recognized (ATS, 2000), followed by the cardiovascular ones (Brook et al., 2010). All these scientific contributions provided a fundamental support for the definition of standardized legislation on air quality in the United States (US EPA, 1996) and in Europe (European Council, 1999; European Parliament and Council, 2008).

In the last years, researchers investigating the health effects of air pollution have achieved important methodological progresses in at least four major fields: (1) an accurate assessment of exposure (although not at the individual level yet) with an almost worldwide coverage (van Donkelaar, Martin, Brauer, & Boys, 2015); (2) a comprehensive picture of the health effects of air pollution, including cancer (Raaschou-Nielsen et al., 2016; Yorifuji & Kashima, 2013), metabolic diseases (Beyerlein et al., 2015), maternal and birth outcomes (Basu et al., 2014; Chen, Zmirou-Navier, Padilla, & Deguen, 2014; Dadvand et al., 2014; Mobasher et al., 2013), development effects (Basagaña et al., 2016), and cognitive impairment (Calderón-Garcidueñas & Villarreal-Ríos, 2017), in addition to usual respiratory and cardiovascular outcomes; (3) an updated definition of "adverse health effect of air pollution" that includes asymptomatic signs of health deterioration, such as biological effects, altered biomarkers, and reduced functions (Liu et al., 2017; Yang, Hou, Wei, Thai, & Chai, 2017); and

(4) a better comprehension of the mechanisms of action (systemic inflammation, oxidative stress, immune modulation, and epigenetic alterations) (Thurston et al., 2017).

Size-fractioned particles have been the most studied pollutants, with research focusing especially on PM $\leq$ 10 μm and PM $\leq$ 2.5 μm, while only recently the interest has spanned across PM size distribution, namely on coarse particles (PM between 2.5 and 10 μm) and ultrafine PM (particles $\leq$ 0.1 μm).

The effects of non-exhaust PM have been less explored compared with those of exhaust PM, despite the relative contribution of non-exhaust PM to the whole traffic particles has been estimated as approximately 50% (Querol et al., 2004). On 2010 the Health Effects Institute's Panel on the Health Effects of Traffic-Related Air Pollution indicated that resuspended road dust, tire wear, and brake wear contain metals and organic compounds that can cause adverse health effects and highlighted the opportunity of considering them in future assessment of motor vehicles impacts on human health (HEI Panel on the Health Effects of Traffic-Related Air Pollution, 2010). Today, the role of metals in inducing oxidative stress provides more relevance to the non-exhaust PM fraction, also in light of the new definition of "health effect" from the European Respiratory Society and the ATS (Thurston et al., 2017), which includes biological processes and functional impairments as essential steps toward symptomatic disease.

From the chemical profile of non-exhaust PM, as previously described in this chapter, it can be observed that (1) metals are the most important components of non-exhaust traffic particles, and oxidative stress is the leading biological mechanism linking non-exhaust PM to health; (2) secondary pollutants, including the inorganic ones, are relevant components of non-exhaust PM because, despite the presumed low toxicity of sulfates and nitrates, they can influence the bioavailability of other components such as metals and enhance their toxicity (Schlesinger & Cassee, 2003; WHO, 2013); (3) although fine particles ($PM_{2.5}$) from road dust represent the most important risk factor to human health, larger particles such as the coarse ones might display a causal role on human health (Kelly & Fussel, 2012).

Typical study designs to investigate acute effects of short-term exposure are time series (Katsouyanni et al. 1997; Samet, Dominici, Curriero, Coursac, & Zeger, 2000) and case-crossover (Janes et al., 2005; Levy, Lumley, Sheppard, Kaufman, & Checkoway, 2001; Lu & Zeger, 2007; Maclure, 1991) approaches, where short-term variations in exposures are related to concurrent variations in health outcomes within a reference population. For long-term exposure, typical studies to investigate long-term effects or life expectancy reduction include cross-sectional or longitudinal approaches, where it is assumed that subpopulations exposed to higher-than-average air pollution levels experience higher mortality/morbidity rates than less exposed subjects, after accounting for differences across subgroups.

More details about the epidemiological study designs are reported in a special section at the end of this chapter: "Epidemiological study designs for the analysis of air pollution and health."

Another possible classification of PM-induced effects distinguishes between changes in health indicators within the general population (e.g., cause-specific mortality, morbidity affecting different organs and apparatus) (see Fig. 3.1); changes in vital functions, such as heart rate (Gold et al., 2000), respiratory volumes (Gehring et al., 2013), and glomerular filtration (Lue, Wellenius, Wilker, Mostofsky, & Mittleman, 2013) (see Fig. 3.1); and early biological effects, i.e., damages of the organs due to either reversible or irreversible impairments at the cellular level. Early biological effects include changes of epigenetic markers such as DNA

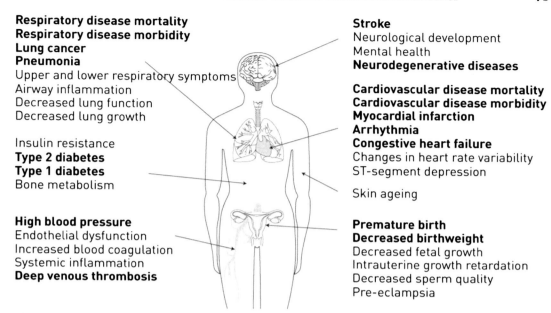

Respiratory disease mortality
Respiratory disease morbidity
Lung cancer
Pneumonia
Upper and lower respiratory symptoms
Airway inflammation
Decreased lung function
Decreased lung growth

Insulin resistance
Type 2 diabetes
Type 1 diabetes
Bone metabolism

High blood pressure
Endothelial dysfunction
Increased blood coagulation
Systemic inflammation
Deep venous thrombosis

Stroke
Neurological development
Mental health
Neurodegenerative diseases

Cardiovascular disease mortality
Cardiovascular disease morbidity
Myocardial infarction
Arrhythmia
Congestive heart failure
Changes in heart rate variability
ST-segment depression

Skin ageing

Premature birth
Decreased birthweight
Decreased fetal growth
Intrauterine growth retardation
Decreased sperm quality
Pre-eclampsia

FIVGURE 3.1   Overview of diseases, conditions, and biomarkers affected by outdoor air pollution. *Acknowledgement Wording: Reproduced with permission from the European Respiratory Society* ©. European Respiratory Journal Jan 2017, 49 (1) 1600419; DOI: 10.1183/13993003.00419-2016.

methylation (Baccarelli et al., 2009) and preclinic damages such as enzyme changes (Thomas, Hodgson, Nieuwenhuijsen, & Jarup, 2009), inflammation (Li, Rittenhouse-Olson, Scheider, & Mu, 2012), and oxidative stress biomarkers (Li et al., 2013; Rücker et al., 2006).

The following sections present a summary of the epidemiological studies on the PM effects reported in people exposed to traffic and the early biologic effects related to the metals mostly identified as tracers of non-exhaust PM. Separate results will be provided for short-term and long-term effects.

3.1 Short-Term Effects

Only few epidemiological studies were able to detect associations between short-term exposure to traffic-related PM components or sources and mortality (either nonaccidental or cause-specific). Most of the studies investigated individual components of PM in relation to health endpoints and argued about the potential role of non-exhaust sources of traffic post hoc. However, few studies tried to derive daily contributions of non-exhaust sources to PM components by using source apportionment methods and applied these estimates to epidemiological studies.

Concerning the first group of studies (those on individual PM components), they provided consistent results from different parts of the world, such as the United States (Ito et al., 2011; Krall, Anderson, Dominici, Bell, & Peng, 2013; Lippmann, 2014; Lippmann, Chen, Gordon, Ito, & Thurston, 2013; Zhou, Ito, Lall, Lippmann, & Thurston, 2011), Chile (Valdés et al., 2012), Europe (Basagaña et al., 2015), and Asia (Son, Lee, Kim, Jung, & Bell, 2012;

Ueda, Yamagami, Ikemori, Hisatsune, & Nitta, 2016). Specifically, metals and secondary pollutants related to non-exhaust PM emissions, such as Cu, Zn, Fe, S, nitrates, and sulfates, were associated with several mortality outcomes (Lippmann, Ito, Hwang, Maciejczyk, & Chen, 2006; Ostro, Feng, Broadwin, Green, & Lipsett, 2007; Ostro, Lipsett et al., 2011; Ostro, Tobias et al., 2011). A more robust evidence emerges on the association between mortality and Ni or V (Bell, Ebisu, Peng, Samet, & Dominici, 2009; Bell & HEI Health Review Committee, 2012; Cao, Xu, Xu, Chen, & Kan, 2012; Franklin, Koutrakis, & Schwartz, 2008; Lippmann et al., 2006) whose presence in the atmosphere is more likely related to exhaust PM (Lippmann et al., 2006). A recent study from China identified K+, Ca++, sulfates, and nitrates present in both $PM_{2.5}$ and PM_{10} to be significantly associated with mortality increments ranging between 0.3% and 1.0% per unit increases in PM compounds and identified resuspended road dust as one of their traffic sources (Li et al., 2015).

Studies on the associations between morbidity outcomes (e.g., hospitalizations) and PM components are as well in the epidemiological literature (Basagaña et al., 2015; Bell et al., 2009, 2014; Li et al., 2015; Lippmann, 2014; Peng et al., 2009; Samoli et al., 2016; Sun, Qiua, Ho, & Tian, 2016; Zanobetti, Franklin, Koutrakis, & Schwartz, 2009). Cardiovascular diseases have been associated with exposure to V and Zn, whereas respiratory exacerbations were found to be related with exposures to Ni, V, Si, and Ti (Bell et al., 2014). Other studies using emergency admissions as morbidity endpoints found associations with elemental carbon and organic carbon (OC) (Peng et al., 2009) and associations between As, Br, Cr, Ni, and OC with cardiac diseases and diabetes (Zanobetti et al., 2009).

Concerning the second group of the few studies that investigated the PM contribution from non-exhaust sources of traffic and daily health outcomes, Ostro, Lipsett et al. (2011) and Ostro, Tobias et al. (2011) applied source apportionment techniques on PM_{10} and $PM_{2.5}$ filters in Barcelona to estimate daily contributions from eight sources, including road dust. They applied the case-crossover design to estimate the association between sources and all-cause and cardiovascular mortality. They found a significant association between road dust $PM_{2.5}$ and all-cause deaths, with mortality increasing by ~4% per 1.8 $\mu g/m^3$ increase in daily road dust contribution to $PM_{2.5}$. The estimate was similar to those found for vehicle exhaust and traffic sources (Ostro, Lipsett et al., 2011; Ostro, Tobias et al., 2011). Similarly, Bell et al. (2014) applied a time series analysis on four counties in Connecticut to investigate the relationship between daily $PM_{2.5}$ sources, as derived from PMF methods, and cardiovascular and respiratory hospitalizations. Of the five sources identified (motor vehicles, road dust/crustal materials, oil combustion, sea salt, and regional sources related to emissions from power plants and other urban areas), road dust was the one mostly associated with hospitalization risk, with excess risks per 1.71 $\mu g/m^3$ increases in road dust equal to 4.51% (95% confidence interval [CI]: 3.30, 6.01% at lag 1) for respiratory admissions and 2.11% (1.09, 3.15% at lag 0) for cardiovascular admissions. Similar results were found by Gent et al. (2009) in New Haven, US, and by Pun et al. (2017) in Hong Kong. They both analyzed source-specific PM daily concentrations in relation to morbidity endpoints, and identified resuspended dust as a significant risk factor for insurgence of symptoms in asthmatic children and respiratory admissions in the general population, respectively.

A few studies found consistent evidence of short-term effects from specific metals, which had been identified as tracer of non-exhaust PM. These include Cu and Zn, which have been identified as possible tracers of brake and tire wears; they have been also found to be significantly associated with oxidative potential in both PM_{10} and bulk road dust (Pant et al., 2015). In addition, they are able to induce a direct production of reactive oxygen species, similarly to other metals with two stable adjacent valence (Fe, Cr, Ni, V), and they can reduce glutathione (an important antioxidant) through indirect mechanisms, similarly to other trace metals such as Cd, Ni, and Pb (Valko, Morris, & Cronin, 2005).

A summary of the main studies investigating the short-term effects of components usually related to traffic non-exhaust PM is reported in Table 3.1.

TABLE 3.1   Summary of the Main Studies[a] Investigating the Short-Term Effects due to Components of Traffic-Related Non-Exhaust Particulate Matter (PM)

Year	First Author, Journal	Setting	Outcomes	Exposures	Main Findings[b] on Non-Exhaust PM-Related Metals	Comments Reported in the Paper
2017	Pun, Atmos Envir	Hong Kong, 2001-2008	Respiratory admissions	19 components of PM10, 8 sources		Resuspended mineral dust
2016	Sun, Env Int	Hong Kong, 1998—2007	Admissions for diabetes, age 65+	17 components of PM_{10}		No association found
2016	Samoli, Occup Environ Med	London, UK, 2011—12	CVD and RESP hospitalizations	5 components of PM_{10} and $PM_{2.5}$	Al, Cu, Zn	
2015	Basagana, Env Int	5 Mediterranean cities, Europe	Cause-specific mortality and admissions	16 components of PM_{10} and $PM_{2.5}$	Ca, Cu, Fe, Mn, Zn, Ni, V, Ti	
2015	Li, Environ Sci Pollut Res Int	Beijing, China, 2005—09	Mortality and Morbidity	9 components of $PM_{2.5}$	Ca++	Resuspended road dust may play a key role
2014	Lippmann, Crit Rev Toxicol	REVIEW	—	$PM_{2.5}$ components	Cu	
2013	Lippmann Res Rep Health Eff Inst	NPACT Initiative, US	—	$PM_{2.5}$ components		
2012	Valdes, Environ Health	Santiago, Chile, 1998—2007	Cause-specific mortality	16 components of $PM_{2.5}$	Cu, Cr, Zn	Cu and Cr showed stronger associations with respiratory and COPD mortality; Zn with cerebrovascular disease

(*Continued*)

TABLE 3.1   Summary of the Main Studies[a] Investigating the Short-Term Effects due to Components of Traffic-Related Non-Exhaust Particulate Matter (PM)—cont'd

Year	First Author, Journal	Setting	Outcomes	Exposures	Main Findings[b] on Non-Exhaust PM-Related Metals	Comments Reported in the Paper
2012	Son, Environ Health Perspect	Seoul, South Korea, 2008–09	All-cause mortality	10 components of $PM_{2.5}$	Mg	
2012	Bell, Res Rep Health Eff Inst	REVIEW of US studies	Morbidity	$PM_{2.5}$ components	Ni, V	
2011	Ostro, Environ Health Perspect	Barcelona, Spain, 2003–07	All-cause and CVD mortality	26 components of $PM_{2.5}$, 8 sources		Road dust
2011	Ito, Environ Health Perspect	New York City, US, 2000–06	CVD mortality, CVD morbidity	10 components of $PM_{2.5}$		
2011	Zhou, Environ Health Perspect	Seattle and Detroit, US, 2002–04	All-cause mortality, CVD, RESP	10 components of $PM_{2.5}$		
2009	Gent, Environ Health Perspect	New Haven County, US, 2000-2004	Symptoms and drug use in asthmatic children	17 components of PM2.5, 6 sources		Road dust associated with shortness of breath
2009	Peng, Environ Health Perspect[c]	119 US communities, 2000–06	CVD and RESP admissions	7 components of $PM_{2.5}$		
2009	Bell, Am J Respir Crit Care Med[c]	106 US counties, 1999–2005	CVD and RESP admissions	20 components of $PM_{2.5}$	Ni, V	
2009	Zanobetti, Environ Health	26 US communities	Admissions for CVD, MI, CHF, RESP, DIAB	18 components of $PM_{2.5}$	Cr, Ni	
2008	Franklin, Epidemiology	25 US communities	All-cause mortality	18 components of $PM_{2.5}$	Al, Ni, S, Si	
2007	Ostro, Environ Health Perspect	6 California counties, US, 2000–03	All-cause mortality, CVD, RESP	19 components of $PM_{2.5}$	K, Cu, Zn	
2006	Lippmann, Environ Health Perspect	NMMAPS cities, US, and Honk Kong	All-cause mortality	16 components of $PM_{2.5}$	Ni, V	

CHF, congestive heart failure; *CVD*, cardiovascular; *DIAB*, diabetes; *MI*, myocardial infarction; *RESP*, respiratory.

[a]*Studies investigating single components are not reported in this review.*

[b]*Findings attributed to non-exhaust PM include only direct contributions from traffic-related sources (brake, tire, road surface). The indirect contributions from resuspended particles are not included.*

[c]*Peng et al. (2009) and Bell et al. (2009) studied county-specific annual concentrations of $PM_{2.5}$ components as effect modifiers of the short-term association between $PM_{2.5}$ and cardiorespiratory hospitalizations.*

3.2 Long-Term Effects

Long-term effects of traffic-related PM components have been extensively investigated in the last few years (Basu et al., 2014; Beelen et al., 2015; Eeftens et al., 2014; Fuertes et al., 2014; Hampel et al., 2015; Ostro et al., 2015; Ostro, Lipsett et al., 2011; Ostro, Tobias et al., 2011; Pedersen et al., 2016; Raaschou-Nielsen et al., 2016; Wang et al., 2017).

The PIAMA study (Bilenko et al., 2015) reported associations of specific components such as Fe, Si, and K with increases in blood pressure among children, which the authors attributed to non-exhaust traffic emissions. A study including 11 European cohorts (Wolf et al., 2015) identified potassium (K) present in both $PM_{2.5}$ and PM_{10} as strongly associated to increases in the incidence of coronary events. The authors acknowledge that K can be considered a tracer of biomass combustion in addition to soil and sea salt. However, they interpret K effects as traffic effects from resuspension of road dust rather than biomass burning effects. In contrast, another study including 19 cohorts from the ESCAPE and TRANSPHORM projects (Wang et al., 2014) did not find any change in cardiovascular mortality as a consequence of long-term exposure to $PM_{2.5}$ or PM_{10} components, including Cu, Fe, K, Ni, S, V, and Zn.

Very recent studies succeeded to identify significant associations between long-term exposure to traffic-related non-exhaust PM and health effects, by combining chemical components of PM to their sources (Basagaña et al., 2016; Crichton et al., 2016; Dadvand et al., 2014; Desikan et al., 2016; Thurston et al., 2016; Tonne et al., 2016).

In a study conducted in London and aimed to quantify the effects of long-term exposure to traffic PM among MI survivors, Tonne et al. (2016) estimated the highest mortality increase in relation to non-exhaust PM_{10} (HR = 1.05; 95% CI = 1.00, 1.10), oxidant gases NO_2 and O_3 (HR = 1.05; 95% CI = 1.00, 1.09), and the coarse fraction of PM (HR = 1.05; 95% CI = 1.00, 1.10), per interquartile range increases of 1.1 $\mu g/m^3$, 3.2 $\mu g/m^3$, and 0.9 $\mu g/m^3$, respectively.

A recent study conducted in Spain (Basagaña et al., 2016) has identified nine sources of $PM_{2.5}$ concentrations, among which three were traffic-related (road dust, secondary sulfates, and secondary nitrates). The authors investigated the relationship between the three traffic-related sources and reductions in cognitive growth in children: they estimated 22% (95% CI: 2%, 42%) reduction of the annual change in working memory, 30% (95% CI: 6%, 54%) reduction of the annual change in superior working memory, and 11% (95% CI: 0%, 22%) reduction of the annual change in the inattentiveness scale. None of the other $PM_{2.5}$ sources was associated with adverse effects on cognitive development.

A summary of the main studies investigating the long-term effects due to components of traffic-related non-exhaust PM is reported in Table 3.2.

3.3 Main Challenges and Novel Approaches for Estimating Effects of Particulate Matter Components

The main challenges in the interpretation of component-specific results are the following ones: (1) the often high correlations between total PM mass and specific PM components do not allow to disentangle the effects of individual species, making even harder to identify possible interventions to reduce emission sources on the basis of the components' levels and their estimated effects on human health; (2) the individual components often share different sources (e.g., traffic-related exhaust and non-exhaust emissions). Source apportionment

TABLE 3.2 Summary of the Main Studies[a] Investigating the Long-Term Effects due to Components of Traffic-Related Non-Exhaust Particulate Matter (PM)

Year	First Author, Journal	Setting	Outcomes	Exposures	Main Findings[b] on Non-Exhaust PM-Related Metals	Comments Reported in the Paper
2017	Wang, Epidemiology	13 million subjects US	Mortality	$PM_{2.5}$ mass; 11 components as effect modifiers	Ca, Cu, Fe	Calcium effect attributed to soil or road dust
2016	Raaschou-Nielsen, Env Int	14 EU cohorts ESCAPE	Lung cancer	8 components of PM_{10} and $PM_{2.5}$	$PM_{2.5}$ Cu; PM_{10} K, Ni, S, Zn	
2016	Pedersen, Environ Health Perspect	8 EU cohorts ESCAPE	Newborn's size different measures	8 components of PM_{10} and $PM_{2.5}$	Ni, S, Zn	
2016	Tonne, Int J Hyg Environ Health	London, UK, 2003–10 cohort	Mortality and admission in MI survivors	Exhaust and non-exhaust primary PM_{10} and $PM_{2.5}$		Largest associations between non-exhaust PM_{10} and mortality
2016	Basagana, Environ Health Perspect	Barcelona, Spain, cohorts of children	Differences in cognitive growth trajectories	9 sources of $PM_{2.5}$ via source apportionment		
2016	Crichton, Sci Tot Environ	London, UK, 2005–12	Stroke incidence	Exhaust and non-exhaust PM_{10} and $PM_{2.5}$		
2016	Desikan, Stroke	London, UK, 2005–12	5-year survival in post-stroke patients	Exhaust and non-exhaust PM_{10} and $PM_{2.5}$		
2016	Thurston, Environ Health Perspect	100 US metropolitan areas, 2000–05	IHD mortality	16 components of $PM_{2.5}$. Sources via APCA		
2015	Bilenko, Env Int	PIAMA cohort, the Netherlands	Blood pressure at age 12	8 components of PM_{10} and $PM_{2.5}$	Fe, Si, and K in PM_{10}	Fe suggests non-exhaust sources
2015	Wolf, Epidemiology	11 EU cohorts ESCAPE	Incidence coronary events	8 components of PM_{10} and $PM_{2.5}$	Fe, Si, and K in PM_{10}	Fe suggests non-exhaust sources
2015	Hampel, Env Int	5 EU cohorts ESCAPE	Biomarkers CRP Fibrinogen	8 components of PM_{10} and $PM_{2.5}$	$PM_{2.5}$ Cu, Zn; PM_{10} Fe	
2015	Beelen, Environ Health Perspect	19 EU cohorts ESCAPE	Natural mortality	8 components of PM_{10} and $PM_{2.5}$	$PM_{2.5}$ S	
2015	Ostro, Environ Health Perspect	California cohorts, US, 2001–07	All-cause, CVD, IHD, RESP mortality in women	CTM from 900 sources		

TABLE 3.2 Summary of the Main Studies[a] Investigating the Long-Term Effects due to Components of Traffic-Related Non-Exhaust Particulate Matter (PM)—cont'd

Year	First Author, Journal	Setting	Outcomes	Exposures	Main Findings[b] on Non-Exhaust PM-Related Metals	Comments Reported in the Paper
2014	Dadvand, Occ Env Med	Barcelona, Spain, 2003-2005	Preeclampsia	8 sources of PM_{10} and $PM_{2.5}$ via PMF		PM10 brake dust associated with preeclampsia
2014	Eeftens, Epidemiology	5 EU birth cohorts ESCAPE	Lung function in children	8 components of PM_{10} and $PM_{2.5}$	PM_{10} Ni and S	PM_{10} Ni and PM_{10} S
2014	Fuertes, Int J Hyg Environ Health	7 EU birth cohorts ESCAPE	Early-life pneumonia	8 components of PM_{10} and $PM_{2.5}$	PM_{10} Zn	Non-tailpipe traffic emissions
2014	Wang, Env Int	19 EU cohorts ESCAPE	CVD mortality	8 components of PM_{10} and $PM_{2.5}$		
2014	Basu, Environ Res	646,296 births in California, US, 2000–06	Birth weight	23 components of $PM_{2.5}$	Cu, Fe, Mn, S, Ti, V, Zn	
2011	Ostro, Environ Health Perspect	California cohorts, US, 2001–07	Natural and cause-specific mortality	8 components of $PM_{2.5}$	Fe, Si, Zn	Some associations of constituents from crustal origin

CRP, c-reactive protein; *CTM*, chemical transport model; *CVD*, cardiovascular; *IHD*, ischemic heart disease; *MI*, myocardial infarction; *RESP*, respiratory.
[a]*Studies investigating single components are not reported in this review.*
[b]*Findings attributed to non-exhaust PM include only direct contributions from traffic-related sources (brake, tire, road surface). The indirect contributions from resuspended particles are not included.*

methods can be extremely useful in these cases; however, they are known to be prone to uncertainty degrees in the source identification and quantification; (3) inside out, many components might coexist within the same PM source influencing each other (e.g., sulfates, nitrates) or their characteristics (e.g., bioavailability of metals). This might give rise to biased interpretations of their respective causal role.

Recent methodological approaches have tried to deal with these challenges. In 2012, Bell and HEI (2012) started a HEI-funded project aimed at (1) characterizing how the chemical composition of $PM_{2.5}$ varied regionally and seasonally in the United States and (2) evaluating whether the seasonal and regional associations between short-term exposure to PM total mass and health effects could be explained by regional and seasonal variations in the chemical composition of $PM_{2.5}$, described in terms of 20 different species. The authors estimated short-term associations between daily $PM_{2.5}$ total mass concentrations and morbidity/mortality outcomes over multiple locations and seasons and then adopted an "effect modification" approach where individual PM components were used to capture differences in regional and seasonal associations.

A different approach, aimed as well at disentangling effects of specific components from those of total PM mass, was suggested by Mostofsky et al. (2012). The authors proposed a method where individual PM components were first regressed against PM total mass, and then the residuals of these models (i.e., the variation in constituent levels independent of $PM_{2.5}$) were associated to daily mortality/morbidity counts. The estimated coefficient for constituent residuals in the health outcome model can be interpreted as "the increase in risk associated with higher levels of the constituent while holding $PM_{2.5}$ constant, that is, higher levels of the constituent (and other constituents that travel with it) and lower levels of other constituents that make up total $PM_{2.5}$ mass" (Mostofsky et al., 2012).

Finally, in 2015 a project was funded by the Health Effects Institute with the declared aim of developing statistical methods for multipollutant research (HEI, 2015). Specifically, in the first part of the report, Coull and colleagues developed a new analysis framework based on methods that simultaneously quantify variability in health outcomes and exposure data for multiple pollutants to identify the mixture profiles (groupings of pollutants and concentrations) most highly associated with the health outcomes (Coull et al., 2015). The second part of the report, instead, developed enhanced statistical methods to jointly assess source factors and health effects using multivariate source characterization and source apportionment models together with a health outcomes analysis. In addition, the investigators' approach incorporated the uncertainty in the source apportionments into the estimation of the source-related health effects (Park, Symanski, Han, & Spiegelman, 2015).

3.4 Health Effects of Non-Exhaust Particles: A Summary

Recent epidemiological studies relating exposure to non-exhaust particles from road dust and health outcomes suggested the possibility for both short-term and long-term health effects. Time series and case-crossover studies estimated increases of daily mortality or hospital admissions of around 4%—5% per ~2 $\mu g/m^3$ increases in road dust contributions to $PM_{2.5}$ concentrations. Such estimates are similar in magnitude to those found for exhaust particles. However, they are based on few studies and locations and are difficult to generalize in other settings. Effects of long-term exposure to non-exhaust particles have been reported in cohort studies; however, the evidence is still inconsistent. Therefore, the plausibility for adverse health effects of long-term exposure to non-exhaust traffic-related sources remain an open issue to be investigated in future studies. Finally, there are no studies showing either short-term or long-term effects of non-exhaust particles different from road dust resuspension, such as tire and break abrasion.

It would be desirable that future campaigns on measurements of non-exhaust PM for health impact evaluations are designed to minimize any sources of bias, possibly by collecting samples in hot spots as well as background sites over multiple seasons and years, to better capture daily, seasonal, and long-term temporal trends in population exposure across wider spatial domains. In addition, data on specific PM components should be complemented, whenever possible, with estimates of their emission sources, via source apportionment techniques. Such database would provide an invaluable contribution to jointly estimate short-term and long-term health effects associated to non-exhaust sources of traffic-related air pollution.

Finally, while the three methodological approaches previously described to catch the effects of single sources (effect modification, residual analysis, and multipollutant modeling) are still quite new and unexplored, we think they can provide an important contribution in the forthcoming years to better elucidate the independent and synergistic health effects of individual components and sources of the air mixture, including those directly related to non-exhaust traffic-related emissions.

EPIDEMIOLOGICAL STUDY DESIGNS FOR THE ANALYSIS OF AIR POLLUTION AND HEALTH

Time series

A time series is most commonly a sequence of discrete time data where observations are taken at successive equally spaced points in time. In air pollution epidemiology, time series analyses are conducted when exposure is measured or estimated for each time unit (for example, day) homogeneously across the study area as well as health outcomes such as the daily counts of natural-cause mortality in the same area (ecological approach), so that temporal variability in exposure is related to temporal variability in outcome. Confounding induced by time-varying covariates (population dynamics, meteorology, etc.) is adjusted for by multivariate regression modeling. In addition, since the interest is in the short-term (within few days) effect, and important confounder adjusted for in time series analysis is long-term and seasonality time trend, this is added to the regression model as a nonlinear term for day (spline, sinusoidal, etc.). The purpose of this term is to remove confounding from all known or unknown variables that might display seasonal or secular trends correlated with trends in both exposure and outcome.

The big advantage of the time series analysis is that it is a quasiexperimental study design, because the exposed population (the whole studied population on a day) is compared with itself over different time points (proximal days). As a consequence, confounding by individual and time-fixed (or slowly varying) characteristics (e.g., age, gender, smoking, diet, comorbidities) is automatically removed, provided that the nonlinear term for time trend is specified in a flexible enough way.

Case-crossover

The case-crossover design was first introduced by Maclure in 1991 to investigate the association between transient exposures and the onset of an acute event, myocardial infarction (Maclure, 1991). The underlying idea is that, when the effect of an exposure is immediate and reversible, the observed cases can be matched with themselves on past periods when the studied outcome had not occurred. Then the exposure on the case day and the exposures on case-free (control) days are contrasted to derive an estimate of the effect of interest. This approach has the advantage of preventing confounding by individual time-fixed characteristics because comparisons are done within subjects. In addition, slowly varying characteristics are controlled by design, provided that control periods are chosen close enough to the case day. Starting from the end of the 1990s, the case-crossover design has been applied also in environmental epidemiology to estimate the short-term (daily) effects of air pollutants and temperature on mortality and morbidity outcomes. It has been shown that, when the

Continued

exposure of interest displays seasonal and long-term time trends, selecting control periods only before the outcome produces bias in the health effect estimates. Therefore, several approaches have been proposed to select controls bidirectionally, either symmetrically before and after the event of interest or randomly within predefined time windows around the case period. The most used strategy for the selection of control periods in case-crossover designs is the "time-stratified" approach (Levy et al., 2001), where control periods are chosen as the same weekdays of the case period within the same month and year. This strategy has been shown to provide unbiased effect estimates under a range of different scenarios of temporal trends in exposure and outcome (Janes et al., 2005).

On 2007, Lu and Zeger (2007) showed that the case-crossover design produces algebraically identical effect estimates as those of time series analyses, if the selection of controls in the former approach resembles the adjustment of time trend in the latter. For example, the "time-stratified" strategy (see above) is analogous to model time trend in time series analysis with a three-way interaction between year, month, and day of the week.

In general, time series analyses are more flexible than case-crossover designs to model time trend, but case-crossover designs are better suited to analyze individual characteristics as effect modifiers, or are the only possible option when the daily exposure of interest is not common to all subject (average across the study area) but is heterogeneous over space (for example, spatiotemporal estimates of PM concentrations from atmospheric or statistical models).

Cohort studies and survival analysis

A cohort is a group of people who share a common exposure within a defined period and whose effect on health may occur even much later. Then, a cohort study is a particular form of longitudinal study that enrolls a cohort and follows it up over time until the occurrence of a specified outcome, end of the study, or lost to follow up. In air pollution epidemiology, cohort studies have been conducted to investigate the relationship between long-term exposure to air pollution and the occurrence of mortality or morbidity endpoints. In environmental epidemiology, exposure is usually assessed at the residential address of cohort participants by land-use regression, dispersion models, or hybrid approaches. Exposure can be "fixed" (when attributed only once at baseline or as a long-term average), or time-dependent (when assessed repeatedly over the follow-up years for each individual). Individuals are then followed over time until the outcome or "censoring", e.g., end of the study or loss to follow-up. Each subject contributes some "person-time" of observation. At any point in time, the "risk set" is made of all subjects at risk of incurring the study outcome at that time, while the cases are the events occurring at that time. The ratio between these two quantities is the "hazard rate," i.e., the instantaneous rate of mortality/morbidity. The "hazard ratio" is the ratio between hazard rates corresponding to different exposure values. In the analysis of long-term effects of air pollution, it is assumed that the risk of mortality/morbidity is higher for increasing levels of long-term exposure at the residence, i.e., higher rates in areas with higher pollution levels. To test this hypothesis, conventional approaches of survival analysis (analysis of time-to-event data) are applied. The most used survival model is the "Cox proportional-hazards model," which assumes that the effect of a unit increase in the exposure is multiplicative with respect to the hazard rate, and it poses no assumption on the shape of the hazard function. In the analysis of the association between long-term exposure to air pollution and mortality/morbidity outcomes, the individual hazard rates are regressed against exposure attributed at the residential address of each subject, while adjusting for individual-level covariates (age, sex, lifestyle variables, comorbidities, etc.) and area-level indicators of socioeconomic status.

Cohort studies require the availability of a cohort of subjects with extensive information at the individual level, including details of the residential addresses and a careful estimation of the exposure to each address. Despite the advantage of being longitudinal (so preserving the temporal sequence exposure ["cause"]/outcome ["effect"]), these studies are not free from confounding induced by unknown or unmeasured individual covariates, and causal interpretation of the results should be taken with caution.

References

Amato, F., Pandolfi, M., Moreno, T., Furger, M., Pey, J., Alastuey, A., et al. (2011). Sources and variability of inhalable road dust particles in three European cities. *Atmospheric Environment, 45,* 6777−6787.

American Thoracic Society. (2000). What constitutes an adverse health effect of air pollution? Official statement of the American Thoracic Society. *American Journal of Respiratory and Critical Care Medicine, 161,* 665−673.

Baccarelli, A., Wright, R. O., Bollati, V., Tarantini, L., Litonjua, A. A., Suh, H. H., et al. (2009). Rapid DNA methylation changes after exposure to traffic particles. *American Journal of Respiratory and Critical Care Medicine, 179,* 572−578.

Basagaña, X., Esnaola, M., Rivas, I., Amato, F., Alvarez-Pedrerol, M., Forns, J., et al. (2016). Neurodevelopmental deceleration by urban fine particles from different emission sources: A longitudinal observational study. *Environmental Health Perspectives, 124,* 1630−1636.

Basagaña, X., Jacquemin, B., Karanasiou, A., Ostro, B., Querol, X., Agis, D., et al. (2015). Short-term effects of particulate matter constituents on daily hospitalizations and mortality in five south-european cities: Results from the MED-particles project. *Environment International, 75,* 151−158.

Basu, R., Harris, M., Sie, L., Malig, B., Broadwin, R., & Green, R. (2014). Effects of fine particulate matter and its constituents on low birth weight among full-term infants in California. *Environmental Research, 128,* 42−51.

Beelen, R., Hoek, G., Raaschou-Nielsen, O., Stafoggia, M., Andersen, Z. J., Weinmayr, G., et al. (2015). Natural-cause mortality and long-term exposure to particle components: An analysis of 19 European cohorts within the multicenter ESCAPE project. *Environmental Health Perspectives, 123,* 525−533.

Bell, M. L., Ebisu, K., Leaderer, B. P., Gent, J. F., Lee, H. J., Koutrakis, P., et al. (2014). Associations of $PM_{2.5}$ constituents and sources with hospital admissions: Analysis of four counties in Connecticut and Massachusetts (USA) for persons $\geq$ 65 years of age. *Environmental Health Perspectives, 122,* 138−144.

Bell, M. L., Ebisu, K., Peng, R. D., Samet, J. M., & Dominici, F. (2009). Hospital admissions and chemical composition of fine particle air pollution. *American Journal of Respiratory and Critical Care Medicine, 179,* 1115−1120.

Bell, M. L., & HEI Health Review Committee. (2012). Assessment of the health impacts of particulate matter characteristics. *Research Report - Health Effects Institute, 161,* 5−38.

Beyerlein, A., Krasmann, M., Thiering, E., Kusian, D., Markevych, I., D'Orlando, O., et al. (2015). Ambient air pollution and early manifestation of type 1 diabetes. *Epidemiology, 26,* e31−e32.

Bilenko, N., Brunekreef, B., Beelen, R., Eeftens, M., de Hoogh, K., Hoek, G., et al. (2015). Associations between particulate matter composition and childhood blood pressure — the PIAMA study. *Environment International, 84,* 1−6.

Brook, R. D., Rajagopalan, S., Pope, C. A., III, Brook, J. R., Bhatnagar, A., Diez-Roux, A. V., et al. (2010). Particulate matter air pollution and cardiovascular disease: An update to the scientific statement from the American Heart association. *Circulation, 121,* 2331−2378.

Brunekreef, B., & Forsberg, B. (2005). Epidemiological evidence of effects of coarse airborne particles on health. *The European Respiratory Journal, 26,* 309−318.

Brunekreef, B., & Holgate, S. T. (2002). Air pollution and health. *Lancet, 360,* 1233−1242.

Bukowiecki, N., Gehrig, R., Lienemann, P., Hill, M., Figi, R., Buchmann, B., et al. (2009). *PM_{10} emission factors of abrasion particles from road traffic (APART)*. Swiss Association of Road and Transportation Experts (VSS).

Calderón-Garcidueñas, L., & Villarreal-Ríos, R. (2017). Living close to heavy traffic roads, air pollution, and dementia. *Lancet, 389*, 675—677. https://doi.org/10.1016/S0140-6736(16)32596-X. pii: S0140—6736(16)32596-X.

Cao, J., Xu, H., Xu, Q., Chen, B., & Kan, H. (2012). Fine particulate matter constituents and cardiopulmonary mortality in a heavily polluted Chinese city. *Environmental Health Perspectives, 120*, 373—378.

Chen, E. K., Zmirou-Navier, D., Padilla, C., & Deguen, S. (2014). Effects of air pollution on the risk of congenital anomalies: A systematic review and meta-analysis. *International Journal of Environmental Research and Public Health, 11*, 7642—7668.

Coull, B. A., Bobb, J. F., Wellenius, G. A., Kioumourtzoglou, M.-A., Mittleman, M. A., Koutrakis, P., et al. (2015). Part 1. Statistical learning methods for the effects of multiple air pollution constituents. In *Development of statistical methods for multipollutant research. Research report 183*. Boston, MA: Health Effects Institute.

Crichton, S., Barratt, B., Spiridou, A., Hoang, U., Liang, S. F., Kovalchuk, Y., et al. (2016). Associations between exhaust and non-exhaust particulate matter and stroke incidence by stroke subtype in South London. *Science of The Total Environment, 568*, 278—284.

Dadvand, P., Ostro, B., Amato, F., Figueras, F., Minguillón, M., Martinez, D., et al. (2014). Particulate air pollution and preeclampsia: a source-based analysis. *Occup Environ Med, 71*, 570—577.

Denier Van der Gon, H., Jozwicka, M., Cassee, F., Gerlofs-Nijland, M., Gehrig, R., Gustafsson, M., et al. (2012). *The policy relevance of wear emissions from road transport, now and in the future*. TNO report. TNO-060-UT-2012—20100732.

Desikan, A., Crichton, S., Hoang, U., Barratt, B., Beevers, S. D., Kelly, F. J., et al. (2016). Effect of exhaust- and nonexhaust-related components of particulate matter on long-term survival after stroke. *Stroke, 47*, 2916—2922.

Dockery, D. W., Pope, C. A., Xu, X., Spengler, J. D., Ware, J. H., Fay, M. E., et al. (1993). An association between air pollution and mortality in six U.S. cities. *The New England Journal of Medicine, 329*, 1753—1759.

van Donkelaar, A., Martin, R. V., Brauer, M., & Boys, B. L. (2015). Use of satellite observations for long-term exposure assessment of global concentrations of fine particulate matter. *Environmental Health Perspectives, 123*, 135—143.

Ebelt, S., Brauer, M., Cyrys, J., Tuch, T., Kreyling, W. G., Wichmann, H. E., et al. (2001). Air quality in post-unification Erfurt, East Germany: Associating changes in pollutant concentrations with changes in emissions. *Environmental Health Perspectives, 109*, 325—333.

EEA (European Environmental Agency). (2016). *Air quality in Europe — 2016 report*. Luxembourg: Publications Office of the European Union.

Eeftens, M., Hoek, G., Gruzieva, O., Mölter, A., Agius, R., Beelen, R., et al. (2014). Elemental composition of particulate matter and the association with lung function. *Epidemiology, 25*, 648—657.

European Council. (1999). Council Directive 1999/30/EC of 22 April 1999 relating to limit values for sulphur dioxide, nitrogen dioxide and oxides of nitrogen, particulate matter and lead in ambient air. *Official Journal of the European Communities L, 163*, 41—60.

European Parliament and Council. (2008). Directive 2008/50/EC of the european parliament and of the council of 21 May 2008 on ambient air quality and cleaner air for Europe. *Off. J. Eur. Union L, 152*, 1—44.

Firket, J. (1936). Fog along the Meuse valley. *Transactions of the Faraday Society, 32*, 1192—1196.

Franklin, M., Koutrakis, P., & Schwartz, J. (2008). The role of particle composition on the association between $PM_{2.5}$ and mortality. *Epidemiology, 19*, 680—689.

Fuertes, E., MacIntyre, E., Agius, R., Beelen, R., Brunekreef, B., Bucci, S., et al. (2014). Associations between particulate matter elements and early-life pneumonia in seven birth cohorts: Results from the ESCAPE and TRANSPHORM projects. *International Journal of Hygiene and Environmental Health, 218*, 656—665.

GBD 2015 Risk Factors Collaborators. (2016). Global, regional, and national comparative risk assessment of 79 behavioural, environmental and occupational, and metabolic risks or clusters of risks, 1990—2015: A systematic analysis for the Global Burden of Disease study 2015. *Lancet, 388*, 1659—1724.

Gehring, U., Gruzieva, O., Agius, R. M., Beelen, R., Custovic, A., Cyrys, J., et al. (2013). Air pollution exposure and lung function in children: The ESCAPE project. *Environmental Health Perspectives, 121*, 1357—1364.

Gent, J. F., Koutrakis, P., Belanger, K., Triche, E., Holford, T. R., Bracken, M. B., et al. (2009). Symptoms and medication use in children with asthma and traffic-related sources of fine particle pollution. *Environ Health Perspect, 117*, 1168—1174.

Gertler, A. W., Gillies, J. A., Pierson, W. R., Rogers, C. F., Sagebiel, J. C., Abu-Allaban, M., et al. (2002). *Real-world particulate matter and gaseous emissions from motor vehicles in a highway tunnel*. Research report 107. Health Effects Institute.

Gold, D. R., Litonjua, A., Schwartz, J., Lovett, E., Larson, A., Nearing, B., et al. (2000). Ambient pollution and heart rate variability. *Circulation, 101*, 1267–1273.

Grigoratos, T., & Martini, G. (2014). *Non-exhaust traffic related emissions. Brake and tyre wear PM. Literature review.* JRC science and policy reports. Luxembourg: Publications Office of the European Union.

Hampel, R., Peters, A., Beelen, R., Brunekreef, B., Cyrys, J., de Faire, U., et al. (2015). Long-term effects of elemental composition of particulate matter on inflammatory blood markers in European cohorts. *Environment International, 82*, 76–84.

Harrison, R. M. (2009). Airborne particulate matter from road traffic: Current status of knowledge and research challenges paper presented at the ETTAO09. In *17th Transport and Air Pollution Symposium and 3rd Environment and Transport Symposium.*

Harrison, R. M., Jones, A. M., Gietl, J., Yin, J., & Green, D. C. (2012). Estimation of the contributions of brake dust, tire wear, and resuspension to non-exhaust traffic particles derived from atmospheric measurements. *Environmental Science & Technology, 46*, 6523–6529.

Health Effects Institute. (2015). *Development of statistical methods for multipollutant research.* Research report 183, Parts 1 & 2. Boston, MA: Health Effects Institute.

HEI Panel on the Health Effects of Traffic-Related Air Pollution (Health Effects Institute). (2010). *Traffic-related air pollution: A critical review of the literature on emissions, exposure, and health effects.* Special report n.17. Health Effects Institute.

IARC (International Agency for Research on Cancer). (2013). *Air pollution and cancer.* Geneva, Switzerland: WHO Press, World Health Organization.

Ito, K., Mathes, R., Ross, Z., Nádas, A., Thurston, G., & Matte, T. (2011). Fine particulate matter constituents associated with cardiovascular hospitalizations and mortality in New York City. *Environmental Health Perspectives, 119*, 467–473.

Janes, H., Sheppard, L., & Lumley, T. (2005). Case-crossover analyses of air pollution exposure data: Referent selection strategies and their implications for bias. *Epidemiology, 16*, 717–726.

Katsouyanni, K., Touloumi, G., Spix, C., Schwartz, J., Balducci, F., Medina, S., et al. (1997). Short-term effects of ambient sulphur dioxide and particulate matter on mortality in 12 European cities: Results from time series data from the APHEA project. Air pollution and health: A European approach. *Bmj: British Medical Journal, 314*, 1658–1663.

Kelly, F. J., & Fussel, J. C. (2012). Size, source and chemical composition as determinants of toxicity attributable to ambient particulate matter. *Atmospheric Environment, 60*, 504–526.

Krall, J. R., Anderson, G. B., Dominici, F., Bell, M. L., & Peng, R. D. (2013). Short-term exposure to particulate matter constituents and mortality in a national study of U.S. urban communities. *Environmental Health Perspectives, 121*, 1148–1153.

Levy, D., Lumley, T., Sheppard, L., Kaufman, J., & Checkoway, H. (2001). Referent selection in case-crossover analyses of acute health effects of air pollution. *Epidemiology, 12*, 186–192.

Li, Y., Nie, J., Beyea, J., Rudra, C. B., Browne, R. W., Bonner, M. R., et al. (2013). Exposure to traffic emissions: Associations with biomarkers of antioxidant status and oxidative damage. *Environmental Research, 121*, 31–38.

Li, Y., Rittenhouse-Olson, K., Scheider, W. L., & Mu, L. (2012). Effect of particulate matter air pollution on C-reactive protein: A review of epidemiologic studies. *Reviews on Environmental Health, 27*, 133–149.

Li, P., Xin, J., Wang, Y., Li, G., Pan, X., Wang, S., et al. (2015). Association between particulate matter and its chemical constituents of urban air pollution and daily mortality or morbidity in Beijing City. *Environmental Science and Pollution Research International, 22*, 358–368.

Lippmann, M. (2014). Toxicological and epidemiological studies of cardiovascular effects of ambient air fine particulate matter ($PM_{2.5}$) and its chemical components: Coherence and public health implications. *Critical Reviews in Toxicology, 44*, 299–347.

Lippmann, M., Chen, L., Gordon, T., Ito, K., & Thurston, G. D. (2013). National particle component toxicity (NPACT) initiative: Integrated epidemiologic and toxicologic studies of the health effects of particulate matter components. *Research Report - Health Effects Institute, 177*, 5–13.

Lippmann, M., Ito, K., Hwang, J. S., Maciejczyk, P., & Chen, L. C. (2006). Cardiovascular effects of nickel in ambient air. *Environmental Health Perspectives, 114*, 1662–1669.

Liu, L., Urch, B., Szyszkowicz, M., Speck, M., Leingartner, K., Shutt, R., et al. (2017). Influence of exposure to coarse, fine and ultrafine urban particulate matter and their biological constituents on neural biomarkers in a randomized controlled crossover study. *Environment International.* https://doi.org/10.1016/j.envint.2017.01.010. pii: S0160–4120(16)30759-0.

Logan, W. P. (1953). Mortality in the London fog incident. *Lancet, 1*, 336—338.

Lu, Y., & Zeger, S. L. (2007). On the equivalence of case-crossover and time series methods in environmental epidemiology. *Biostatistics, 8*, 337—344.

Lue, S. H., Wellenius, G. A., Wilker, E. H., Mostofsky, E., & Mittleman, M. A. (2013). Residential proximity to major roadways and renal function. *Journal of Epidemiology and Community Health, 67*, 629—634.

Maclure, M. (1991). The case-crossover design: A method for studying transient effects on the risk of acute events. *American Journal of Epidemiology, 133*, 144—153.

Mobasher, Z., Salam, M. T., Goodwin, T. M., Lurmann, F., Ingles, S. A., & Wilson, M. L. (2013). Associations between ambient air pollution and hypertensive disorders of pregnancy. *Environmental Research, 123*, 9—16.

Mostofsky, E., Schwartz, J., Coull, B. A., Koutrakis, P., Wellenius, G. A., Suh, H. H., et al. (2012). Modeling the association between particle constituents of air pollution and health outcomes. *American Journal of Epidemiology, 176*, 317—326.

Ostro, B., Feng, W. Y., Broadwin, R., Green, S., & Lipsett, M. (2007). The effects of components of fine particulate air pollution on mortality in California: Results from CALFINE. *Environmental Health Perspectives, 115*, 13—19.

Ostro, B., Hu, J., Goldberg, D., Reynolds, P., Hertz, A., Bernstein, L., et al. (2015). Associations of mortality with long-term exposures to fine and ultrafine particles, species and sources: Results from the California Teachers study cohort. *Environmental Health Perspectives, 123*, 549—556.

Ostro, B., Lipsett, M., Reynolds, P., Goldberg, D., Hertz, A., Garcia, C., et al. (2011). Long-term exposure to constituents of fine particulate air pollution and mortality: Results from the California Teachers study. *Environmental Health Perspectives, 118*, 363—369.

Ostro, B., Tobias, A., Querol, X., Alastuey, A., Amato, F., Pey, J., et al. (2011). The effects of particulate matter sources on daily mortality: A case-crossover study of Barcelona, Spain. *Environmental Health Perspectives, 119*, 1781—1787.

Pant, P., Baker, S. J., Shukla, A., Maikawa, C., Godri Pollitt, K. J., & Harrison, R. M. (2015). The PM_{10} fraction of road dust in the UK and India: Characterization, source profiles and oxidative potential. *Science of The Total Environment, 530—531*, 445—452.

Pant, P., & Harrison, R. M. (2013). Estimation of the contribution of road traffic emissions to particulate matter concentrations from field measurements: A review. *Atmospheric Environment, 77*, 78—97.

Park, E. S., Symanski, E., Han, D., & Spiegelman, C. (2015). Part 2. Development of enhanced statistical methods for assessing health effects associated with an unknown number of major sources of multiple air pollutants. In *Development of statistical methods for multipollutant research. Research report 183*. Boston, MA: Health Effects Institute.

Pedersen, M., Gehring, U., Beelen, R., Wang, M., Giorgis-Allemand, L., Andersen, A. M., et al. (2016). Elemental constituents of particulate matter and Newborn's size in eight European cohorts. *Environmental Health Perspectives, 124*, 141—150.

Peng, R. D., Bell, M. L., Geyh, A. S., McDermott, A., Zeger, S. L., Samet, J. M., et al. (2009). Emergency admissions for cardiovascular and respiratory diseases and the chemical composition of fine particle air pollution. *Environmental Health Perspectives, 117*, 957—963.

Pope, C. A., III (1989). Respiratory disease associated with community air pollution and a steel mill, Utah valley. *American Journal of Public Health, 79*, 623—628.

Pope, C. A., III, & Dockery, D. W. (1999). Epidemiology of particle effects. In S. T. Holgate, H. Koren, R. Maynard, & J. Samet (Eds.), *Air pollution and health* (pp. 673—705). London, England: Academic Press.

Pope, C. A., III, & Dockery, D. W. (2006). Health effects of fine particulate air pollution: Lines that connect. *Journal of the Air & Waste Management Association, 56*, 709—742.

Pun, C. V., Tian, L., & Ho, K. (2017). Particulate matter from re-suspended mineral dust and emergency cause-specific respiratory hospitalizations in Hong Kong. *Atmos Environ, 165*, 191—197.

Querol, X., Alastuey, A., Ruiz, C. R., Artiñano, B., Hansson, H. C., Harrison, R. M., et al. (2004). Speciation and origin of PM_{10} and $PM_{2.5}$ in selected European cities. *Atmospheric Environment, 38*, 6547—6555.

Raaschou-Nielsen, O., Beelen, R., Wang, M., Hoek, G., Andersen, Z. J., Hoffmann, B., et al. (2016). Particulate matter air pollution components and risk for lung cancer. *Environment International, 87*, 66—73.

Rückerl, R., Ibald-Mulli, A., Koenig, W., Schneider, A., Woelke, G., Cyrys, J., et al. (2006). Air pollution and markers of inflammation and coagulation in patients with coronary heart disease. *American Journal of Respiratory and Critical Care Medicine, 173*, 432—441.

Samet, J. M., Dominici, F., Curriero, F. C., Coursac, I., & Zeger, S. L. (2000). Fine particulate air pollution and mortality in 20 U.S. cities, 1987—1994. *The New England Journal of Medicine, 343*, 1742—1749.

Samoli, E., Atkinson, R. W., Analitis, A., Fuller, G. W., Green, D. C., Mudway, I., et al. (2016). Associations of short-term exposure to traffic-related air pollution with cardiovascular and respiratory hospital admissions in London, UK. *Occupational and Environmental Medicine, 73*, 300–307.

Schlesinger, R. B., & Cassee, F. (2003). Atmospheric secondary inorganic particulate matter: The toxicological perspective as a basis for health effects risk assessment. *Inhalation Toxicology, 15*, 197–235.

Son, J. Y., Lee, J. T., Kim, K. H., Jung, K., & Bell, M. L. (2012). Characterization of fine particulate matter and associations between particulate chemical constituents and mortality in Seoul, Korea. *Environmental Health Perspectives, 120*, 872–878.

Stafoggia, M., Samoli, E., Alessandrini, E., Cadum, E., Ostro, B., Berti, G., et al. (2013). Short-term associations between fine and coarse particulate matter and hospitalizations in southern Europe: Results from the MED-PARTICLES project. *Environmental Health Perspectives, 121*, 1026–1033.

Stafoggia, M., Zauli-Sajani, S., Pey, J., Samoli, E., Alessandrini, E., Basagaña, X., et al. (2016). Desert dust outbreaks in southern Europe: Contribution to daily PM_{10} concentrations and short-term associations with mortality and hospital admissions. *Environmental Health Perspectives, 124*, 413–419.

Stanek, L. W., Sacks, J. D., Dutton, S. J., & Dubois, J. J. B. (2011). Attributing health effects to apportioned components and sources of particulate matter: An evaluation of collective results. *Atmospheric Environment, 45*, 5655–5663.

Sun, S., Qiua, H., Ho, K., & Tian, L. (2016). Chemical components of respirable particulate matter associated with emergency hospital admissions for type 2 diabetes mellitus in Hong Kong. *Environment International, 97*, 93–99.

Thomas, L. D., Hodgson, S., Nieuwenhuijsen, M., & Jarup, L. (2009). Early kidney damage in a population exposed to cadmium and other heavy metals. *Environmental Health Perspectives, 117*, 181–184.

Thurston, G. D., Burnett, R. T., Turner, M. C., Shi, Y., Krewski, D., Lall, R., et al. (2016). Ischemic heart disease mortality and long-term exposure to source-related components of U.S. Fine particle air pollution. *Environmental Health Perspectives, 124*, 785–794.

Thurston, G. D., Kipen, H., Annesi-Maesano, I., Balmes, J., Brook, R. D., Cromar, K., et al. (2017). A joint ERS/ATS policy statement: What constitutes an adverse health effect of air pollution? An analytical framework. *The European Respiratory Journal, 49*(1). https://doi.org/10.1183/13993003.00419-2016. pii: 1600419.

Tonne, C., Halonen, J. I., Beevers, S. D., Dajnak, D., Gulliver, J., Kelly, F. J., et al. (2016). Long-term traffic air and noise pollution in relation to mortality and hospital readmission among myocardial infarction survivors. *International Journal of Hygiene and Environmental Health, 219*, 72–78.

Ueda, K., Yamagami, M., Ikemori, F., Hisatsune, K., & Nitta, H. (2016). Associations between fine particulate matter components and daily mortality in Nagoya, Japan. *Journal of Epidemiology, 26*, 249–257.

US EPA. (1996). *Air quality criteria for particulate matter.* Washington, DC: U.S. Environmental Protection Agency. EPA/600/P-95/001Cf.

US EPA. (2009). *Final Report: Integrated science assessment for particulate matter.* Washington, DC: U.S. Environmental Protection Agency. EPA/600/R-08/139F.

US EPA (United States Environmental Protection Agency). (2017a). *Air pollutant emissions trends data.* Available at https://www.epa.gov/air-emissions-inventories/air-pollutant-emissions-trends-data.

US EPA (United States Environmental Protection Agency). (2017b). *Emission standards reference guide for on-road and nonroad vehicles and engines.* Available at https://www.epa.gov/emission-standards-reference-guide.

Valdés, A., Zanobetti, A., Halonen, J. I., Cifuentes, L., Morata, D., & Schwartz, J. (2012). Elemental concentrations of ambient particles and cause specific mortality in Santiago, Chile: A time series study. *Environmental Health: a Global Access Science Source [electronic Resource], 11*, 82.

Valko, M., Morris, H., & Cronin, M. T. (2005). Metals, toxicity and oxidative stress. *Current Medicinal Chemistry, 12*, 1161–1208.

Wang, F., Ketzel, M., Ellermann, T., Wahlin, P., Jensen, S. S., Fang, D., et al. (2010). Particle number, particle mass and NOx emission factors at a highway and an urban street in Copenhagen. *Atmospheric Chemistry and Physics, 10*, 2745–2764.

Wang, M., Beelen, R., Stafoggia, M., Raaschou-Nielsen, O., Andersen, Z. J., Hoffmann, B., et al. (2014). Long-term exposure to elemental constituents of particulate matter and cardiovascular mortality in 19 European cohorts: Results from the ESCAPE and TRANSPHORM projects. *Environment International, 66*, 97–106.

Wang, Y., Shi, L., Lee, M., Liu, P., Di, Q., Zanobetti, A., et al. (2017). Long-term exposure to $PM_{2.5}$ and mortality among older adults in the Southeastern US. *Epidemiology, 28*, 207–214.

Watson, J. G., & Chow, J. C. (2007). Receptor models for source apportionment of suspended particles. In B. Murphy, & R. Morrison (Eds.), *Introduction to environmental forensics* (2nd ed., vol. 2, pp. 279–316). New York, NY: Academic Press.

Westerdahl, D., Wang, X., Pan, X., & Zhang, M. K. (2009). Characterization of on-road vehicle emission factors and microenvironmental air quality in Beijing, China. *Atmospheric Environment, 43,* 697–705.

Wichmann, H. E., Spix, C., Tuch, T., Wölke, G., Peters, A., Heinrich, J., et al. (2000). *Daily mortality and fine and ultrafine particles in Erfurt, Germany. Part I: Role of particle number and particle.* Research report 98. Health Effects Institute.

Wolf, K., Stafoggia, M., Cesaroni, G., Andersen, Z. J., Beelen, R., Galassi, C., et al. (2015). Long-term exposure to particulate matter constituents and the incidence of coronary events in 11 European cohorts. *Epidemiology, 26,* 565–574.

World Health Organization. (2013). *Review of evidence on health aspects of air pollution — REVIHAAP Project.* Technical report. Copenhagen, Denmark: WHO Regional Office for Europe.

Yang, L., Hou, X. Y., Wei, Y., Thai, P., & Chai, F. (2017). Biomarkers of the health outcomes associated with ambient particulate matter exposure. *Science of The Total Environment, 579,* 1446–1459.

Yorifuji, T., & Kashima, S. (2013). Air pollution: Another cause of lung cancer. *The Lancet Oncology, 14,* 788–789.

Zanobetti, A., Franklin, M., Koutrakis, P., & Schwartz, J. (2009). Fine particulate air pollution and its components in association with cause-specific emergency admissions. *Environmental Health: a Global Access Science Source [electronic Resource], 8,* 58.

Zhou, J., Ito, K., Lall, R., Lippmann, M., & Thurston, G. (2011). Time-series analysis of mortality effects of fine particulate matter components in Detroit and Seattle. *Environmental Health Perspectives, 119,* 461–466.

Regulation on Brake/Tire Composition

Theodoros Grigoratos

Environmental Chemist, Ranco (VA), Italy

O U T L I N E

1. Brakes Regulation	89	
1.1 Asbestos-Related Regulation	89	
1.2 Trace Elements and Heavy Metals	92	
1.3 REACH and REACH-Like Regulations	94	
1.4 CLP Regulation	96	
2. Tires Regulation	97	

2.1 REACH Regulation	97
2.2 Global Automotive Declarable Substances List and International Material Data System	99
3. Summary	99
References	100

1. BRAKES REGULATION

1.1 Asbestos-Related Regulation

Asbestos was historically used as a friction material in brake linings, disc brake pads, and clutch facings in vehicles because of its unique fire resistance and wear properties. Asbestos has been proven dangerous to work with and is associated to occupational diseases, including respiratory problems (Stayner, Kuempel, Gilbert, Hein, & Dement, 2008), mesothelioma (Lemen, 2004), and lung cancer (NTIS, 2005). For that reason the manufacturing of asbestos containing friction materials—including brake pads—has been ceased in many places worldwide.

In Europe, **the asbestos-free directive** (Council Directive 83/477/EEC on the protection of workers from the risks related to exposure to asbestos at work) was introduced with the aim of protecting workers against risks to their health arising—or likely to arise—from exposure to asbestos at work. For the purposes of the Council Directive 83/477/EEC, the term "asbestos" includes the following fibrous silicates: actinolite (CAS No 77536-66-4), asbestos grunerite (amosite—CAS No 12172-73-5), anthophyllite (CAS No 77536-67-5), chrysotile (CAS No 12001-29-5), crocidolite (CAS No 12001-28-4), and tremolite (CAS No 77536-68-6). The limit values pertaining to in-air concentrations are as follows:

- Chrysotile: 0.60 fibers/cm^3 calculated or measured for an 8-h reference period.
- Other forms of asbestos: 0.30 fibers/cm^3 calculated or measured for an 8-h reference period.

Among the amending acts that include **Directive 91/382/EEC**, **Directive 98/24/EC**, **Directive 2003/18/EC**, and **Directive 2007/30/EC**, the Directive 2003/18/EC reduced the limit value for occupational exposure of workers to asbestos by setting a single maximum limit value for airborne concentration of asbestos of 0.1 fibers/cm^3 as an 8-h time-weighted average and prohibited activities exposing workers to asbestos fibers, with the exception of the treatment and disposal of products resulting from demolition and asbestos removal. The Directive and its amendments shall not prejudice the right of Member States to apply or introduce laws, regulations, or administrative provisions, ensuring greater protection for workers, in particular with regard to the replacement of asbestos by less-dangerous substitutes. In that context many European Member States have introduced total bans in the use of asbestos for the production of brake pads, while there are still some countries where the compliance with this directive has not been verified (http://ibasecretariat.org/alpha_ban_list.php).

In Canada the Ministry of Labour considers all potential worker exposures to asbestos to be a serious workplace hazard. Asbestos is a designated substance under Ontario's Occupational Health and Safety Act and there are three regulations addressing occupational exposures to asbestos:

- Regulation 490/09—Designated Substances
- Regulation 278/05 Designated Substances—Asbestos on Construction Projects and in Buildings and Repair Operations
- Regulation 833—Control of Exposure to Biological or Chemical Agents

The **regulation respecting Asbestos on Construction Projects and in Buildings and Repair Operations** sets out the requirements for repair operations involving vehicles and brake repair operations. This regulation prescribes protective measures and procedures, which among others include the removal of any visible dust on the pad or drum with a damp cloth or a vacuum equipped with a high efficiency particulate air filter before beginning the repair work; controlling the spread of asbestos dust by using measures that are appropriate to the repair work; not using compressed air to clean up or remove dust from any surface; and not permitting eating, drinking, chewing, or smoking in the work area.

However, no direct ban of the use of asbestos in brake linings has been applied. Instead, according to Statistics Canada figures, the imports of asbestos-related items rose in 2014 compared with 2013 and reached a 7-year high with the bulk of these goods consisting of asbestos brake linings and pads. Recently, the Canadian Science Ministry announced that the government is moving to ban asbestos by 2018. The ban will apply to the manufacture of any products containing asbestos, as well as imported products containing the mineral.

Similarly in the United States, brake friction materials are not covered by US Environmental Protection Agency asbestos regulations in a federal level. Legislation has been enacted in California **(Senate Bill (SB) 346)** and Washington **(Senate Bill (SB) 6557)**, eliminating the use of several toxic chemicals, including asbestos, in brake pads, drum linings, and heavy-duty brake block for over-the-road vehicles from 2014. In California the SB346, which is effective since January 2017, prohibits the sale of any motor vehicle brake friction materials exceeding the following concentrations:

- Cadmium exceeding 0.01% by weight
- Chromium (VI) salts exceeding 0.1% by weight
- Lead exceeding 0.1% by weight
- Mercury exceeding 0.1% by weight
- Asbestiform fibers exceeding 0.1% by weight

In Washington the SB6557, among others, foresees the following provisions:

- Brake pads and shoes manufactured after January 2015 must not contain asbestos, hexavalent chromium, mercury, cadmium, or lead. Auto shops and other distributors of brakes will be able to sell any existing inventory for 10 years.
- Brake pads manufactured after January 2021 must not contain more than 5% copper by weight.
- Beginning in 2015, Ecology reviewed relevant information and consulted with a committee of experts to determine that alternative brake friction materials containing less than 0.5% copper were available, meaning that a complete phaseout of copper was feasible.
- In 2025, 8 years after Ecology made the determination that alternative brake friction materials were available, brake pads containing more than 0.5% copper may not be sold in Washington.
- Brake manufacturers will use accredited laboratories and certify to Ecology that their brake pads and shoes comply with the law and will mark proof of certification on all pads and packaging offered for sale in Washington.
- Ecology will track data provided by manufacturers to ensure that concentrations of nickel, zinc, and antimony in automobile brake pads do not increase by more than 50%.

In Japan a total ban of asbestos in all products, including brake systems, has been applied by the Ministry of Health, Labour and Welfare since 2005. More specifically, the production, import, transfer, provision, or use of asbestos or any material containing more than 0.1% asbestos by weight is totally prohibited. Any material containing more than 0.1% by weight of the following silicates: actinolite, asbestos grunerite, anthophyllite, chrysotile, crocidolite, and tremolite is illegal. The Japanese government requests that enterprises importing machinery or other products into Japan confirm that the packing, gaskets, etc., of such products contain no asbestos, based on supporting documents or analytical results before importing such products. In practice the case of Japan could be considered as a total asbestos ban because 0.1% is below the detection limit of currently existing methodologies.

Despite the global trend for a total asbestos ban, according to the International Ban Asbestos Secretariat, developed countries that still permit asbestos use in some ways are China, Russia, India, Brazil, and the United States (http://ibasecretariat.org/alpha_ban_list.php).

1.2 Trace Elements and Heavy Metals

Several regulative actions related to the use of trace elements and heavy metals in vehicle-related industry have been put into effect worldwide. As a consequence, brake production industry has also been affected. In Europe, the **End of Life Vehicle directive (Directive 2000/53/CEE)** requires that "Member States shall ensure that materials and components of vehicles put on the market after 1 July 2003 do not contain **Lead (Pb)**, **Hexavalent Chromium (Cr VI)**, **Cadmium (Cd)** and **Mercury (Hg)** other than in cases listed in Annex II under the conditions specified therein." The vehicle manufacturer shall be required to demonstrate, through contractual arrangements with his suppliers, that compliance with Article 4(2a) of Directive 2000/53/EC is ensured.

The objective of the directive is the prevention of waste from vehicles and also reuse, recycle, and recovery of end-of-life vehicles and components, thus improving environmental performance for all operators involved.

Despite there is no federal regulation regarding the usage of heavy metals in brake friction materials in the United States, several states have adopted such measures. Legislation has been enacted in California **(Senate Bill (SB) 346)**, Washington **(Senate Bill (SB) 6557)**, Rhode Island State **(Senate Bill (SB) H7997)**, and New York State **(Senate Bill (SB) A10871)**, eliminating the use of several toxic chemicals in brake pads, drum linings, and heavy-duty brake block for over-the-road vehicles.

These states have passed laws that limit the weight percentage of specific constituents within a friction material as given in Table 4.1. Exemptions may only be issued for small volume motor vehicle manufacturer's specific motor vehicle models, or special classes of vehicles, such as fire trucks, police cars, and heavy or wide-load equipment hauling, provided the manufacturer can demonstrate that complying with the requirements of the legislation is not feasible (Rhode Island State and New York State).

TABLE 4.1 Heavy Metals Restrictions Enforced in California and Washington and Date of Implementation for Each Level of Regulation

Material	First Implementation Date		
	January 2014 (Level A)	January 2021 (Level B)	January 2025 (Level N)
Cu	No limit	5.0% wt.	0.5% wt.
Cr (VI), Pb, Hg	0.1% wt.	0.1% wt.	0.1% wt.
Cd	0.01% wt.	0.01% wt.	0.01% wt.
Ni, Sb, Zn	Currently none—monitored and maybe regulated in the future		

Further to heavy metals restricted in Europe (Cr (VI), Pb, Hg, Cd), there is a special focus on the use of **Copper (Cu)** in brake systems. The issue started in the early 1990s, when cities south of San Francisco were having trouble meeting Clean Water Act requirements to reduce copper in urban runoff flowing into San Francisco Bay. The Brake Pad Partnership, a cooperative effort among representatives of the auto industry, brake pad manufacturers, environmental groups, storm water agencies, and coastal cities, found that brakes account for 35%—60% of Cu in California's urban watershed runoff, leading the group's members to propose that the most effective action would be to pursue legislation mandating the phased reduction of copper used in brakes (California Stormwater Quality Association, 2016).

As shown in Table 4.1, it is foreseen that Cu shall be largely replaced by other elements in all brake pads and discs produced in these states by 2025. States require manufacturers to test their friction materials for these constituents using appropriately accredited third-party labs. The labs should then communicate the results to a third party so that information on each friction material is posted on a public website and shared with the appropriate government agencies. The laws require that friction materials are marked with their level (A, B, or N) of compliance with the law. Marks used in the certification scheme include **A** that translates to asbestos, Cr (VI), Pb, Hg with concentrations lower than 0.1% wt. as well as Cd lower than 0.01% wt., **B** that translates to A plus Cu with concentration lower than 5% wt., and **N** that translates to A plus Cu with concentration lower than 0.5% wt. Since 2015 (Washington) and 2014 (California), each manufacturer must:

- test friction materials with a third party lab to determine compliance level;
- submit signed self-certification documentation to NSF International registrar;
- mark friction material with compliance level (A, B, or N) and the year of manufacture;
- mark friction material's packaging with trademark showing compliance level; and
- retest and recertify friction materials every 3 years.

It is recommended that all brake pads intended for sale in California follow all of Washington's marking requirements. Packaging of friction materials must have marked proof of certification with mark as shown in Fig. 4.1.

Finally, special attention is paid on potentially hazardous elements such as **Nickel (Ni)**, **Antimony (Sb)**, and **Zinc (Zn)**. Despite these elements are not currently regulated in any state, they are being monitored and they may be regulated in the future.

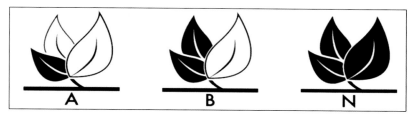

FIGURE 4.1   Marks used in certification (A: Asbestos, Cr (VI), Pb, Hg <0.1% and Cd <0.01%; B: A + Cu <5%; N: A + Cu <0.5%).

1.3 REACH and REACH-Like Regulations

The **REACH** regulation **(Regulation 1907/2006)** aims at improving the protection of human health and the environment through the better and earlier identification of the intrinsic properties of chemical substances. This is done by the four processes of REACH, namely the Registration, Evaluation, Authorization, and Restriction of chemicals. REACH is applicable in all 28 EU countries.

The general principle of REACH regulation is "No data, no market." This practically means that if a substance has not been registered for a specific application, it must not be used. Manufacturers and importers are required to gather information on the properties of their chemical substances, which will allow their safe handling, and to register the information in a central database in the European Chemicals Agency (ECHA) in Helsinki. This regulation also calls for the progressive substitution of the most dangerous chemicals (referred to as "substances of very high concern") when suitable alternatives have been identified. In practice, market surveillance includes any necessary action (e.g., bans, withdrawals, recalls) to stop the circulation of products that do not comply with all the requirements set out in the relevant EU harmonization legislation, to bring the products into compliance, and to apply sanctions.

Regarding brake systems, the Automotive Industry Guideline on REACH clearly defines the brake friction pad as "an article." According to **Article 3.3** of the REACH regulation, an article is defined as "an object which during production is given a special shape, surface or design which determines its function to a greater degree than does its chemical composition." On the other hand, **Article 7.1** of the REACH regulation defines that a registration of substances in articles is mandatory **IF** the substances are intended to be released from the produced or imported article(s) during normal and reasonable foreseeable conditions of use. Release of particles or wear debris from tires or rubber belts, brake linings, and discs is not considered to be intended, thus no registration is required. Exception to the general rule is an article that contains **Substances of Very High Concern (SVHC)** in the quantity of 0.1% wt. and in which case they should be reported. A Member State or ECHA at the request of the European Commission can propose a substance to be identified as an SVHC. If identified, the substance is added to the Candidate List, which includes candidate substances for possible inclusion in the Authorization List (Annex XIV).

The publication of intentions of proposing a substance to be identified as an SVHC is submitted so as to give advanced information to industry and other stakeholders. After the publication of the proposal, anyone can comment on it or add further information, for example, related to the properties, uses, and risks of the proposed substance or its alternatives. If no comments are received, the substance is included in the Candidate List. The proposals and the comments will be forwarded to the Member State Committee to agree on the identification as an SVHC. The Candidate List can be accessed at https://echa.europa.eu/candidate-list-table.

Apart from the 28 EU countries, **REACH-like** regulations have been applied in countries such as China, Malaysia, and Korea. Despite the similar concept of **REACH-like** and **REACH** regulations, there are several differences in the timing for notifications/registrations, the threshold values applied, the availability of exemptions, and the communication in the supply chain.

In China the Order No. 7 of Ministry of Environmental Protection (MEP) came in 2010 and stipulates that new chemical substances have to be notified to the Chemical Registration Centre of the MEP, irrespective of annual tonnage. The notification applies not only to new substance on its own, in preparation or articles intended to be released, but also to new substances used as ingredients or intermediates for pharmaceuticals, pesticides, veterinary drugs, cosmetics, food additives, and feed additives. Exemptions to the notification are chemicals subject to other existing laws and regulations, substances existing in nature (unprocessed or processed only physically), substances of noncommercial purpose or unintentionally produced, and some special categories. The regulation applies to manufacturers of new substances in China, importers of new substances in China, and foreign companies selling new substances to China.

A new chemical is defined as any substance not listed in the Inventory of Existing Chemical Substances produced or imported in China (IECSC). There were 45,612 substances in IECSC in 2013 when the Chinese government updated the inventory. Someone could check for substances listed in IECSC by using CIRS's free Chinese Chemical Inventory Search System.

Updated in June 2013, the Act on the Registration and Evaluation of Chemicals (known as Korea REACH) passed the plenary session of the National Assembly in Korea on April 30, 2013, and came into force on Jan 1, 2015. The purpose of this act is to protect public health and the environment through these provisions:

- Registration of chemical substances
- Screening of hazardous chemical substances
- Hazard and risk assessment of products containing chemical substances and hazardous substances
- Sharing information of chemical substance

The Ministry of Environment (MoE) is responsible for the registration and evaluation of chemical substance under this act. The previously existing current Toxic Chemical Control Act was divided into Korea REACH and Chemicals Control Act (CCA) in 2015, with K-REACH focusing on registration and evaluation of substance while CCA focusing on the control of hazardous substance and response to chemical accidents. The full English text of Korea REACH is available free of charge from International Zinc Association's website (International Zinc Association, 2016).

Japanese **Chemical Substances Control Law (CSCL)** was initially implemented in 1974, amended in 2009, and entered into force in 2011. **Japan CSCL** covers general industrial chemical products, which include both new chemicals and existing chemicals. Food or feed additives, pharmaceuticals, cosmetics, agricultural chemicals, and fertilizers are subject to different laws and acts. General industrial chemical products might also be subject to the requirements of the Industrial Safety and Health Law, the Poisonous and Deleterious Substances Control Act, the Air Pollution Control Law, and the Water Pollution Control Act. The three government bodies responsible for the implementation of **CSCL** are the Ministry of Economy, Trade and Industry, Labor and Welfare, and the MoE. Manufacturers and importers in Japan shall submit new chemical notification to all three authorities to obtain approval notice from the governments before the manufacture or import of the new substance. Foreign manufacturers exporting new chemicals to Japan may also submit new chemical notifications by themselves.

The following categories of chemicals are regulated by Japan CSCL:

- New Chemical Substances
- General Chemicals
- Priority Assessment Chemical Substances
- Monitoring Chemical Substances
- Class II Specified Chemical Substances
- Class I Specified Chemical Substances

1.4 CLP Regulation

The CLP Regulation **(Regulation 1272/2008)** is the European regulation on **Classification, Labeling** and **Packaging** of chemical substances and mixtures. The CLP Regulation ensures that the hazards presented by chemicals are clearly communicated to workers and consumers in the European Union. The CLP Regulation entered into force in January 2009 and replaced two previous pieces of legislation, the Dangerous Substances Directive and the Dangerous Preparations Directive.

According to the CLP Regulation before placing chemicals on the market, the industry must establish the potential risks to human health and the environment of such substances and mixtures, classifying them in line with the identified hazards. The hazardous chemicals also have to be labeled according to a standardized system so that workers and consumers know about their effects before they handle them. Thanks to this process, the hazards of chemicals are communicated through standard statements and pictograms on labels and safety data sheets. For example, when a supplier identifies a substance as "acute toxicity category 1 (oral)," the labeling will include the hazard statement "fatal if swallowed," the word "Danger," and a pictogram with skull and crossbones.

The obligations of a downstream user under CLP can be summarized to the following:

- Take over the classification for a substance or mixture derived already by another actor in the supply chain, provided that they do not change the composition of this substance or mixture.
- If there is new information that may lead to a change of the harmonized classification and labeling elements of a substance, they should submit a proposal to the Competent Authority in one of the Member States in which the substance is placed on the market (CLP Article 37(6)).
- They should assemble and keep available all the information required for the purposes of classification and labeling under CLP for a period of at least 10 years after they have last supplied a substance or mixture.

The CLP Regulation mostly affects brake fluids and not the main body of the brake system (i.e., pads and disc).

2. TIRES REGULATION

2.1 REACH Regulation

The **REACH** regulation **(Regulation 1907/2006)** aims at improving the protection of human health and the environment through the better and earlier identification of the intrinsic properties of chemical substances. This is done by the four processes of REACH, namely the Registration, Evaluation, Authorization, and Restriction of chemicals. REACH is applicable in all 28 EU countries.

According to the REACH regulation, tires fall under the definition of an article; therefore manufacturers and importers of tires have certain obligations under REACH. An EU company is defined as a tire "importer" if it imports tires from outside Europe, while any European company supplying tires to the European market is classed as a tire "supplier."

Tire importers are obliged to ask information on the chemicals included into the tires during their manufacture from non-EU tire suppliers. In addition, the amount of each chemical in the tire should be known. When necessary, importers should notify the ECHA for tires that contain a substance of very high concern at a concentration above 0.1% in quantities exceeding 1 t/year as well as provide their customers with information that will allow for the safe use of the tire, including, as a minimum, the name of the Candidate List substance. Suppliers of tires containing any Candidate List substance at a concentration above 0.1% are also obliged to provide customers with information that will ensure the safe use of the tire, including, as a minimum, the name of the Candidate List substance.

At the moment there is one group of substances specifically relevant to the manufacture, import, supply, and use of tires, which is common to both the Candidate List and Restrictions under REACH, namely the **polycyclic aromatic hydrocarbons (PAHs)**. PAHs are a group of over 100 different chemicals that are formed during the incomplete burning of coal, oil and gas, garbage, or other organic compounds. PAHs are known for their carcinogenic, mutagenic, and teratogenic properties (Kim, Jahan, Kabir, & Brown, 2013). PAHs have been used in extender oils used in the production of tires. REACH Annex XVII has placed a restriction on the use of eight PAHs in tires and extender oil. The eight PAHs are given in Table 4.2.

The restriction on PAHs prohibits the placing on the market or use of extender oils containing certain PAHs for the production of tires or parts of tires if they contain:

- more than 1 mg/kg (0.0001% by weight Benzo[a]pyrene—BaP) and
- more than 10 mg/kg (0.001% by weight) of the sum of all listed PAHs.

The standard EN 16143:2013 (petroleum products—determination of content of benzo(a) pyrene (BaP) and selected PAHs in extender oils—procedure using double liquid chromatography cleaning and gas chromatography/mass spectroscopy analysis) shall be used as the test method for demonstrating conformity with the limits referred above (British Institute of Petroleum, 2013).

For the purpose of the described legislation, "tires" shall mean tires for vehicles covered by

- Directive 2007/46/EC of the European Parliament and of the Council of September 5, 2007, establishing a framework for the approval of motor vehicles and their trailers,

TABLE 4.2   PAHs Restricted Under the REACH Regulation

Name	CAS	Structure
Benzo[a]pyrene	50-32-8	
Benzo[e]pyrene	192-97-2	
Benzo[a]anthracene	56-55-3	
Chrysene	218-01-9	
Benzo[b]fluoranthene	205-99-2	
Benzo[k]fluoranthene	207-08-9	
Benzo[j]fluoranthene	205-82-3	
Dibenzo[a,h]anthracene	53-70-3	

- Directive 2003/37/EC of the European Parliament and of the Council of May 26, 2003, on type approval of agricultural or forestry tractors, their trailers, and interchangeable towed machinery, together with their systems, components, and separate technical units, and
- Directive 2002/24/EC of the European Parliament and of the Council of March 18, 2002, relating to the type approval of two- or three-wheel motor vehicles and repealing Council Directive 92/61/EEC.

2.2 Global Automotive Declarable Substances List and International Material Data System

More than direct regulation both International Material Data System (IMDS) and Global Automotive Declarable Substance List (GADSL) are tools developed by the automotive industry to support among others regulations, such as the End of Life Vehicle directive (Directive 2000/53/EC) and REACH directive concerning obligations for substances in articles.

The **IMDS** is an online system for collecting information on substances used in the automobile industry in Japan, United States, and Europe to comply with End of Life Vehicle directives. The system was designed to collect, maintain, analyze, and archive information on materials used for automotive components. Over time, it has been adapted to meet the obligations placed on automobile manufacturers, and thus on their suppliers, by national and international standards, laws and regulations, scientific findings, and risk assessments, according to the GADSL. Suppliers are required to enter parts and materials on individual request from automotive producer into IMDS website (http://www.mdsystem.com/). Depending on the complexity, there are between 4000 & 9000 main components contained in a vehicle platform. IMDS entry is obligatory for recycling rate management, management of substances of concern, and REACH compliance. Vehicle manufacturers can ask to the system in which components a substance has contained.

The **GADSL** is the result of a year-long global effort of representatives from the automotive, automotive parts supplier, and chemical/plastics industries who have organized the Global Automotive Stakeholders Group (GASG). It is supported by both the European and American car manufacturers and their suppliers. The Japanese and Korean car manufacturers joined the initiative in 2006. The GASG's purpose is to facilitate communication and exchange of information regarding the use of certain substances in automotive products throughout the supply chain. The GADSL only covers substances that are expected to be present in a material or part that remains in a vehicle. This list (http://www.gadsl.org) shows regulated substances used in the automobile and chemical industries in Japan, Europe, and the United States (substances prohibited and regulated under laws and regulations worldwide). Suppliers are required to comprehensively manage substances of concern and to make continuous efforts to reduce use in accordance with GADSL.

3. SUMMARY

The most important regulations affecting brake and tire production, worldwide, have been described in detail. Brake pads composition is mainly affected from the worldwide asbestos ban, REACH and REACH-like regulations, the European regulation on Classification, Labeling and Packaging of chemical substances and mixtures (CLP Regulation), and regulations related to restrictions on the use of particular trace elements and heavy metals. Table 4.3 gives a brief overview of brake composition related regulation and the countries where each regulation has been applied. Some countries are given with an asterisk due to either not full implementation of the regulation or due to other restrictions. Table 4.3 shall serve only as a basis as new countries continuously adopt some of the regulations.

TABLE 4.3   Brake Composition—Related Regulation Worldwide

Regulation	Country
Asbestos ban	Algeria, Argentina, Australia, Bahrain, Brunei, Chile, Egypt, EU*, FYROM, Gabon, Gibraltar, Honduras, Iceland, Iraq, Israel, Japan, Jordan, Kuwait, Mauritius, Monaco, Mozambique, New Caledonia, New Zealand, Norway, Oman, Qatar, Saudi Arabia, Serbia, South Africa, South Korea, Switzerland, Turkey, Uruguay, USA*
REACH regulation	EU
REACH-like regulations	Albania, Australia, Brazil*, Canada, China, FYROM, Hong Kong, Iceland, India, Indonesia*, Israel, Japan, Malaysia, Mexico, Norway, Philippines, Serbia, South Africa*, South Korea, Switzerland, Taiwan, Thailand, Turkey*, UAE, USA*, Vietnam*
End of Life Vehicle directive	EU
Senate Bill (SB) regulations	California (Senate Bill (SB) 346), Washington (Senate Bill (SB) 6557), Rhode Island State (Senate Bill (SB) H7997), and New York State
CLP Regulation	EU

Not applied in every member state or applied with some modifications.

On the other hand, tire composition is driven mainly from REACH and REACH-like regulations (Table 4.3 shows where it applies), while it is affected also from the Global Automotive Declarable Substances List and International Material Data System.

References

British Institute of Petroleum. (2013). *Petroleum products. Determination of content of Benzo(a)pyrene (BaP) and selected polycyclic aromatic hydrocarbons (PAH) in extender oils. Procedure using double LC cleaning and GC/MS analysis.* Standard Number BS EN 16143:2013, BS 2000-605:2013.

California Stormwater Quality Association. (2016). *Estimated urban runoff copper reductions resulting from brake pad copper restrictions.* Available on-line at https://www.casqa.org/sites/default/files/library/technical-reports/estimated_urban_runoff_copper_reductions_resulting_from_brake_pad_copper_use_restrictions_casqa_4-13.pdf.

International Zinc Association's website. (2015). *Act on registration and evaluation, etc. of chemical substance.* Available on-line at http://www.lawbc.com/uploads/docs/English_translation_-_Korea_REACH__Act_on_Registration_and_Evaluation_e.pdf.

Kim, K. H., Jahan, S. R., Kabir, E., & Brown, R. (2013). A review of airborne polycyclic aromatic hydrocarbons (PAHs) and their human health effects. *Environment International, 60,* 71—80.

Lemen, R. A. (2004). Asbestos in brakes: Exposure and risk of disease. *American Journal of Industrial Medicine, 45,* 229—237.

NTIS, National Toxicology Program. (2005). Asbestos. In *Report on carcinogens* (11th ed.). U.S. Department of Health and Human Services, Public Health Service, National Toxicology Program.

Stayner, L., Kuempel, E., Gilbert, S., Hein, M., & Dement, J. (2008). An epidemiological study of the role of chrysotile asbestos fibre dimensions in determining respiratory disease risk in exposed workers. *Occupational and Environmental Medicine, 65,* 613—619.

European Emission Inventories and Projections for Road Transport Non-Exhaust Emissions
Analysis of Consistency and Gaps in Emission Inventories From EU Member States

Hugo Denier van der Gon[1], Jan Hulskotte[1], Magdalena Jozwicka[1,2], Richard Kranenburg[1], Jeroen Kuenen[1], Antoon Visschedijk[1]

[1]TNO, Department Climate, Air and Sustainability, Utrecht, The Netherlands;
[2]European Environment Agency, Copenhagen, Denmark

O U T L I N E

1. Introduction	102	3.1 Resuspension in the Reporting Requirements for the EU Member States	108
2. Trends in Road Transport Exhaust and Wear Emissions	104	4. Analysis and Comparison of Official Reported Emissions Data	109
2.1 Projected Road Transport Wear Emissions	106	4.1 Implied Emission Factors per Vehicle Kilometer	109
3. Emission Reporting in Europe for Brake and Tire Wear and Road Abrasion	108	4.2 Size Distribution of Reported Wear Particles	113

5. Resuspension due to Road Transport 114

6. Tracers to Verify Brake, Tire, and Road Wear 116

7. Conclusions 118

References 120

1. INTRODUCTION

An emission inventory is the foundation for air quality management because from the inventory we understand the relation between (human) activities, in-use technologies, and the related release of pollutants. Moreover, from the inventory we understand the relative importance of sources for the overall release of a pollutant and the possibilities for mitigation. Technical guidance to prepare national emission inventories is provided by, for example, the joint EMEP/EEA air pollutant emission inventory guidebook (EEA, 2016), which supports the mandatory reporting of emissions data under the UNECE Convention on Long-range Transboundary Air Pollution (CLRTAP) and the EU National Emission Ceilings Directive (UNECE, 2009). It provides expert guidance on how to compile an atmospheric emission inventory. Similar guidance is also provided by the US Environmental Protection Agency and other environment agencies.

Emission inventories are typically developed by using a bottom-up approach, i.e., combining available statistics on fuel combustion, industrial production, animal husbandry, etc., with the most appropriate emission factors. A typical emission inventory is compiled by collecting activity data and appropriate emission factors and subsequently calculating the total emission by pollutant (Eq. 5.1).

$$\text{Emission}_{\text{pollutant}} = \sum_{\text{activities}} \text{Activity rate}_{\text{activity}} \times \text{Emission factor}_{\text{activity, pollutant}} \qquad (5.1)$$

Taking the example of direct brake wear, the activity rate could be the number of vehicle kilometers (vkm) driven by a particular vehicle class on a particular road class, e.g., vkm by passenger cars (PCs) on urban roads. In the nomenclature for reporting (NFR), wear emissions involve the following categories: NFR 1.A.3.b.vi—road transport: automobile tire and brake wear and NFR 1.A.3.b.vii—road transport: automobile road abrasion (Table 5.1).

In this chapter we will mostly focus on the categories direct brake wear, direct tire wear, and direct road wear, but we will also address the importance of road dust resuspension. Direct brake, tire, and road wear are explicitly treated by Ntziachristos and Boulter (2016) in the EMEP/EEA emission inventory guidebook (EEA, 2016). Unfortunately, in the reporting to UNECE-CLRTAP, direct brake and tire wear are combined in one category (Table 5.1), which may not be optimal for transparency as we will see in Section 6.6.

For the direct brake, tire, and road wear, uncertainties in the emission inventories are substantial because there is considerable uncertainty about the type of activity as well as the related emission factors. This is illustrated by an example of wear factors for light-duty vehicle tires (Fig. 5.1).

TABLE 5.1 The Nomenclature for Reporting 2014 (NFR14) Categories to be Used for Reporting of Road Transport Emissions by Parties to the Long-range Transboundary Air Pollution Convention (EEA, 2016)

NFR Code	Description
N14 1A3bi	Passenger cars
N14 1A3bii	Light commercial trucks
N14 1A3biii	Heavy-duty vehicles including buses
N14 1A3biv	Motorcycles
N14 1A3bv	Evaporation
N14 1A3bvi	Brake and tire wear
N14 1A3bvii	Road surface abrasion

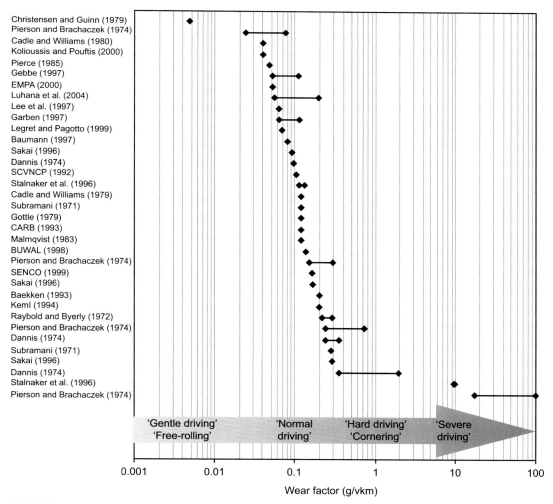

FIGURE 5.1 Wear factors for light-duty vehicle tires (Boulter, 2005; Ntziachristos & Boulter, 2016). *vkm*, vehicle-km.

Even the classification "normal driving" has a considerable range in associated emission factors (note the logarithmic scale in Fig. 5.1). This will, at least partly, be due to the fact that the definition of "normal driving" is quite arbitrary. Recently, the UNECE Informal Working Group on the Particle Measurement Programme further highlighted this problem concluding that an extremely wide range of driving conditions as well as different sampling techniques has been applied in non-exhaust–related studies, which has often led to different results and/or contradictory conclusions (PMP IWG, 2016).

We refer to Ntziachristos and Boulter (2016) for further detailed background information on inventory compilation for these sources, the processes involved, calculation methods proposed, and overall data quality. In this chapter we investigate how the reported emissions for brake, tire, and road wear have changed over time, may change in the future, and the consistency of the reported emissions. The official reported data discussed in this chapter were downloaded on February 15, 2017. Those are the data reported until 2016 with emissions for 2014 as the most recent year. The emission data for the historic years are annually revised. Hence, during 2017 and beyond, the official reported emissions for 2014 may deviate from the data presented here. This process is, however, transparent as the CEIP website (http://www.ceip.at/) allows selection of emission year as well as reporting year.

2. TRENDS IN ROAD TRANSPORT EXHAUST AND WEAR EMISSIONS

Up until the early 1990s, road transport emissions were dominated (80%–90%) by exhaust emissions (Denier van der Gon et al., 2013); but over time, this proportion has gradually decreased to less than 50% (the Netherlands; Fig. 5.2A). The trend as shown for the Netherlands applies also to the EU15, with the brake-even point between exhaust and wear emissions around the year 2012 (Fig. 5.2B). As can be seen for the Netherlands in Fig. 5.2A, since 1990 the exhaust emission reduction strategies have been extremely successful and currently wear emissions dominate over combustion emissions. Therefore, further progress in emission reduction by addressing combustion emissions only will be more limited. For the EU new Member States (NMS) the development since 2000 is different with, at first, an increasing trend of combustion emissions, but since 2008 this trend reversed (Fig. 5.2C). Currently (year 2014 emissions), the wear emissions (i.e., excluding resuspension) for the EU-NMS are about 31% of the total road transport PM_{10} emissions. The trend of increasing relative importance of wear emissions in the EU-NMS is likely to continue. This is due to two processes: (1) further exhaust emission reduction due to fleet renewal and (2) continuing growth in vkm. Because wear emissions are proportional to vkm and current legislation does not require a reduction in wear emissions per vkm traveled, the wear particulate matter (PM) emissions will increase also in absolute terms.

While for combined EU15 road transport PM_{10} emissions, the contribution from wear emissions now exceeds exhaust PM_{10} emissions (Fig. 5.2B); this is not the case for $PM_{2.5}$. About 50% of the wear emissions is assumed to be $PM_{2.5}$ (Ntziachristos & Boulter, 2016), whereas for exhaust emissions that is almost 100%. Therefore, exhaust still dominates $PM_{2.5}$ emissions and will remain to do so for a number of years (Fig. 5.3). Fig. 5.3 largely

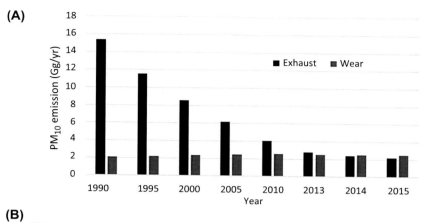

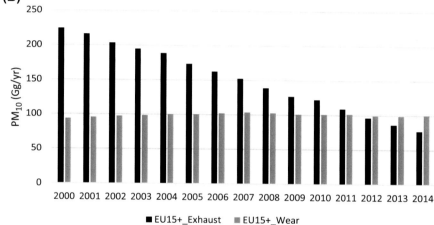

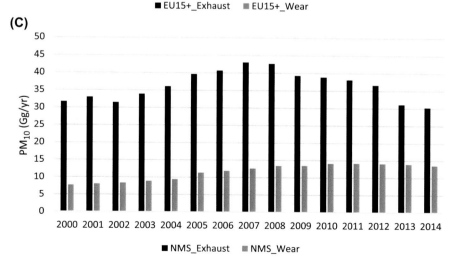

FIGURE 5.2 PM$_{10}$ emission from road transport exhaust and wear in the Netherlands 1990−2015 (A) (data NL PRTR, 2017); 2000−2014 for EU15, Norway and Switzerland (B) and the EU new Member States (C) (data TNO-MACC (Kuenen, Visschedijk, Jozwicka, & Denier van der Gon, 2014) and CEIP).

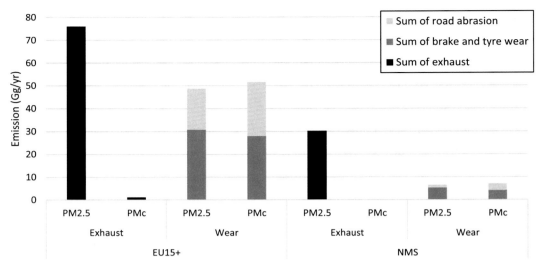

FIGURE 5.3 Road transport $PM_{2.5}$ and PMc ($PM_{2.5-10}$) emissions of the EU15 + Norway + Switzerland and the EU new Member States for 2014. Wear emissions are split in brake and tire wear and road abrasion.

reflects the results of using the guidebook (EMEP/EEA, 2016) for splitting PM_{10} in a fine fraction ($PM_{2.5}$) and a coarse fraction (PMc or $PM_{2.5-10}$). Data on size distribution of wear emissions are scarce, often contradictory and therefore uncertain. This topic deserves more attention in future research. However, in this chapter we focus on the country-reported emissions, as presented in the overview given in Fig. 5.3.

2.1 Projected Road Transport Wear Emissions

Rexeis and Hausberger (2009) predicted, using a detailed emission model for the Austrian fleet, that the percentage of PM non-exhaust of the total PM emissions will increase from about 50% between 2005 and 2010 up to some 80%–90% by 2020. Hulskotte and Jonkers (2012) modeled the effect of a range of possible emission factors on the local air quality in three different urban situations with respect to the traffic pattern but with more or less the same traffic intensity. The dominance of wear-related processes in the contribution of traffic on the local concentrations of PM_{10} was illustrated for all scenarios and even more strongly in the near future (2020). To get insight in the level of concentrations of wear dust to which citizens potentially could be exposed, a traffic influenced route through the center of Rotterdam was selected. Within the selected route, different traffic flow patterns are present while the traffic intensity (vehicles/day) is more or less the same (Fig. 5.4). The traffic flow patterns considered by Hulskotte and Jonkers (2012) are as follows:

- Urban free flow
- Urban intermediate flow
- Urban traffic junction (specially defined situation)

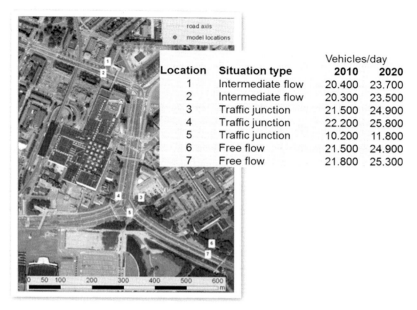

Location	Situation type	Vehicles/day 2010	2020
1	Intermediate flow	20.400	23.700
2	Intermediate flow	20.300	23.500
3	Traffic junction	21.500	24.900
4	Traffic junction	22.200	25.800
5	Traffic junction	10.200	11.800
6	Free flow	21.500	24.900
7	Free flow	21.800	25.300

FIGURE 5.4 Aerial photo of sample location with model receptor points and applied model traffic intensity and traffic type nearby receptor points indicated.

Along this selected route, emission factors of wear processes were differentiated according to three scenarios: (1) the standard CAR (Calculation of Air pollution by Road transport) model emission factors (Jonkers, 2007), (2) elevated $PM_{2.5}$ and PM_{10} fractions in the total PM from wear processes, and (3) elevated emission factors due to deceleration and acceleration at junctions. The results are summarized in Table 5.2. In the high PM fraction scenario (2) and the elevated emissions at junctions scenario (3), the share of wear in PM_{10} emission is

TABLE 5.2 Modeled Road Traffic Contribution to Ambient Concentration of PM_{10} ($\mu g/m^3$ and Between Brackets Wear Share %) in the Street With Standard CAR Emission Factors and With Alternative Wear Scenarios

	Wear Scenario					
Standard Emission Factors for Tailpipe in CAR[a] Model	Standard		High PM_{10} Fraction		Elevated Emission at Traffic Junction: "Heavy Braking"	
Year	2010	2020	2010	2020	2010	2020
Urban free flow	3.5 (49%)	2.8 (76%)	5.4 (66%)	4.9 (86%)		
Urban intermediate flow	3.0 (48%)	2.2 (74%)	4.5 (65%)	4.0 (85%)		
Congested traffic	4.2 (38%)	2.7 (66%)	5.7(54%)	4.6 (80%)	7.2 (64%)	6.3 (86%)

[a]Jonkers (2007).
CAR, Calculation of Air pollution by Road transport

already more than 50% in 2010, whereas the shares of wear in 2020 grow to more than 80% under all traffic flow patterns.

The emission data for the EU15 suggest that the overall break-even point (highway, rural, and urban roads) was around 2011–12 (Fig. 5.2). For the NMS this will take considerably longer (Fig. 5.2), but the trend is unmistakable; for the total European Union domain, including Norway and Switzerland, the wear contribution to PM_{10} (114 Gg/year) exceeded the exhaust contribution (107 Gg/year) in 2014 (based on data reported until 2016). From projection studies modeling the future wear contributions, e.g., Rexeis and Hausberger (2009) and Hulskotte and Jonkers (2012), the share of wear emissions in total PM from road transport will increase further. Hence, in the European urban environment, where most Europeans live and spend their time, the contribution of wear in total exposure to urban road transport PM_{10} emissions already reaches 50% and is predicted to grow to about 80%–90% after 2020. In the following section, we analyze the available emission data for non-exhaust since 2000 in more detail.

3. EMISSION REPORTING IN EUROPE FOR BRAKE AND TIRE WEAR AND ROAD ABRASION

Fifty-one countries in Europe and North America, including the EU as a whole, have to annually submit their emissions of air pollutants for the latest year and all historic years to EMEP (cooperative program for monitoring and evaluation of long-range transmission of air pollutants in Europe) as laid out in the UNECE CLRTAP, which was signed in 1979. The reporting follows well-defined guidelines and asks countries to complete a predefined template with emissions by year, pollutant, and sector (defined by the NFR; see also Table 5.1). Countries are encouraged to set up their own inventory system and choose the best methodologies for emission estimation, which fit their national situation. Although the motivation is to get the best, most appropriate, and accurate emission estimates, this results in substantial variation between countries, which is not always well understood. Nevertheless, substantial progress has been made in recent years (EMEP, 2016). The official submitted data for all countries are collected by the Centre for Emission Inventories and Projections (CEIP) and made available online (http://www.ceip.at/). Because of the more detailed methodologies included in most inventories and the national focus of each of the inventories, the reported emissions often provide the most accurate estimate for a country. Nevertheless, gaps and errors do exist in the reported emission data. Especially, the consistency in emissions reporting for consecutive years is problematic.

3.1 Resuspension in the Reporting Requirements for the EU Member States

Traffic-induced resuspension of deposited road dust is not represented in the official emission inventories. Current NFR source categorization includes tire wear together with brake abrasion as one source category (NFR 1.A.3.b.vi) and road surface abrasion emissions as a second source category (NFR 1.A.3.b.vii); the latter however does not include resuspension but only primary particle generation. Resuspension, strictly speaking, is not a primary emission but a reemission. This is the most important argument why it is not included in the reporting requirements for the EU Member States under the CLRTAP and EMEP. Another consideration is the lack of a level playing field; resuspension emission is highly dependent on climatic conditions (Amato et al., 2014, 2012; see also Chapter 1). Due to the dependencies on climate, countries may not be fully in control of their resuspension emissions, which would make demands for emission reduction difficult. However, traffic-induced resuspension of deposited road dust may dominate non-exhaust PM emissions, especially in cities (Denier van der Gon et al., 2013; Denier van der Gon, Jozwicka, Hendriks, Gondwe, & Schaap, 2010). Therefore, resuspension emissions have to be taken in consideration. Once resuspension emission is included in the official emission reporting, the estimated PM emissions will substantially change (see also Section 6.5). Including resuspension is necessary to understand the origin of real-world ambient PM concentrations currently observed in European cities and their streets. For an estimation of resuspension emissions using derived emission factors, we refer to Section 6.5 and Chapter 11. For the uncertainties and shortcomings in emission inventories in relation to resuspension, we refer to Denier van der Gon et al. (2013, 2010), Amato et al. (2014), and Chapter 11.

4. ANALYSIS AND COMPARISON OF OFFICIAL REPORTED EMISSIONS DATA

The total emissions by country are related to the size of the country and the number of inhabitants. Large countries with a large population, such as Germany and France, will have higher emissions than small countries, such as Belgium or Denmark. Therefore comparing emissions by country provides little insight in the underlying data, similarities, and/or discrepancies between country emission reporting. It is more informative to normalize the emission estimates and make them comparable between countries. This is achieved here by expressing the emissions per vkm.

4.1 Implied Emission Factors per Vehicle Kilometer

Wear emissions differ between vehicle types and road types (Denier van der Gon et al., 2013). Emission factors per vkm are higher for heavy-duty vehicles (HDV) than for PCs. Within the category of HDV, the wear emissions are related to the size of the vehicle and the load factor, as explained in the Tier 2 methodology described by Ntziachristos and Boulter (2016). The underlying specification of the emissions shown in Fig. 5.2, by vehicle type and country, may not be publicly accessible, but we can compare more in-depth by a normalization of the emissions using an approximation of the activity data. For this purpose we derive the vkm data by vehicle type from the TRACCS database (Papadimitriou et al., 2013). The normalized share of different vehicle types for selected countries is presented in Fig. 5.5.

The overall picture is rather consistent with about 75% of the vkm being driven by PCs and the remainder being dominated by light-duty vehicles (12%) and HDV (8%). The most deviating country is Greece, but because no emissions are reported by Greece for the wear categories, we will not use those data.

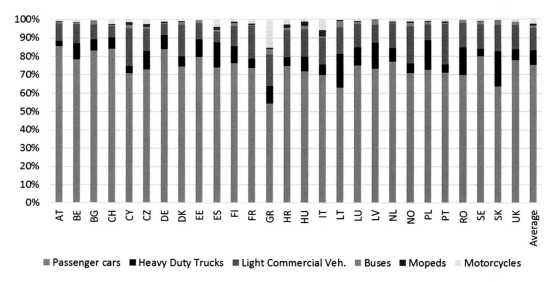

■ Passenger cars ■ Heavy Duty Trucks ■ Light Commercial Veh. ■ Buses ■ Mopeds ■ Motorcycles

FIGURE 5.5 Share of vehicle types to total vehicle kilometers driven by country in 2010 based on Papadimitriou et al. (2013).

When we compare the results (Table 5.3, Fig. 5.6), several observations can be made. A number of countries do not report road abrasion emissions: Bulgaria, Switzerland, Italy, Lithuania, Luxembourg, and Romania (see BG, CH, CY, IT, LT, LU, RO in Fig. 5.6). Some of these countries may have included road abrasion emissions under brake and tire wear. This appears to be the case, for example, for Switzerland, Luxembourg, and Romania because their brake and tire wear implied emission factors are well above the average for this category (Table 5.3). Information about emissions being included elsewhere (IE) can often be found on the CEIP website (http://www.ceip.at/). However, disaggregation is not possible and one cannot be sure what "including" exactly means. Hence, we do not try to elucidate this further. Instead, we group the countries according to their reporting to make comparisons. Note that brake and tire wear is one category in the reporting requirements; countries do not report brake wear and/or tire wear individually. The country grouping is based on completeness (which categories are reported) and the information we obtain from the implied emission factors. Groups I and II are mutually exclusive, but a country of Group I will also belong to either Group III or IV.

- Group I = countries that report both brake and tire wear and, separately, road abrasion
- Group II = countries that only report brake and tire wear
- Group III = countries reporting road abrasion in the range up to 5 mg/vkm
- Group IV = countries reporting road abrasion above 5 mg/vkm due to road sanding and/or use of studded tires

The results are visualized in Fig. 5.6 for an overview, revealing some relative extreme values.

TABLE 5.3 Implied Emission Factors for PM_{10} and $PM_{2.5}$ for Brake and Tire Wear and for Road Abrasion for European Countries That Report at Least One of the Categories.

Country[a]	Brake and Tire Wear (mg/vkm)		Road Abrasion (mg/vkm)	
	PM_{10}	$PM_{2.5}$	PM_{10}	$PM_{2.5}$
Finland	7.6	4.2	49.3	26.7
Sweden	6.6	3.6	26.2	2.5
Norway	4.0	2.8	10.2	1.7
Austria			8.6	2.6
Slovakia	7.9	4.2	4.7	2.5
Latvia	7.6	4.0	3.7	2.0
Estonia	6.3	3.4	3.5	1.9
Germany	6.0	3.3	3.4	1.8
Belgium	6.0	3.2	3.3	1.8
Czech Republic	5.6	3.0	3.3	1.8
Hungary	6.7	3.6	3.2	1.8
Denmark	6.0	3.4	3.2	1.7
Spain	5.2	3.0	3.1	1.7
United Kingdom	5.7	3.2	3.0	1.6
Poland	4.3	2.8	3.0	0.0
Croatia	6.1	3.2	2.7	1.5
France	4.9	2.7	2.7	1.4
Netherlands	3.2	0.6	2.6	0.4
Bulgaria	7.1	3.8		
Switzerland	14.3	2.1		
Cyprus	6.2	3.3		
Italy	5.5	3.0		
Lithuania	6.9	3.7		
Luxembourg	11.2	6.2		
Romania	9.1	4.8		
Average Group I	5.9	3.2		
Average Group II	8.5	3.7		
Average Group III			3.2	1.6
Average Group IV			23.6	8.4

[a]Group I = all countries that report both brake and tire wear; Group II = countries that only report brake and tire wear; Group III = countries that report road abrasion in the range up to 5 mg/vkm; Group IV = countries that report road abrasion above 5 mg/vkm due to road sanding and/or use of studded tires.
Underlying Emission Data Downloaded From Centre for Emission Inventories and Projections (http://www.ceip.at/) on February 15, 2017.

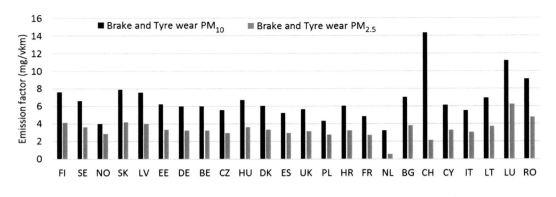

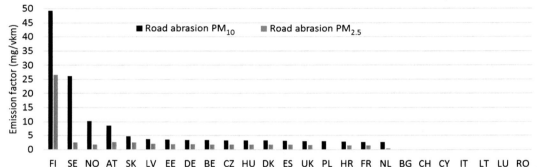

FIGURE 5.6 Calculated implied emission factors for PM_{10} and $PM_{2.5}$ for brake and tire wear (top) and for road abrasion (bottom) for European countries that report for at least one of the categories.

It should be noted that Group IV is probably not only reporting primary road abrasion but also, to varying degrees, resuspension of abraded dust and/or sand. The current emission reporting format does not allow a separation between primary road abrasion and resuspension (see also Section 6.5). For break and tire wear Group I and Group II the average PM_{10} emission factors are 5.9 and 8.5 mg/vkm, respectively, while the average $PM_{2.5}$ emission factors are 3.2 and 3.7 mg/vkm, respectively (Table 5.3). For Group I, a considerable range is observed of 3.2–7.9 mg/vkm for PM_{10} and especially for $PM_{2.5}$ with 0.6–4.2 mg/vkm. The Netherlands represents the low end, whereas Finland and Slovakia represent the high end. The range of a factor of 2 or more suggests that the emission calculations, although referring to guidebooks, are not yet harmonized.

For road abrasion, the major difference between Groups III and IV is the inclusion of road abrasion due to road sanding and/or the use of studded tires. This is applicable for Scandinavian and Alpine countries. For group IV it is difficult to draw conclusions about consistency because studded tires and road sanding lead to different wear patterns and one needs to understand the local situation thoroughly. Moreover, the period in which the antiskid measures occur may differ, as well as the duration. Nevertheless, the discrepancies between, for example, Norway and Finland for $PM_{2.5}$ (Table 5.3, Fig. 5.6) cannot be easily understood from a process understanding. For Group III, road abrasion excluding road sanding and/or studded tires, the implied emission factors for PM_{10} appear quite harmonized. For $PM_{2.5}$, however, substantial differences exist varying from 0 (Poland) to 2.5 (Slovakia) mg/vkm $PM_{2.5}$.

The data compiled in Fig. 5.6 and Table 5.3 illustrate that in-depth comparisons and a harmonization exercise between countries could significantly influence the emission estimates by country. It should be stressed that a decade ago comparing implied emission factors for wear emissions for various countries would have been impossible due to a lack of data. So much progress has been made. On the other hand, the analysis here suggests that currently a level playing field for reduction strategies is not yet reached, with countries using significantly varying emission factors and $PM_{2.5}/PM_{10}$ fractions.

4.2 Size Distribution of Reported Wear Particles

Because the regulated size fractions are PM_{10} and $PM_{2.5}$, we do not discuss a further breakdown to PM_1 and $PM_{0.1}$ (ultra fine particles). For these size classes it is better not to rely on official emission reporting, as they are not compulsory and therefore receive limited attention. For particle numbers including $PM_{0.1}$ we refer to the recently published inventory by Paasonen et al., (2016). The share of the fine fraction ($PM_{2.5}$) and the coarse fraction (PMc; $PM_{2.5-10}$) for brake and tire wear and for road wear for the EU15, Norway, and Switzerland and the EU NMS is presented in Figs. 5.7 and 5.8, respectively. It can be seen that Austria and Portugal do not report brake and tire wear. Road surface wear is not reported by Switzerland, Italy, Luxemburg, Portugal (Fig. 5.7) and Bulgaria, Cyprus, Lithuania, and Romania (Fig. 5.8). Moreover, in some cases the emissions are presented in the reporting as "not estimated" or IE. For the latter category it is usually not known where included exactly, and what the specific contribution elsewhere is. This illustrates that it is complicated to compare properly between countries. Nevertheless, useful insight in the status of emission estimations in the European countries can be derived from Figs. 5.7 and 5.8.

The fraction of $PM_{2.5}$ in PM_{10} for both road surface wear and the combined brake and tire wear category according to the EMEP/EEA guidebook is 54% (Table 5.4). This ratio

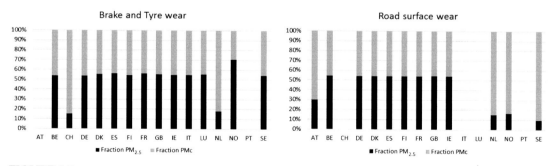

FIGURE 5.7 Fraction of $PM_{2.5}$ and PMc ($PM_{2.5-10}$) for the EU15, Norway, and Switzerland in brake and tire wear (left) and road surface wear (right). Greece is not included because it reports incomplete for both wear and exhaust emissions; this excludes the possibility that the emissions are accounted for but included elsewhere, for example Portuguese states.

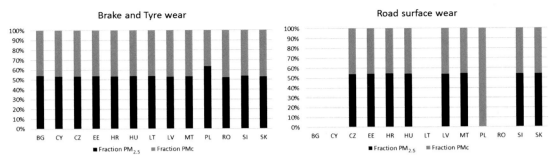

FIGURE 5.8   Fraction of $PM_{2.5}$ and PMc ($PM_{2.5-10}$) for the EU new Member States in brake and tire wear (left) and road surface wear (right).

TABLE 5.4   Emission Factors for Total Suspended Particulates (TSP), PM_{10} and $PM_{2.5}$ for Passenger Cars (PC) From the EMEP/EEA Emission Inventory Guidebook Tier 1 and Tier 2 Methodology

Source	TSP[a] Emission (mg/km)	Mass Fraction of TSP		PM_{10} (mg/km)	$PM_{2.5}$ (mg/km)	Fraction $PM_{2.5}$ in PM_{10} (%)
		PM_{10}	$PM_{2.5}$			
PC tire wear	10.7	0.60	0.42	6.4	4.5	70
PC brake wear	7.5	0.98	0.39	7.4	2.9	40
Brake and tire wear aggregated	18.2			13.8	7.4	54
PC road surface wear	15.0	0.50	0.27	7.5	4.1	54

[a]Total suspended particles.
Compiled From Ntziachristos and Boulter (2016).

is recognized in Figs. 5.7 and 5.8 for the vast majority of countries. Notable exceptions using a different methodology are Switzerland, Netherlands, Austria, Norway, Sweden, and Poland. Brake and tire wear are caused by the same activity and this, to some extent, justifies grouping and reporting them as one category for simplicity and compactness. A complicating issue, however, is that the $PM_{2.5}$ fraction in PM_{10} is different for brake or tire wear (Ntziachristos & Boulter, 2016; PMP IWG, 2016). Therefore, for a good understanding of $PM_{2.5}$ levels and the contribution of different processes in the street environment, separate reporting would be preferred.

5. RESUSPENSION DUE TO ROAD TRANSPORT

Resuspension of dust particles smaller than 10 microns due to vehicle-generated turbulence is widely recognized as a significant source of PM_{10}, especially in the urban environment. The material that is being resuspended varies with the nature of the local circumstances. It typically includes particles from vehicle tire wear, brake wear, primary exhaust emissions that have settled on road surfaces (perhaps adhering to larger particles), and environmental dust from many sources, such as pollen, sea salt, construction work, and wind-blown soil. In addition, the contribution from resuspension to total PM depends on meteorological conditions; during episodes of rain and road surface wetness, resuspension is reduced (Amato et al., 2012; Denby et al., 2013).

The mix of particles with varying chemical composition and the effects from local and climatic conditions make resuspension a difficult source to quantify. Various studies have suggested that emissions from resuspension are of the same order of magnitude as primary emissions from road traffic, relating to vehicle exhaust fumes, tire wear, and brake wear (Gehrig et al., 2004; Harrison et al., 2001; Lenschow et al., 2001; Thorpe, Harrison, Boulter, & McCrae, 2007). Furthermore, a number of studies have reported that resuspension due to HDV is higher by about a factor of 8—10 compared with light-duty vehicles (e.g., Abu-Allaban, Gillies, Gerler, Clayton, & Proffitt, 2003; Gehrig et al., 2004; Thorpe et al., 2007). An extensive review of the literature on mineral dust and resuspension due to road traffic is made by Denier van der Gon et al. (2010). Based on this review, a selection of emission factors for resuspension expressed in mg/vkm is proposed (Table 5.5).

Special correction factors are developed to accommodate adjustment of calculated emissions for climate (southern Europe) and for nonskid measures, such as studded tires and road gritting (northern Europe), which are known to increase resuspension emissions (Denier van der Gon et al., 2010; Schaap et al., 2009). The result of this approach is that estimated resuspension emissions will show variation from year to year because climate variables such as rainfall, ice, and snow vary from year to year. Here we use the modeled resuspension emissions for an average year and compare with reported total wear emissions for 2010 (Fig. 5.9). The comparison suggests that resuspension emissions can be substantially larger than primary wear emissions. The resuspension is a function of the number of vkm driven; hence, large countries with more vkm also have higher emissions. The countries are grouped in Fig. 5.9 according to their climate and geographic location. Resuspension emissions in the southern European countries stand out due to their arid climate and related large dust reservoir. The northern countries have relatively large emissions, despite their wet climate due to the road sanding and/or use of studded tires. Finland is the only country that reports higher wear emissions than our model-calculated resuspension emission. A reason for this may be that Finland includes an estimate for resuspension in their wear emission reporting (see also the discussion in Section 6.4.1). The modeled resuspension emissions presented in

TABLE 5.5 $PM_{2.5-10}$ Emission Factor (mg/vkm) for Traffic-Related Resuspension as a Function of Road Type for Light- and Heavy-Duty Traffic, for Application in the LOTOS-EUROS Air Quality Model (Denier van der Gon et al., 2010). In the Emission Estimation, the Emissions are Modified by the Model for Climatic and Specific Country Conditions.

	Road Type		
	Highway	Rural	Urban
Heavy-duty vehicles	198	432	432
Passenger cars and light-duty vehicles	22	48	48

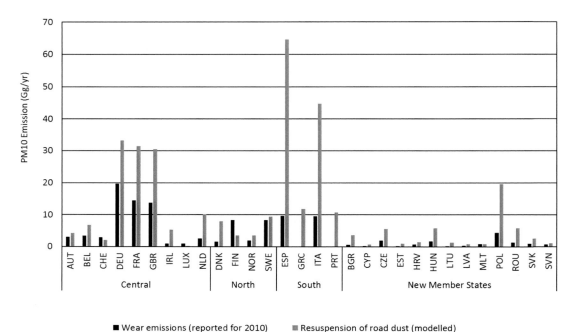

FIGURE 5.9 Total PM_{10} wear emissions (brake, tire, road) reported by countries to EMEP and modeled PM_{10} emissions due to resuspension by road transport.

Fig. 5.9 are only indicative and are highly uncertain. Nevertheless, they convincingly make the point that resuspension should be quantified and properly understood, as it may well be the dominant source of especially coarse PM_{10} ($PM_{2.5-10}$) (Denier van der Gon et al., 2013, 2010; Amato et al., 2014).

6. TRACERS TO VERIFY BRAKE, TIRE, AND ROAD WEAR

Tracers have been identified that can help to quantify the road transport wear emissions in ambient PM (Denier van der Gon et al., 2013; Denier van der Gon, Hulskotte, Visschedijk, & Schaap, 2007), but generally these substances are not reported separately in emission inventories. An exception is copper (Cu), a good tracer for brake wear (Denier van der Gon et al., 2007). Heavy metals such as copper are persistent in the environment and are known to have adverse effects on the environment and human health (WHO, 2007). The Protocol for Heavy Metals under the UNECE CLRTAP articulates the importance of heavy metals for air quality, including copper (UNECE, 1998). Although reporting for heavy metals is notoriously incomplete, there are enough Cu emission data for the EU15 to investigate consistency between brake wear PM reporting and Cu emission reporting for the sector road transport. A complication is that the reporting guidelines combine road transport brake and tire wear (Table 5.1 and EEA, 2016). Brake wear is a dominant source of copper, but the tire wear contribution to copper emissions is relatively negligible (Denier van der Gon et al., 2007; Ntziachristos and Boulter, 2016). A good approximation of the contribution by brake wear only can be made by using the ratio's between brake and tire wear shown in Table 5.4; the result is presented in the first column of Table 5.6.

Assuming that brake wear is responsible for the copper emission in the category brake and tire wear, the fraction of copper in reported PM_{10} emissions can be calculated (Fig. 5.10). This reveals some surprising extremes: 0.5% Cu (UK) up to 30% Cu (Germany) in brake wear PM_{10}, but the majority of the values is between 2% and 7%. The outlier data for the UK and Germany are most likely mistakes or leftover from a previous estimation procedure. The only other explanation would be that in those countries the brake pad and disc compositions would be significantly different from other European countries. This example shows how a "normalization" can help to quickly flag reporting that needs more attention; when only looking at total national emissions, these outliers would not have been identified. How can we support our hypothesis that these outliers are likely to errors? Hulskotte, Roskam, and Denier van der Gon (2014) provide an elemental composition profile of brake pads and brake discs as used in the Netherlands in 2012. Because car, engine, and safety regulations are not nationally determined but controlled by European legislation, the resulting profiles may be representative for the European PC fleet. The average brake pad profile

TABLE 5.6 PM_{10} From Brake Wear Derived From Reported Emissions, Reported Copper Emissions, and the Fraction of Copper in Brake Wear PM_{10} for EU15 Countries Reporting Both Substances and the Value Published by Hulskotte et al., (2014) Based on Analysis of Brake Pads and Brake Discs for Comparison

	Brake Wear PM_{10}[a] (Gg/year)	Copper From Brake and Tire Wear[b] (Gg/year)
Belgium	1.18	0.02
Denmark	0.56	0.039
Finland	0.61	0.044
France	5.05	0.133
Germany	6.83	2.066
Ireland	0.35	0.01
Italy	5.15	0.098
Netherlands	0.76	0.021
Spain	3.27	0.142
Sweden	0.91	0.046
United Kingdom	4.87	0.023
Norway	0.29	0.015

[a]To derive the brake wear only emissions, the ratio between brake wear and tire wear contribution in Table 5.4 is used, which allocates 55% of the combined PM_{10} emissions to brake wear.
[b]Reported jointly, but all Cu assumed to be released from brake wear based on Denier van der Gon et al. (2007) and Ntziachristos and Boulter (2016).

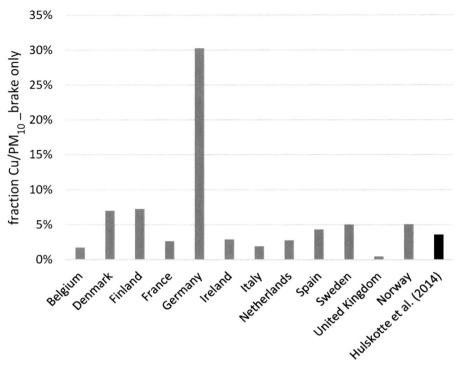

FIGURE 5.10 PM_{10} from brake wear derived from reported emissions, reported copper emissions, and the fraction of copper in brake wear PM_{10} for EU15 countries reporting both substances and the value published by Hulskotte et al. (2014) for comparison.

contained about 50% of nonmetal material, and 20% Fe, 10% Cu, 4% Zn, and 3% Sn as the dominant metals (Hulskotte et al., 2014). One of the remarkable findings of Hulskotte et al. (2014) is that 70% of the brake wear originates from the discs and only 30% from the brake pads. This has important consequences for the elemental composition of overall brake wear as the discs almost entirely consist of Fe (>95%), but the brake pads contain a wide variety of metals with Fe, Cu, Zn, and Sn together making up about 80%—90% of the metals present. The new brake wear profile derived by Hulskotte et al. (2014) suggests a copper fraction in brake wear material (discs and pads combined) of about 3.6%. Several countries report close to this value (Fig. 5.10), but overall the conclusion is that consistency and harmonization of emissions by source sectors across pollutants is limited and needs further attention. Moreover, a value of 1.8% or 7% may seem close to 3.6%; it still results in a factor of 2—4 different emissions for a health relevant heavy metal such as copper.

7. CONCLUSIONS

Stringent EU policies succeed in reducing the road transport exhaust PM emissions, but do not address "non-exhaust" emissions from brake wear, tire wear, road abrasion, and road dust suspension. This is reflected in the trends of reported PM_{10} and $PM_{2.5}$ emission from road transport by European countries. The engine exhaust emissions dominated the road transport emissions pre-2000, but since about 2012, wear emissions contribute more to road transport PM_{10} emissions than exhaust emissions in the EU15, Norway, and Switzerland. This is not yet the case in the EU NMS where exhaust emissions are still dominant. The main reason for this difference is a relatively older vehicle fleet with higher exhaust emissions per vkm. The trend, however, is similar to the EU15 and in the near future wear emissions may also become the dominant road transport PM_{10} emissions source in the EU NMS.

In the comparison between exhaust emissions and wear emissions, it is important to distinguish between PM_{10} and $PM_{2.5}$. Exhaust emissions still dominate the road transport $PM_{2.5}$ emissions in Europe because exhaust emissions consist almost entirely of $PM_{2.5}$, whereas only about 50% of the PM_{10} wear emissions are thought to consist of $PM_{2.5}$.

When intercomparing the emission reported by countries in Europe, we conclude that harmonization and consistency have much improved over the past 10 years. This is at least partly, if not foremost, due to wide adoption of the EEA/EMEP emission inventory guidebook (EEA, 2016). It stresses the importance of critically reviewing and maintaining the emission inventory guidebook. However, consistency in itself does not imply that the most consistently used emission factors are also the most accurate. Because wear emissions are (becoming) the dominant road transport PM_{10} source, more research is needed to establish a large data set to underpin, and possibly adjust, the guidebook emission factors. Recently, this was also stressed by the UNECE Informal Working Group on the Particle Measurement Programme, stating that an extremely wide range of driving conditions as well as different sampling techniques have been applied in non-exhaust—related studies, which has often led to different results and/or contradictory conclusions (PMP IWG, 2016).

When we compare and analyze the official reported emissions, it is concluded that a substantial number of countries still report incomplete, despite the improvements over the years. These countries do one or more of the following: not reporting any wear emissions, no separation between exhaust and wear emissions, and not separate brake and tire wear from road abrasion or report only one of the two wear categories (brake and tire wear or road abrasion). Moreover, when the reported emissions are normalized by expressing them per vkm driven, more discrepancies are observed. The implied emission factors per vkm should not necessarily be the same for each country, but the ranges are much larger than can be understood due to different climatic conditions or vehicle fleets. The calculated PM_{10} brake and tire wear implied emission factor is 6.7 mg/vkm (range 3.2—14.3 mg/vkm). For $PM_{2.5}$ this is 3.4 mg/vkm (range 0.6—6.2 mg/vkm). A possible reason for some of the high-end values can be the inclusion of road abrasion, as was discussed, but differences up to a factor of 2—4 persist even if the extremes are excluded.

For road abrasion the situation is even more complex, partly because country climatic conditions differ. Scandinavian and Alpine countries use road sanding and/or studded tires to ensure road safety in winter months. This is known to generate intense road abrasion. For the Scandinavian and Alpine countries the implied PM_{10} emission factor for road abrasion varies between 8.6 and 49.3 mg/vkm ($PM_{2.5}$: 1.7—26.7 mg/vkm). This range is extremely large and poorly understood. For the non-Scandinavian and non-Alpine countries that do report road abrasion, the ranges are more comparable: for PM_{10} 3.2 (2.6–4.7) mg/vkm and for $PM_{2.5}$ 1.6 (0–2.5) mg/vkm.

Copper can be used as a tracer to identify brake wear. It was investigated if the reported copper emissions due to brake wear are consistent with the reported PM_{10} emissions from brake wear. This revealed some significant outliers that should probably be corrected. The purpose of this comparison is merely to illustrate that consistency in the reporting within a country is not guaranteed. If we are to use a tracer, such as Cu for brake wear, to identify the contribution of the brake, tire or road abrasion—the inventory for the tracer needs to be made independently and bottom-up for the entire domain.

An important conclusion is that resuspension of road dust may well be the dominant source of road transport PM_{10} in many cities but is currently excluded from official emission reporting. It would be highly recommendable to include road resuspension as a separate category. The lack of a clear location for reporting resuspension, in the future, may lead to a further increase of inconsistency between countries. For transparency and intercomparison, it would be advantageous to have a separate NFR category for resuspension in the official reporting, even if reporting would be on a voluntary basis, e.g., as a memo item.

Finally, the overall conclusion is that in the European urban environment, where most Europeans live and spend their time, the contribution of wear emissions to urban road transport PM_{10} emissions already reaches more than 50% in many countries and is predicted to grow to about 80%–90% after 2020, even when resuspension is not or only incompletely taken into account.

References

Abu-Allaban, M., Gillies, J. A., Gerler, A. W., Clayton, R., & Proffitt, D. (2003). Tailpipe, resuspended road dust, and brake-wear emission factors from on-road vehicles. *Atmospheric Environment, 37*, 5283–5293. https://doi.org/10.1016/j.atmosenv.2003.05.005.

Amato, F., Cassee, F. R., Denier van der Gon, H. A. C., Gehrig, R., Gustafsson, M., Hafner, W., et al. (2014). Urban air quality: The challenge of traffic non-exhaust emissions. *Journal of Hazardous Materials, 275*, 31–36.

Amato, F., Schaap, M., Denier Van Der Gon, H. A. C., Pandolfi, M., Alastuey, A., Keuken, M., et al. (2012). Effect of rain events on the mobility of road dust load in two Dutch and Spanish roads. *Atmospheric Environment, 62*, 352–358.

Boulter, P. G. (2005). *A review of emission factors and models for road vehicle non-exhaust particulate matter. TRL Report PPR065.* Wokingham, UK: TRL Limited.

Denby, B. R., Sundvor, I., Johansson, C., Pirjola, L., Ketzel, M., Norman, M., et al. (2013). A coupled road dust and surface moisture model to predict non-exhaust road traffic induced particle emissions (NORTRIP). Part 2: Surface moisture and salt impact modelling. *Atmospheric Environment, 81*, 485–503.

Denier van der Gon, H. A. C., Gerlofs-Nijland, M. E., Gehrig, R., Gustafsson, M., Janssen, N., Harrison, R. M., et al. (2013). The policy relevance of wear emissions from road transport, now and in the future — an international workshop report and consensus statement. *American Journal of the Air & Waste Management Association, 63*, 136–149.

Denier van der Gon, H. A. C., Hulskotte, J. H. J., Visschedijk, A. J. H., & Schaap, M. (2007). A revised estimate of copper emissions from road transport in UNECE-Europe and its impact on predicted copper concentrations. *Atmospheric Environment, 41*, 8697–8710.

Denier van der Gon, H. A. C., Jozwicka, M., Hendriks, E., Gondwe, M., & Schaap, M. (2010). *Mineral dust as a component of particulate matter.* PBL report Report 500099003, ISSN: 1875–2322 (print) ISSN: 1875-2314 (on line). Bilthoven, The Netherlands: Netherlands Environmental Assessment Agency.

EEA (European Environment Agency). (2016). *EMEP/EEA air pollutant emission inventory guidebook, 2016 edition.* http://www.eea.europa.eu/publications/emep-eea-guidebook-2016. EEA Report No 21/2016.

EMEP. (2016). *Transboundary particulate matter, photo-oxidants, acidifying and eutrophying components.* Joint MSC-W & CCC & CEIP Report, EMEP Status Report 1/2016.

Gehrig, R., Hill, M., Buchmann, B., Imhof, D., Weingarter, E., & Baltensprenger, U. (2004). Separate determination of PM_{10} emission factors of road traffic for tailpipe emissions and emissions form abrasion and resuspension processes. *International Journal for Environment and Pollution, 22,* 312–332.

Harrison, R. M., Yin, J., Mark, D., Stedman, J., Appleby, R. S., Booker, J., et al. (2001). Studies of the coarse particle (2.5-10μm) component in UK urban atmospheres. *Atmospheric Environment, 35,* 3667–3679.

Hulskotte, J. H. J., & Jonkers, S. (2012). *Exploration of uncertainty in wear emission parameters and related bandwidth in the contribution of traffic to future kerbside concentrations of PM, TNO-060-UT-2012–20100585.* Utrecht: TNO.

Hulskotte, J. H. J., Roskam, G. D., & Denier van der Gon, H. A. C. (2014). Elemental composition of current automotive braking materials and derived air emission factors. *Atmospheric Environment, 99,* 436–445.

Jonkers, S. (2007). *Handleiding car II.* Versie 6.1, TNO-rapport 2007-A-R0788/B.

Kuenen, J. J. P., Visschedijk, A. J. H., Jozwicka, M., & Denier van der Gon, H. A. C. (2014). TNO-MACC_II emission inventory: A multi-year (2003–2009) consistent high-resolution European emission inventory for air quality modelling. *Atmospheric Chemistry and Physics, 14,* 10963–10976. https://doi.org/10.5194/acp-14-10963-2014.

Lenschow, P., Abraham, H. J., Kutzner, K., Lutz, M., Preu, J. D., & Reichenbacher, W. (2001). Some ideas about the sources of PM_{10}. *Atmospheric Environment, 35,* 23–33.

Ntziachristos, L., & Boulter, P. (2016). Chapter 1.A.3.b.vi Road transport: Automobile tyre and brake wear and 1.A.3.b.vii Road transport: Automobile road abrasion. In *EMEP/EEA air pollutant emission inventory guidebook 2016.* EEA Report No 21/2016.

Paasonen, P., Kupiainen, K., Klimont, Z., Visschedijk, A., Denier van der Gon, H. A. C., & Amann, M. (2016). Continental anthropogenic primary particle number emissions. *Atmospheric Chemistry and Physics, 16,* 6823–6840. https://doi.org/10.5194/acp-16-6823-2016.

Papadimitriou, G., Ntziachristos, L., Wuetrich, P., Notter, B., Keller, M., Fridell, E., et al. (2013). *TRACCS - transport data collection supporting the quantitative analysis of measures relating to transport and climate change.* EMISIA Report No. 13.RE.025.V1 http://traccs.emisia.com.

PMP IWG (PMP – Particle Measurement Program UNECE Informal Working Group). (2016). *Non-exhaust traffic related particle emissions, (brake and tyre/road wear) summary report from the PMP IWG Informal document GRPE-73-14, 73rd GRPE, 6–10 June 2016.* https://www.unece.org/fileadmin/DAM/trans/doc/2016/wp29grpe/GRPE-73-14.pdf.

Rexeis, M., & Hausberger, S. (2009). Trend of vehicle emission levels until 2020 – Prognosis based on current vehicle measurements and future emission legislation. *Atmospheric Environment, 43,* 4689–4698.

Schaap, M., Manders, A. M. M., Hendriks, E. C. J., Segers, A. J. S., Denier van der Gon, H. A. C., Jozwicka, M., et al. (2009). *Regional modelling of particulate matter for The Netherlands: Technical background report BOP, ECN, TNO, PBL report 5000990088* (Bilthoven, Netherlands).

Thorpe, A., Harrison, R. M., Boulter, P. G., & McCrae, I. S. (2007). Estimation of particle resuspension source strength on a major London Road. *Atmospheric Environment, 41,* 8007–8020. https://doi.org/10.1016/j.atmosenv.2007.07.006.

UNECE. (1998). *The 1998 Aarhus protocol on heavy metals.* Available at http://www.unece.org/env/lrtap/hm_h1.html.

UNECE. (2009). *Guidelines for estimating and reporting emission data under the convention on long-range transboundary air pollution (ECE/EB.AIR/97).*

WHO (World Health Organization). (2007). *Health risks of heavy metals from long range transboundary air pollution.* Copenhagen, Denmark: WHO Regional Office for Europe, ISBN 978 92 890 7179 6.

Review of Brake Wear Emissions
A Review of Brake Emission Measurement Studies: Identification of Gaps and Future Needs

Jana Kukutschová[1], Peter Filip[1, 2]

[1]VSB – Technical University of Ostrava, Ostrava, Czech Republic; [2]Southern Illinois University Carbondale, Carbondale, United States

O U T L I N E

1. Introduction	123	4.2 Brake Dynamometer Studies	131
		4.3 Field Studies	138
2. Automotive Friction Brake Materials	125	5. Identification of Gaps	141
3. Transformations of Friction Brake Materials and Release of Wear Emissions	126	6. Future Trends in Development of Materials for Automotive Brakes	142
3.1 Formation of Friction Layer	126	7. Summary	143
3.2 Wear and Wear Products	127	Acknowledgments	143
4. Brake Wear Emission Studies	129	References	144
4.1 Pin-On-Disc Studies	129		

1. INTRODUCTION

Every day millions of automobiles worldwide need to decelerate and stop every time the driver depresses the brake pedal. Every application of a friction brake leads to the release of direct wear products, either gaseous or particulate matter. Depending on driving conditions and formulation of brake materials, the released wear debris can be partially attracted to the vehicle hardware and fall on road surfaces, but a considerable fraction is released as airborne matter. Because the number of vehicles in operation worldwide is increasing (over 1 billion units in 2010) (Sousanis, 2011), the contribution of road transportation to environmental pollution is identified as significant. Exhaust and non-exhaust sources were estimated to contribute approximately equal to traffic-related PM_{10} emissions (Grigoratos & Martini, 2015). Over the last decades, the automotive industry has also made enormous efforts to minimize exhaust emissions because of strict exhaust emission control, whereas emissions related to direct brake wear are not addressed as strictly by governmental agencies. Brake wear is considered to be one of the most important non-exhaust traffic-related sources of particulate emissions. Garg, Cadle, Mulawa, and Groblicki (2000) estimated total brake wear of nonasbestos organic (NAO) brake pads for a small passenger car ranging between 3.2 and 8.8 mg/km. Abu-Allaban, Gillies, Gertler, Clayton, and Proffitt (2003) quantified brake wear contribution to emissions per light-duty vehicle to PM_{10} and $PM_{2.5}$ as 0—80 and 0—5 mg/km, respectively. Hagino, Oyama, and Sasaki (2016) observed airborne brake wear emissions related to NAO pads ranging from 0.04 to 1.4 mg/km/passenger car for PM_{10} and from 0.04 to 1.2 mg/km/car for $PM_{2.5}$. Sanders, Xu, Dalka, and Maricq (2003) reported that approximately 50% of brake wear was emitted as airborne in both field and dynamometer tests irrespectively of a formulation of brake material tested and PM_{10} accounted for 63%—85%. However, most of these studies were focused on estimation and characterization of PM_{10} or $PM_{2.5}$ emissions only. In the last decade, more attention is also being paid to generation, monitoring, and characterization of nanoparticulate emissions (Alemani, Nosko, Metinoz, & Olofsson, 2016; Hagino, Oyama, & Sasaki, 2015, 2016; Kukutschová et al., 2011; Nosko & Olofsson, 2017; Wahlström, Söderberg, Olander, Jansson, & Olofsson, 2010).

Several review papers tried to summarize the current knowledge on automotive brake friction materials and their wear emissions. Chan and Stachowiak (2004) published a review only on automotive brake friction materials as a summary of components used in brake pad formulations with no emphasis on wear and wear particles. A review by Thorpe and Harrison (2008) was focused on non-exhaust particulate emissions from road traffic in general, with a partial focus on brake wear emissions, their physical properties, and chemical composition. A specific composition-related issue such as the role of copper in automotive brake materials, its environmental impact, and potential substitutes was summarized in the review of Straffelini, Ciudin, Ciotti, and Gialanella (2015). The most detailed review was published by Grigoratos and Martini (2015) summarizing published information on particulate emissions from automotive friction brakes. Special focus was put on methods for generation, quantification, and characterization of particulate emissions from brake pad materials. An overview of the most important studies investigating the mass and number distribution of airborne brake wear particles and characterization of their chemical composition is also provided. To quantify a contribution of brake wear emissions, a topic of emission factors (EF) is included. Perricone et al. (2016) published a review focused on the issue of quantification of airborne brake wear emissions via estimation of EF of different brake materials based on particle mass and number. The review by Peikertová and Filip (2016) summarized different approaches adapted when addressing the detection of brake wear in the environment (air pollution, road dust, soil and sediments, water runoffs) and the possible impact to plants, animals, and human health as well. Guevara (2016) provided an overview on emissions of primary PM with a special focus on source categories, size distribution, and speciation source profiles and inconsistently with other studies reported that wear of brakes is contributing only to the emissions of coarse particles (>2.5 μm). Since 2015, numerous research papers dealing with methods for simulation of wear and particle emissions (Wahlström, 2015), laboratory generation of wear particles and their chemical composition (Verma et al., 2016, 2015), online monitoring of particle size distribution (Alemani et al., 2016; Hagino et al., 2015, 2016; Perricone et al., 2016), chemical analyses and testing of wear particles (Beddows, DallÓsto, Olatunbosun, & Harrison, 2016; Kazimirová et al., 2016; Malachová et al., 2016), and monitoring of traffic tracers for brake wear in urban areas and road dust (Megido, Negral, Castrillón, Maranón, & Fernandez-Nava, 2016) were issued.

2. AUTOMOTIVE FRICTION BRAKE MATERIALS

Friction brakes in passenger vehicles use friction between a brake material and a rotating part (drum or disc) to slow down or stop a vehicle by transforming kinetic energy mostly into frictional heat. Design principles of automotive brake systems are thoroughly discussed in Day (2014) and Limpert (2011). In general, there are two main groups of brake systems: disc and drum. Disc brakes typically have a rotating disc made of a pearlitic gray cast iron (in some cases made of steel, carbon—carbon, ceramic, or aluminum matrix composites). Brake lining materials underwent a distinct development in the past three decades, which is related to an effort to develop optimal materials, which would reflect the requirements determined by transportation safety, economy, environmental concerns, comfort, and increased performance demand. As a rule, the brake lining material should have (1) appropriate mechanical and thermal properties, (2) adequate and stable coefficient of friction in a wide range of operating conditions (temperature, pressure, environment, e.g., dust, water, deicing agents), and (3) high resistance to wear and good compatibility with the rubbing counterpart. Ideally, they should operate reliably under hot, dry, wet, or cold conditions, and without pollution and noise, and should be easily manufactured at low costs. Brake systems represent a relevant element in transportation safety and, in spite of replacement of asbestos and gradual elimination of copper, also in environmental loading. Over the last decade, several research teams were dealing with the development of eco-friendly brake pad formulations via replacement of some potentially hazardous ingredients such as Cu, Sn, antimony trisulfide, and whisker materials (Aranganathan & Bijwe, 2016; Lee & Filip, 2013; Qi et al., 2014; Yun, Filip, & Lu, 2010).

According to the terminology accepted in the automotive industry, typically used brake materials are roughly categorized as NAO, low metallic (LM) or low steel, and semimetallic (SM) (Table 6.1).

These groups can be considered organic because the matrix of these complex composites is formed by one or more organic polymers. The friction materials usually contain four classes of ingredients (Filip, 2013):

- *Reinforcing constituents* bring strength to the entire composite. Characteristic examples could be metallic fibers and particulates (steel, Fe, Cu, Al), ceramic and mineral fibers and particulates (e.g., wollastonite, $ZrSiO_4$), and polymer fibers or pulp (e.g., Kevlar, acrylic fiber).

TABLE 6.1   Main Groups of Brake Pad Formulations

Formulation	Main Ingredients	Characteristics
NAO	Organic compounds, mineral fibers, graphite, titanates, etc.	Relatively soft, low brake noise, lower friction performance, sensitive to elevated temperature
LM	Mixture of organic and metallic components	Higher friction performance, higher durability
SM	Predominantly metallic—steel fibers, over 30, typically less then 60 wt.% of low carbon steel fiber and/or iron powder	Highest durability, good heat transfer, higher wear of rotor

LM, low-metallic; NAO, nonasbestos organic; SM, semimetallic.

- *Friction modifiers* influence the level of friction. They are either solid lubricants (such as graphite and different metal sulfides), lowering the friction level, or abrasives (e.g., Al_2O_3, SiC) increasing the friction level.
- *Fillers* are thermally stable inexpensive ingredients occupying the volume and having a minimum impact on the performance. The most frequently used filler is barium sulfate.
- *Binders* hold all ingredients together. The most frequently used is phenolic resin or modified phenolic resin, combined with different rubbers (styrene butadiene rubber and nitrile butadiene rubber) to improve the ability of a material to absorb energy.

A characteristic brake pad material is a multicomponent composite, typically formulated of more than 10 constituents bound in a polymer matrix. Several thousand different raw materials and their allotropic modifications have been used in formulations of brake pads by different brands (Filip, Kovarik, & Wright, 1997). Hulskotte, Roskam, and Denier van der Gon (2014) analyzed the composition of 65 automotive brake pads and 15 rotors used in passenger cars, collected from maintenance facilities in the Netherlands, to obtain a representative picture of chemical composition of brake materials currently on roads. All pads were LM or SM with Fe as the dominant element, followed by Cu, Zn, Sn, Al, Si, S, Zr, Ti, Sb, Cr, Mo, and Mn. Some of the pads were free of Cu and Sb.

3. TRANSFORMATIONS OF FRICTION BRAKE MATERIALS AND RELEASE OF WEAR EMISSIONS

A friction process is always accompanied with the formation of friction products, which includes the development of friction debris with subsequent formation of a friction layer, release of brake wear particles, and gaseous emissions.

3.1 Formation of Friction Layer

Generated friction debris adheres to the friction surface of brake pads and forms a friction layer. This layer is crucial for friction performance of each brake pad. Because of the complexity of friction phenomena in polymer matrix composites, and despite the great importance of the structures existing on the surface of brake pads, the mechanisms of friction layer formation is still not fully understood (Filip, Weiss, & Rafaja, 2002; Neis, Ferreira, Fekete, Matozo, & Masotti, 2017). Friction layer is not continuous and differs according to temperature and general conditions experienced during friction processes. Lower energy dissipation during less severe brake applications, typically at lower temperatures, is in general accompanied with lower wear and friction debris release. A thick friction layer is developed predominantly in scratches, which are filled with generated wear debris (Filip et al., 2002).

The friction debris and friction layer represent newly formed matter. Direct wear debris adheres to the bulk material of a brake pad and rotating counterpart. The layer is heterogeneous and consists of several different phases, e.g., fragments of carbonaceous matter, oxidized metals, various newly formed materials, and degradation products of phenolic resin (Filip et al., 2002). Composition of the friction layer also differs for various conditions. More thermally stable components, such as graphite, zircon ($ZrSiO_4$), and barite, do not significantly interact with other components or with the environment and can be detected in friction layer and wear debris regardless of testing conditions. In contrast to the stable components, vermiculite is prone to a dehydration process, and exposure to 200°C and higher temperatures leads to loss of water and transfers into micaceous or talc-type mineral (Filip et al., 2002). Metal sulfides (Sb_2S_3 and MoS_2), which serve as solid lubricants, undergo an oxidation process to form not only various oxides but also elemental antimony responsible for alloying processes of Cu or Fe (Filip et al., 2002; Matějka et al., 2011). Very important from the environmental point of view are the changes in Fe and Cu metals. It has been proven that iron oxides typically represent the major component of the friction layer (Österle et al., 2014) and are also dominant compounds in airborne and nonairborne wear debris (Kukutschová et al., 2011; Verma et al., 2016; Wahlström, Söderberg, Olander, Olofsson, & Jansson, 2010). Filip et al. (2002) conducted dynamometer testing of LM brake materials at different temperatures and detected magnetite (Fe_3O_4) and maghemite (γ-Fe_2O_3) origination at temperatures above 400°C, while hematite (α-Fe_2O_3) formation above 600°C. Verma et al. (2015) performed pin-on-disc testing of an LM brake material, where maximum achieved contact temperature was below 50°C, and reported hematite (Fe_2O_3), as a product of tribooxidation of cast iron disc, to be the major iron form on the friction layer. Copper is providing good thermal conductivity and is linked to the capacity to eliminate thermal fade of brakes, act as a solid lubricant at high temperatures, and influence general tribological performance (Aranganathan & Bijwe, 2016; Lee & Filip, 2013; Österle, Prietzel, Kloß, & Dmitriev, 2010). At elevated temperatures between 400 and 600°C, Cu is being oxidized to Cu_2O and to CuO above 600°C (Filip et al., 2002). In addition, elemental copper (Cu^0) in the form of recrystallized copper nanosized particles with diameters about 100 nm (after moderate testing conditions) and about 10 nm (after severe conditions) were observed in the friction layer (Österle et al., 2010). Thus, various newly formed chemical phases and particle fractions, often significantly differing from the bulk material, are being released into the environment.

3.2 Wear and Wear Products

Wear of friction brakes represents a relevant performance parameter, as it dictates their life span. Brake performance typically decreases with increasing temperature, in extreme situations, leading to thermal fade. Because the friction process takes place on asperities, and the real contact area is significantly smaller than the apparent surface area, temperature and pressure on the friction surface are higher than values indicated by thermocouples. The stochastic nature of the braking process, the high-energy conditions, and complex mechano-chemical interactions on the friction surface during braking make it difficult to predict the chemistry and particle size of newly formed species. Even if the raw materials selected for manufacturing brake pads are in conformation with the environmental requirements, it is always seen that the newly formed wear particles will have a different chemistry and structure (Filip et al., 2002; Kukutschová et al., 2009). Similar to friction performance, wear of brake materials depends on temperature, sliding speed, pressure applied, and the chemistry and the structure of friction couple and surrounding environment. Typically, wear increases with increasing temperature (Filip, 2013; Filip et al., 2002). There is no simple and linear relationship that describes general wear dependence on distance or time when the brakes were applied. Although there is increased tendency to design one friction couple (pads and rotor) for the entire life span of a vehicle, for practical purposes, one member of the friction couple typically wears faster and can easily be replaced at a reasonably low cost (Filip, 2013).

Wear products can be generated by different wear mechanisms during friction processes. At temperatures over 300°C, wear of polymer matrix pads is accompanied with oxidative processes associated with considerable mass loss due to polymer degradation and formation of volatiles. This is characteristic for high-temperature braking scenarios, when numerous thermally less stable components (e.g., phenolic resin, rubber, graphite, coke) interact with available gases and oxygen from the ambient air (Kukutschová et al., 2009) or undergo a pyrolysis (Plachá et al., 2016). This degradation of organic components is associated with emissions of very fine amorphous carbon particles and volatile organic compounds (Kukutschová et al., 2011; Plachá et al., 2017). A combination of adhesive, abrasive, fatigue, and oxidative wear can also be observed at high-temperature conditions. At lower temperatures, "mechanical" wear dominates. In general, the adhesive wear mechanism is combined with abrasive wear, fatigue wear mechanisms, and oxidative wear as well. Thus, the produced brake wear debris is a complex mixture containing particles with sizes ranging from several nanometers to millimeters and the chemistry of wear debris is significantly different compared with the original pad material constituents, but typically comparable with the chemistry of a friction layer (Kukutschová et al., 2009; Roubíček, Raclavská, Juchelková, & Filip, 2008). Wear of pads and rotors generate particles of various sizes and morphology, and each combination of speed, pressure, and temperature leads to a different amount of wear (Kukutschová et al., 2009). Oxidative wear can generate very fine (submicron-sized), typically round-shaped particles, by degradation of organic binder with subsequent condensation and agglomeration of these finest fractions. Iron oxide particles, produced not only by oxidative mild wear of the cast iron disc but also from oxidation of Fe-based ingredients of metallic pads, are typically present in friction layer and together with elemental carbon from resin and other organic pad constituents represent one of the main components of airborne wear debris. Mechanical wear (abrasive and fatigue wear) typically leads to the release of larger particles, belonging mainly to PM_{10} or $PM_{2.5}$ fractions. These particles usually have sharper edges and irregular morphology (Kukutschová et al., 2011).

The majority of studies deal with experimental measurements of wear and detection of wear emissions; however, a full factorial design developed by Wahlström (2016) allows for prediction of the effect of wear parameters on friction performance of pad material and wear emissions. Although Wahlström suggested that a stable third body, a low specific wear, a stable resin, and a relatively high amount of metal fibers result in low wear emissions, this is not necessarily always true, particularly when adhesive mechanisms of wear are employed significantly. Nevertheless, it is possible to argue that the extent of emissions produced can be mitigated by a proper brake pad formulation strategy.

While the contribution of automotive brake wear particulate emissions to environmental pollution is being addressed to a considerably larger extent recently (Beddows et al., 2016; Gietl, Lawrence, Thorpe, & Harrison, 2010; Hagino et al., 2016; Kukutschová et al., 2011; Perricone et al., 2016; Wahlström, Olofsson et al., 2010), only limited attention is paid to the released organic compounds, despite the fact that organic components represent a significant part of brake pad formulations (Plachá et al., 2016). Because the friction processes related to braking are always associated with relatively high temperatures and pressures on the friction surface (at least in localized regions), a thermal decomposition of the organic and carbonaceous components and thermal fade of brake pads may occur (Yun et al., 2010). Thermal degradation of organic (polymer matrix) brake pads starts at approximately 150°C; however, this temperature is not sufficient for considerable oxidation of metallic particles (Fe, Sn, Cu). The maximum mass loss related to oxidative thermal degradation of organic binder and the release of volatiles typically occurs above 300°C (Křístková, Weiss, Filip, & Peter, 2004; Kukutschová et al., 2010). Metals and their oxides can catalyze the degradation process of phenol-formaldehyde resin (Křístková et al., 2004). These organic compounds may be released directly to the atmosphere or be adsorbed on solid wear particles (Plachá et al., 2016). Release of volatile organic compounds obviously happens also in the manufacturing process during hot pressing and postcuring processes.

4. BRAKE WEAR EMISSION STUDIES

Operation of brake systems is quite complex and often stochastic in nature. It is therefore not possible to simulate all braking scenarios. It is well known that friction and wear are system properties; nevertheless, researchers and manufacturers typically perform series of tests starting with small laboratory tests (e.g., pin-on-disc testing), followed by subscale and full-scale dynamometers and real field tests, demonstrating how the tested brakes perform in a wide range of conditions (Filip, 2013). Studies dealing with automotive brake wear emissions also use these methods to generate brake wear emissions for their subsequent evaluation. Advantage of small testers is typically related not only to cost but also to a considerably higher accuracy of detected physical variables compared with large-scale and field tests and often allows for a better understanding of wear mechanisms. Nevertheless, these tests have to be well understood and hardly can predict the wear behavior of real brake systems (Lee & Filip, 2013).

4.1 Pin-On-Disc Studies

Several studies generated brake wear particulate emissions using pin-on-disc tribometer (Mosleh, Blau, & Dumitrescu, 2004; Nosko & Olofsson, 2017; Verma et al., 2016; Wahlström,

Jansson et al., 2010), where pin samples made of real friction brake pads were pressed against a rotating cast iron disc. Mosleh et al. (2004) generated wear particles from commercial metallic truck brake pads and focused on settled particles only. The study addressed effects of contact pressure, sliding speed, and continuity of sliding contact on particle size distribution and the chemistry of collected wear particles. The generated particles had bimodal distribution, with first peak at approximately 350 nm and composition of these submicron particles was corresponding with the cast iron disc. The testing conditions did not affect the size distribution. The second peak was between 2 and 15 μm, depending on the pressure and sliding speed and particles having high content of C, Si, Al, Fe, Mo, Sb, Si, S, and Cu. When the motion was discontinuous at a repeated brake action, smaller wear particles were generated. However, the study evaluated nonairborne emissions only and contribution to airborne fraction is not clear.

Wahlström, Jansson et al. (2010) generated airborne wear particles by pin-on-disc tribometer equipped with particle counting instruments and tested NAO and LM brake pads against cast iron disc. Their results indicated that the LM pads showed higher friction performance and caused more wear to the rotor than the NAO pads, which was resulting in higher mass losses and higher number concentrations of airborne wear particles. Although there were variations in the measured particle concentrations, similar size distributions of airborne wear particles were obtained regardless of the pad material. The characteristic number distributions were with maxima at about 100, 280, 350, and 550 nm. Verma et al. (2016) conducted pin-on-disc tests with LM brake pad material with a real-time monitoring, size-resolved sampling of the airborne wear particles from 6 nm to 10 μm and nonairborne fraction and their subsequent characterization. The mass distribution showed the dominant of airborne fractions around 5 μm, while number distribution revealed concentrations of nanosized particles (<100 nm) by three orders of magnitude higher compared with microsized particles. Both airborne and nonairborne fractions had Fe as the major element consistently with studies where chemistry of the friction layer was studied (Österle et al., 2014). Submicron particles were present in all size fractions of airborne particles, as well as in the settled fraction. Airborne particles were based on hematite (Fe_2O_3), wüstite (FeO), and elemental Fe, while the settled fraction contained magnetite (Fe_3O_4) together with Fe and Fe_2O_3. It is evident that wear particles have different phase compositions according to different testing conditions and also for different size fractions.

Nosko and Olofsson (2017) quantified ultrafine (<100 nm), submicron (0.1–1.0 μm), and microsized (1–10 μm) airborne wear particles generated by pin-on-disc testing of several LM representatives for the European market and one NAO typical for the US market against cast iron disc.

Mass and mainly number of ultrafine particles were strongly influenced by rotor temperature achieved during the tests. Significant increase in the number of ultrafine particles was observed at temperatures above 200°C (Fig. 6.1) with dominant particle sizes between 11 and 29 nm. These findings are consistent with results of the study by Alemani et al. (2016) investigating airborne brake wear emissions generated by pin-on-disc machine testing similar formulations (LM and NAO) in terms of size distribution and elemental composition of size fractions from 6 nm to 10 μm. They found that with temperature increasing from 100 to 300°C, the ultrafine particle emissions intensifies, while the coarse particle emission decreases. Number concentration of ultrafine wear particles between 11 and 35 nm in

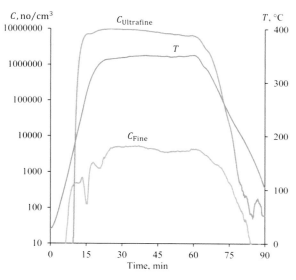

FIGURE 6.1 Example of particle number concentrations of fraction 0.1—10 µm (C_{Fine}) and below 100 nm ($C_{Ultrafine}$) versus time and temperature (T) for a low-metallic material. *Reprinted with permission from O. Nosko, U. Olofsson: Quantification of ultrafine airborne particulate matter generated by the wear of car brake materials. Wear, vol. 374—375, p. 92—96, Elsevier, 2017. Copyright (2017) Elsevier.*

diameter significantly increases at temperatures above 165—190°C (according to a formulation tested). Above the temperature of 190°C, ultrafine particles constitute almost 100% of the total particle number. Fine particles (<2.5 µm) dominate in number over the coarse and ultrafine particles at temperatures below 165°C. In accordance with data of Verma et al. (2016), all particle fractions consist of Fe and Cu, with iron varying from 20 to 66 wt.%, where contribution of pad wear, same as discs, has to be taken into account. However, the authors did not estimate carbon forms, which have to be present as well, because the binder in most of the brake materials is a phenolic resin. The authors recognize that the testing conditions are considerably different from the real conditions of the automotive brakes operation, including size of the tested pad samples, adopted sliding velocity and deceleration, particle generation rate, and airflow regime. These differences result in lower energy generated per area and mass and per unit time to heat up the tested materials, which can also influence parameters of wear particles. Obviously, the small size noninertial friction tester cannot generate corresponding temperatures, nor vibrational conditions, typically seen in real brake systems.

4.2 Brake Dynamometer Studies

Numerous studies generated brake wear emissions using dynamometer testing with a real hardware and brake materials having real size and trying to simulate conditions close to real urban or suburban driving.

Iijima et al. (2007) measured particle size distributions of direct brake wear particles generated by open disc brake dynamometer (without an environmental chamber) and using three commercially available NAO brake pads typically used in Japanese passenger cars and cast iron disc. Wear test procedure was based on JASO-C427. The particle size distribution was measured at disc temperatures of 200, 300, and 400°C and initial speed was 50 km/h at

200 and 300°C, while at 400°C it was 80 km/h. However, concentrations of emitted airborne wear particles were measured only by aerodynamic particle sizer spectrometer detecting particles from 500 nm to 20 µm and thus information on real-time concentrations of particles <500 nm was not obtained. Size-resolved sampling presented in this study enabled capturing of particles in nine stages (<430 nm up to >11 µm) and their subsequent elemental analysis by ICP-AES and ICP-MS. The authors observed that particle size distributions of number concentrations obtained for all tested NAO brake pads had a peak between 1 and 2 µm; however, in distributions by mass concentration the peak was shifted to 3–6 µm. With respect to quantity, Iijima et al. (2007) estimated that fine particles ($PM_{2.5}$) dominate, i.e., approximately 90% of particles by number and 30% by mass is emitted as $PM_{2.5}$.

Garg et al. (2000) determined mass emission rates from seven brake pad materials in high-volume use on new vehicles of General Motors in the 1998 production year, including semi-metallic and NAO formulations. Direct wear emissions were generated with brake assembly enclosed in a chamber and using wear test procedure BSL-035 simulating a regular on-road braking event, which is similar to JASO-C427. All braking events were done from 50 to 0 km/h at a deceleration rate of 2.94 m/s^2 with temperatures 100, 200, 300, and 400°C. Particle mass distribution and particle number distribution from nano-to microsized particles were estimated and elemental composition of the collected particles and gases was analyzed. Lowest wear rate was at temperatures below 300°C and highest mass wear at 400°C, which is higher than the data published by Alemani et al. (2016) and Nosko and Olofsson (2017) using the pin-on-disc experiment, where the maximum quantities of wear particles were emitted below 200°C. One reason for this relevant difference between the brake dynamometer and the pin-on-disc based experiments is because the pin-on-disc testing is not "able" to generate sufficient power per unit area or unit mass of tested brake material samples to increase temperature on the friction surface. In brake dynamometer tests, the mass of the airborne emissions and the total mass loss of brake composites were increasing with temperature exhibiting the maximum values at 400°C. However, the percentage of the airborne PM was higher for 100°C tests than for 400°C tests (Alemani et al., 2016). As discussed earlier, different wear mechanisms of particle formation occur at different temperatures and lead to generation of various size fractions and amounts of brake wear (Kukutschová et al., 2011; Verma et al., 2016). The average airborne fractions found by Garg et al. (2000) at 100 and 200°C were approximately 30%. Particle size concentrations of airborne wear particles were expressed by mass only; however, the size fraction below 1 µm had minimum mass, while the mass of microsized fractions dominated. Usage of particle size distributions, expressed by number of particles instead of their mass, would probably show opposite tendency in amount of particles in the different detected size fractions, as later demonstrated by Kukutschová et al. (2011). Chemical composition of the particles captured by Garg et al. (2000) consisted dominantly of carbonaceous and metal-based components. The average total carbon (TC) emission rates were approximately 0.15 mg/stop and, on average, 18% of the airborne PM captured was carbonaceous material. In accordance with several other studies (Alemani et al., 2016; Kukutschová et al., 2011; Österle et al., 2014; Verma et al., 2016), iron was the most dominant metal detected in all studies addressing the brake wear debris.

Sanders et al. (2003) and Garg et al. (2000) also conducted dynamometer measurements with SM, LM, and also NAO brake materials typically used for midsize and full-size cars as well as the full-size trucks. Sanders et al. (2003) used an open disc system, while the measurements by Garg et al. (2000) were carried out in a closed chamber. Dynamometer testing simulated regular on-road braking events "typical for urban driving," i.e., speed from 90 to 0 km/h (Sanders et al., 2003) and from 50 to 0 km/h (Garg et al., 2000), deceleration below 1.6 m/s^2 (Sanders et al., 2003) and 2.94 m/s^2 (Garg et al., 2000) and, importantly, the data generated for the particle mass and the particle number distributions, as well as the chemical composition of direct wear debris, were compared with the on-vehicle brake particle measurements (Sanders et al., 2003). The number distribution was dominated by particles larger than 200 nm in diameter, which is not in agreement with the discussed pin-on-disc studies (Alemani et al., 2016; Nosko & Olofsson, 2017) and later dynamometer study (Kukutschová et al., 2011), which detected particles down to 10 nm in diameter. Sanders et al. (2003) estimated the major difference between LM and NAO pad formulations in number concentration of wear particles with LM formulation generating two to three times larger number of wear particles than SM and NAO (Fig. 6.2). By comparison of final pad and rotor mass, Sanders et al. (2003) observed that when LM brake pads are used, 60% of the wear debris comes from the rotor and 40% from the pads. Because rotors are contributing to direct brake wear, observations regarding the presence of iron in wear debris, as the most dominant metallic element, are consistent in numerous studies (Garg et al., 2000; Kukutschová et al., 2011; Österle et al., 2014; Sanders et al., 2003; Wahlström, Olander, & Olofsson, 2010). The elements detected by Sanders et al. (2003) with the highest frequency and in the highest amounts were Fe, Cu, Si, Ba, K, and Ti. However, they did not observe any difference in distribution of elements among studied size fractions, i.e., coarser and finer fractions were of similar elemental composition.

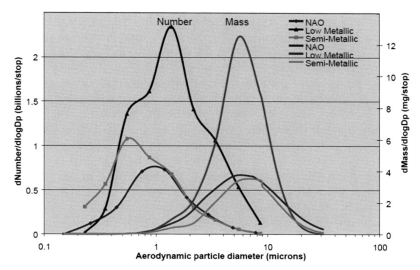

FIGURE 6.2 Number versus mass distributions of brake wear particles generated from low-metallic, semi-metallic, and nonasbestos organic (NAO) brake pad materials. *Adapted with permission from Sanders P.G. et al.: Airborne brake wear debris: size distributions, composition, and a comparison of dynamometer and vehicle tests. Environ. Sci. Technol. (2003) 37:4060—4069). Copyright (2003) American Chemical Society.*

Because the airflow around the brake hardware mounted on a dynamometer is typically not representing airflow of a real vehicle, Sanders et al. (2003) performed particle measurements with a real passenger vehicle in a wind tunnel and onboard measurements on the same vehicle using a testing track. An estimated release of airborne wear particles was up to 21 mg/stop with "aggressive LM pads" in the wind tunnel. In addition, it was found that approximately 50% of the brake wear was emitted as airborne particulate matter (Sanders et al., 2003). This is consistent, although not exactly equal, with their data obtained in the brake dynamometer tests, which indicated 50%—70% of wear particles being released as airborne. Importantly, since the friction and wear are system properties, the difference between on-road and dynamometer generated particles was also in elemental composition. On-road particles were typified by a significantly lower content of Fe and Cu; however, Ba content was higher (Sanders et al., 2003). The authors (Sanders et al., 2003) recognized the resuspended dust as a potential source of some elements, while high content of Ba is unclear.

Wahlström, Olander et al. (2010) used a laboratory stand with front right assembly enclosed in a chamber with a controlled environment and generated airborne wear particles emitted by NAO and LM pads and cast iron disc under unrealistic conditions of constant brake pressure and speed. Nevertheless, they believe that the testing conditions simulated light braking starting at a vehicle speed of 55 km/h. Online monitoring, sampling, and chemical analysis of the generated wear particles were carried out. The characteristic size distribution had distinct maxima similar for NAO and LM materials at sizes of approximately 280 and 350 nm and the majority of particles were of submicron size. No significant contribution of ultrafine particles (<100 nm) was detected. The total concentrations were 6.3×10^{10} particles/m^3 for LM and 7.7×10^{10} particles/m^3 for NAO, however, without detailed information on contribution of single size fractions. The captured particles of ultrafine, fine, and coarse fractions contained Fe, Ti, Zn, Ba, Mn, and Cu.

The brake dynamometer study by Kukutschová et al. (2011) focused on generation of airborne wear particles from a commercial original equipment LM brake material for a midsize passenger car and approved for the EU market with subsequent characterization of number distribution, morphology, and chemical composition. The entire front wheel with the brake system was encapsulated in the environmental chamber with controlled filtered air and the simulation representing a suburban driving segments (i.e., cycles from 73 to 67 km/h followed by idle run). Online monitoring of wear particles with diameter from 10 nm to 20 μm and sampling of particles with mean aerodynamic diameter between 37 nm and 9.5 μm were performed. Particle number and mass distribution, morphology of particles, and elemental and phase composition of thus obtained particles were estimated. It was found that emissions of the finest airborne particles depend on the rotor temperature. Testing conditions of a relatively cold rotor (below and around 200°C) led to negligible emissions of submicron particles. After the rotor temperature reached 300°C, a gradual increase of the finest fractions, including the nanosized particles (<100 nm) reaching concentration up to 10^6 per cm^3, was detected. From the shape of the particle size distributions and their variation with time, it could be assumed that submicron particles are formed by the evaporation/ condensation process with a subsequent aggregation of the primary nanosized particles. In contrast, the detected concentrations of larger microsized particles were not so strongly affected by increasing rotor temperature, and it can be assumed that these larger particles are generated by mechanical wear, not the oxidative wear mechanisms, as are the finest fractions. The maximum amount of the finest airborne particles was released when the rotor temperature was over 300°C, which is consistent with thermogravimetric measurements of the initial brake pad material (Kukutschová et al., 2011). In both an inert and oxidative atmosphere, a significant mass loss was observed at temperature above 300°C when oxidation or pyrolysis of volatile organic compounds occurs. The finest fractions dominated and the detected quantities of the finest particles were by three orders of magnitude higher compared with the microsized particles, when number distribution was used. Interestingly, when the interpretation of the airborne emissions was based on the mass distribution, the microparticles represented the dominant fraction, which is in agreement with other dynamometer as well as pin-on-disc studies (Nosko & Olofsson, 2017; Perricone et al., 2016; Sanders et al., 2003; Verma et al., 2016; Wahlström, Olander et al., 2010). Because the reactivity and the environmental risks of the nanosized particles are much higher compared with the microparticles, obviously, the usage of mass distribution for the quantification of the released direct brake wear particles for further evaluation of their effects on the environment and health may lead to underestimation. In the study by Kukutschová et al. (2011), consistently with other dynamometer and pin-on-disc studies, the dominant metallic element in all captured size fractions was Fe. This is not an unexpected result because all tests were performed with the cast iron or steel counterparts and many of brake formulations contained Fe-based materials. Other metals detectable even in the submicron and nanofractions were Cu, Sn, and Zn. As mentioned above, the Cu-, Sn-, and Zn-containing compounds have been used for brake pad formulations and the rotors could be treated by these elements and their compounds to mitigate corrosion. Nevertheless, a significant difference was detected in morphology of the captured fractions, when the referenced studies are compared. While the microsized fractions contained particles with sharp edges (Fig. 6.3A and B), in the finest nanosized fractions, clusters of spherical particles only were detected (Fig. 6.4A). Moreover, these finest particles were observed to be attached to the surface of larger particles as well (Fig. 6.4B). TEM analysis

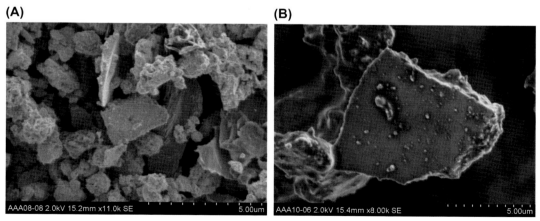

FIGURE 6.3 SEM images of the fine <2.5 μm (A) and the coarse >2.5 μm (B) brake wear particle fractions. *Adapted with permission from J. Kukutschová et al.: On airborne nano/micro-sized wear particles released from low-metallic automotive brakes. Environ Poll, vol. 159 (4), p. 998—1006, Elsevier, 2011. Copyright (2011) Elsevier.*

(A)   **(B)**

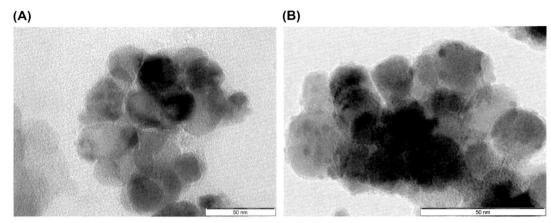

FIGURE 6.4   Bright field TEM images of direct brake wear nanosized particles detected in fraction with particle size <56 nm (A) and <2.5 μm (B). *Adapted with permission from J. Kukutschová et al.: On airborne nano/micro-sized wear particles released from low-metallic automotive brakes. Environ Poll, vol. 159 (4), p. 998–1006, Elsevier, 2011. Copyright (2011) Elsevier.*

revealed that these particles have similar size to the particles captured in the nanofractions (Fig. 6.4C). This finding clearly shows that precise quantification of nanosized particles by an impactor sampling technique is incorrect. Several analytical techniques identified phase composition of the airborne wear particles captured and found that the submicron and also microsized fractions consisted of amorphous carbon, graphitic particles (Fig. 6.5), and various metal-based components dominated by magnetite and hematite.

Hagino et al. (2015; 2016) studied particulate emissions generated by dynamometer testing of NAO pads for passenger cars and a middle-class truck typical for the Japanese market. Simulations of low-temperature braking conditions typical for urban districts (deceleration from 60 to 0 km/h) were performed. In the study (Hagino et al., 2015) they focused on quantification of PM_{10} and $PM_{2.5}$ emissions by mass and evaluation of resuspended particles. Airborne wear particles were not only detected at deceleration as a direct brake wear but also occurred at acceleration (nonbraking event), suggesting resuspension of wear particles. These particles had increasing concentration during acceleration with an increasing initial speed. Based on their findings, the resuspended particles should be included in emission measurements. This study was focused on larger particle size fractions quantified by mass only and thus potential contribution of finest nanosized particles was neglected. The study (Hagino et al., 2016) used the same experimental setup and NAO pads but focused more on measurements and chemical composition of emitted airborne particles (PM_{10} and $PM_{2.5}$) and nonairborne fraction as well. The observed airborne wear emissions ranged from 0.04 to 1.4 mg/km/vehicle for PM_{10} and from 0.04 to 1.2 mg/km/vehicle for $PM_{2.5}$. The proportion of brake wear debris emitted as airborne wear particles was 2%–21% of the mass of wear. The dominating size fractions were <10 μm, with a unimodal distribution between 0.68 and 3.5 μm and no contribution of ultrafine particles (<100 nm). These results may suggest three eventualities: (1) there was a contribution of ultrafine particles but mass distribution did underestimate the number of "weightless particles," (2) the testing conditions were

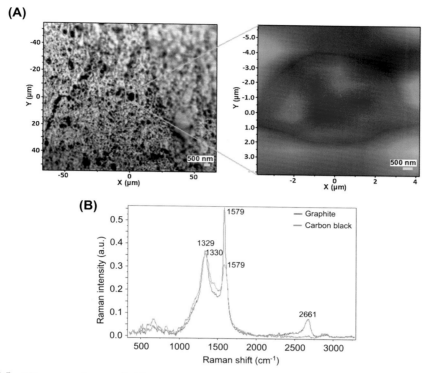

FIGURE 6.5 Microscopy image of airborne wear particles with corresponding Raman map (A) and Raman spectra of two components detected (B). *Adapted with permission from J. Kukutschová et al.: On airborne nano/micro-sized wear particles released from low-metallic automotive brakes. Environ Poll, vol. 159 (4), p. 998–1006, Elsevier, 2011. Copyright (2011) Elsevier.*

too mild to initiate nanoparticulate emissions, or (3) brake materials tested have an eco-friendly character and do not release nanosized particles. TC emissions of PM_{10} and $PM_{2.5}$ were accounted for two tested passenger cars as 71 and 110 µgC/km/vehicle and 67 and 83 µgC/km/vehicle, respectively. TC emissions in PM_{10} and $PM_{2.5}$ for the midclass truck were lower, 66 and 63 µgC/km/vehicle, respectively. Oxygenated carbonaceous components were observed in the airborne PM, which indicated that the oxidation occurred during the friction process. The most abundant inorganic element in the wear particles was Fe (from 12% to 26% in PM_{10} and ranging between 10% and 28% in $PM_{2.5}$). The majority of this Fe containing particles came from wear of the rotor because the NAO brake pad formulations without steel or iron constituents were used. In studies testing SM brake pad formulation, the contribution of rotor and pads wear to Fe emissions are difficult to estimate precisely because the pads typically contain more than 50 wt.% of iron-based constituents. Other key tracers identified by this study (Hagino et al., 2016) at comparable emission levels with traffic-related atmospheric environments (tunnels, street canyons, or roadside environment) were Cu, Ba, and Sb. All these tracers will be inapplicable in future because the brake pad manufacturers are introducing Cu- and Sb-free formulations.

Perricone et al. (2016) used a dynamometric inertia bench stand adapted for airborne PM measurements with hardware of medium-size European passenger car and generated wear particles from pairs of NAO and LM pads rubbed against gray cast iron discs. The brake assembly was enclosed in an overpressurized box with a controlled supply of clean air and implementation of isokinetic sampling with a high sampling efficiency. Testing procedures were based on a modified wear test (SAE J 2707) using only the sections representative for urban driving. Particulate emissions were measured online with subsequent sampling by a cascade impactor to determine mass and number EF. Both determined factors strongly depended on pad formulation. A ranking of the pad–rotor combinations revealed NAO pads to have the lowest mass EF, but highest number EF, whereas LM pads had higher mass EF and lower number EF. This finding is inconsistent with the full-scale brake dynamometer and real vehicle study of Sanders et al. (2003), where LM formulations had higher emissions in mass and number as well.

Overview of the published brake wear studies trying to quantify direct brake emissions generated mainly by a brake dynamometer (Table 6.2) clearly shows impossibility to make a comparison of quantity of brake wear emissions, as there are several variables (testing device, testing procedure, sampling conditions, brake materials, etc.) and no recommended approach for generation, measurement, and expression of brake wear emissions is clearly defined.

A comparison of experimental data addressing the number and volume distributions of airborne wear particles generated by LM brake pads and cast iron discs for passenger cars and measured online in field tests, disc brake assembly test stand, and a pin-on-disc machine was performed by Wahlström, Olofsson et al. (2010). A pin-on-disc machine and disc brake assembly (or dynamometers) allows measurement of airborne wear particles with low background concentrations. In addition, a contamination of the collected particles by particles from other road traffic sources can be completely eliminated, thus chemical analyses more accurately reflect on the contribution of brake materials wear. Vehicle field tests are limited by the background particle concentration in the surrounding air and it could be possible to distinguish brake wear particles measured near the disc brake. A correlation between the three different testing methods was observed. When the number distribution was used, submicron particles dominated in all testing methods with maximum at about 300 nm. Particles below 100 nm were not detected due to the size interval of the aerosol spectrometer used (0.25–32 μm). Obviously, this correlation can change with different materials and it is safe to suggest that further work is necessary to develop a representative testing methodology well reflecting the contribution of brake materials to environmental pollution.

4.3 Field Studies

Field studies can be divided into two main categories: (1) on-vehicle measurements (i.e., wind tunnel, test track measurements, and measurements in traffic) and (2) sampling of airborne traffic-related particles. When measuring, sampling and analyzing airborne brake wear particles in field tests, it is difficult to clearly distinguish direct brake wear from other sources, such as resuspended road dust, tire, pavement wear, and similar other sources of pollutants. Therefore, it is suitable to perform these simulations in a controlled environment with a contribution of the tested brake pads and rotors only.

TABLE 6.2 Overview of Studies Quantifying the Airborne Fraction of Direct Brake Wear Particulate Emissions

Reference	Pads	Type of Testing	Testing Procedure	Particle Quantification	Output
Garg et al. (2000)	SM NAO	BD	Wear Test—General Motors BSL-035 (deceleration 2.94 m/s², max. speed 50 km/h, temperatures 100, 200, 300, and 400°C)	Mass of filters (TSP sampling, size-resolved by MOUDI)	Overall average: 4.6−12.1 mg of PM_{10}/mile/vehicle 3.4−8.9 mg of $PM_{2.5}$/mile/vehicle 1.9−5.0 mg of $PM_{0.1}$/mile/vehicle
Sanders et al. (2003)	LM SM NAO	BD	Urban driving program (deceleration < 1.6 m/s², max. speed 90 km/h)	Mass of filters (TSP sampling, size-resolved by MOUDI and ELPI)	~ 8 mg of TSP/stop/brake (LM) ~ 2 mg of TSP/stop/brake (SM) ~ 2 mg of TSP/stop/brake (NAO)
Hagino et al. (2015)	NAO	BD	Own urban driving program (deceleration < 3.0 m/s², max. speed 60 km/h)	Mass of filters (DustTrak + impactor)	0.006−0.016 mg of PM_{10}/braking/wheel
Hagino et al. (2016)	NAO	BD	• Wear test—JASO C427 (deceleration 2.94 m/s², max. speed 50 km/h) • Japanese exhaust emission/fuel economy tests (JC08/JE05) (max. speed 90 km/h)	Mass of filters (DustTrak + impactor)	0.04−1.4 mg of PM_{10}/km/vehicle 0.04−1.2 mg of $PM_{2.5}$/km/vehicle
Perricone et al. (2016)	LM NAO	BD	Modified wear test (SAE J 2707) (max. deceleration 3.92 m/s, max. speed 100 km/h, initial rotor temperature 100 −200°C)	Mass of filters Number concentration (ELPI + cascade impactor)	14.5−46.4 mg of PM_{10}/stop/brake (LM) 8.5−9.2 mg of PM_{10}/stop/brake (NAO) 8-91 × 10^{10}# of PM_{10}/stop/brake (LM) 153 × 10^{10}# of PM_{10}/stop/brake (NAO)
Nosko and Olofsson (2017)	LM	PoD	Own testing program (contact pressure 0.5−1.5 MPa, deceleration 0.8−1.6 m/s)	Calculation of volume concentration based on size distribution (FMPS, OPS)	~200 μg of PM_{10}/m³

BD, brake dynamometer; *LM*, low-metallic; *NAO*, nonasbestos organic; *PoD*, pin-on-disc; *SM*, semimetallic; *TSP*, total suspended particles.

Few field studies focusing on brake wear particles have been reported in the literature. Sanders et al. (2003) performed measurements in a wind tunnel, test track, and field test in traffic with a full-size car and LM brake pads to make a comparison with brake dynamometer testing. The wind tunnel measurements were performed at speeds from 96 km/h, an initial brake temperature of 130°C, and constant wind speed at 64 km/h. In fact, these measurements are free from road dust because the entire vehicle remains stationary and also free of tire wear particles due to low wear from smooth steel roller that supplies the load to the wheels. A high-speed track and also public roads were used for generation of wear particles from LM and NAO brake materials. Particle size distributions were measured in all tests performed. All stops on the test track were made from 60 km/h at different deceleration. Filters were installed in the vehicle to capture the generated airborne particles and to perform chemical analysis for a comparison with composition of wear debris collected during the dynamometer tests. The dynamometer tests were consistent with field tests in particle mass distributions. However, the mass weighted mean diameter decreases from approximately 6 μm for the dynamometer tests to about 3 μm for the test track data. A negligible contribution of the submicron fraction to the total mass of wear particles was observed in all tests and for all materials. This is given by the selected parameter for particle quantification and the fact that the finest, especially nanosized, particles are "weightless." The number distributions had maxima about 1 μm, with 10^9 of released particles per stop for LM and NAO materials. The mass of PM_{10} wear particles generated by dynamometer tests were ~8 mg/stop/brake for LM pads and ~2 mg/stop/brake for NAO, while after the wind tunnel tests with LM pads the mass of airborne wear particles was ~21 mg/stop/brake. A series of field tests using a passenger car with LM brake pads were conducted also in studies of Wahlström, Olofsson et al. (2010), Wahlström and Olofsson (2015). The tests were performed on a test track simulating urban driving, country roads, and a highway. To reduce the influence of resuspended traffic-generated particles, the tests were conducted on days when it rained. The measured number concentration was 1.7×10^7 particles/m^3 and the peak in number concentration was around 410 nm (Wahlström, Olofsson et al., 2010). In the study (Wahlström et al., 2015), there was a clear difference between the concentrations measured near the brake and at the front of the car. The maximum particle size concentration registered was at approximately 350 nm.

Numerous studies were focused on sampling and characterization of PM from roadside environments, but only a few addressed indirect detection of brake wear emissions based on searching for a suitable tracer. Detailed overview of recent findings related to brake wear particles in the environment can be found in the review published by Peikertová and Filip (2016). Within a search for the most suitable tracer of brake wear, Gietl et al. (2010) recommended Ba as a quantitative tracer of brake wear emissions based on sampling and analyses of size-segregated aerosols in London urban and suburban areas. They found Fe, Cu, Ba, and Sb to be indicative of a common traffic-related source. Ba was found to comprise 1.1% of brake wear PM_{10} fraction from the traffic fleet as a whole. This tracer seems promising for the future, as barium sulfate is a component in brake materials, which is not hazardous, and no strong efforts to replace it are planned. Although not all brake friction materials contain $BaSO_4$, Sanderson et al. (2016) collected and characterized Fe-rich airborne submicron particles in the roadside environment of Birmingham and New Castle, United Kingdom. The particles were classified into high Fe content (c.90 wt.%) and moderate Fe content

(<75 wt.%) with high contents of Mn and Si and also a mix of Zn, Cu, Ba, Al, and Ca. The Fe-rich nanosized particles exhibited sizes ranging from 20 nm and also some larger particles up to 100 nm could be observed. These primary particles tend to form clusters ranging from 200 nm to 1 μm in diameter. The iron-rich particles were observed in the form of Fe(II) and Fe(III) oxides (FeO, Fe_3O_4, α-Fe_2O_3, and γ-Fe_2O_3) with Fe(III) being much more frequently observed. This observation of nanoparticulate emissions is consistent with characteristics of submicron particles released by wear of LM pads published earlier by Kukutschová et al. (2011). However, iron is an element released by a range of sources and therefore it is not a suitable tracer of brake wear emissions. Another study (Megido et al., 2016) analyzed PM_{10} collected in suburban area of northern Spain with special attention to a relationship between carbonaceous fraction and metals. On contrary to the results of Gietl et al. (2010), the authors suggested Cu, Sb, and Sn as the elements, having most likely originated from brake wear and be the suitable brake wear tracers. Because the brake pad manufacturers are introducing Sb- and Cu-free formulation, these tracers will not be suitable in future.

5. IDENTIFICATION OF GAPS

Based on the current literature review, there are several critical issues in the area of generation, detection, quantification, and characterization of airborne brake wear emissions, which make a comparison of single studies difficult (Peikertová & Filip, 2016). The following topics are still to be defined:

- *Testing devices for generation of representative wear emissions* (tribometers, dynamometers, or field testing)—e.g., how the dynamometer testing is far from reality and whether these emissions are representative for real driving. The field testing on real vehicles can generate representative emissions; however, a contamination by resuspended particles and emissions from other sources make the precise description of brake wear emissions almost impossible.
- *Testing procedures representing typical driving conditions*—currently, there is a wide range of testing conditions used in the studies published which lead to different results. There are numerous recommended practices for friction performance and wear testing accepted by industry, but these procedures mostly do not represent conditions which can contribute to brake wear emissions most (i.e., city driving, suburban, highway). Thus, these commonly used procedures can generate emissions with significantly different size distributions and chemistry. There are several testing procedures, such as Mojacar test, Los Angeles City Traffic test, or Detroit City Traffic test, representing different driving conditions, which differ in deceleration rates, brake phase duration, vehicle speed, and number of stops. Currently, there is no testing procedure widely accepted for generation of brake wear emissions.
- *Best available techniques for particle quantitative measurements and sampling*—numerous previously published studies quantified brake wear particulate emissions based on mass distribution obtained by size-resolved sampling of wear particles. Size-resolved sampling of airborne particles often requires various techniques to be employed. There

would be a significant benefit in a sampler that could reliably collect size resolved samples across the entire size range of airborne particles. In addition, quantification of submicron or ultrafine particles based on weight of these size fractions can lead to underestimation of these emissions because the finest fractions are often attached to the surface of larger particles and in fact are present in the majority of size fractions of airborne particles captured. As there are many technical difficulties, especially in the case of submicron particles, these size fractions cannot be precisely quantified by mass. Therefore, number distributions are more suitable for more accurate quantification. Also, an issue regarding how to express brake wear emissions (mg/stop, mg/km, number of particles/stop, etc.) is still unresolved.

- *Characterization of brake wear particulate emissions*—single size fractions of airborne particles captured are in most of the published studies analyzed by elemental analyses. Only few studies evaluated detailed phase composition of emitted wear particles, although this information is also relevant for interpretation of toxicological results. Moreover, a majority of studies deal with analyses of metals and elemental carbon, while organic compounds could also be present in considerable amounts.

- *Impact on environment and health*—since there are numerous issues how to obtain representative samples of direct brake wear particles, it is unclear what environmental and health impact has and will have the brake wear debris. Several studies (Kazimirová et al., 2016; Kukutschová et al., 2009; Malachová et al., 2016; Peikertová & Filip, 2016; Shupert, Ebbs, Lawrence, Gibson, & Filip, 2013) previously indicated/discussed potential negative impact based on simplified studies of model brake materials; considerably more work is necessary to understand this complex topic.

6. FUTURE TRENDS IN DEVELOPMENT OF MATERIALS FOR AUTOMOTIVE BRAKES

The future of friction brake materials for the automotive industry is very dependent on the speed of development of efficient "regenerative braking" technology. When effectiveness of regenerative braking increases to the level of maturate inexpensive and reliable technology, it seems probable that the friction brake will be used as an emergency/parking brake system only.

On the other hand, the rapid development of autonomous vehicles, with optimized response to driving speed and stopping distance, typically requires more frequent engagement of friction brakes, leading to higher demands put on friction materials. It will be necessary to further improve their wear resistance when the environmental aspects are addressed. This challenge is complex, however, and it does not only include the development of new brake friction materials but also the entirely improved brake design with a more efficient heat dissipation/management, vibrational response, and related noise, while still lowering the mass of braking system.

It is also possible to expect an increased demand for shorter traveling times and a related increase in the maximum speed of vehicles and related reinforcement of safety measures. Hence, the abovementioned "overlapping period" between the dominant use of friction brakes and the use of mature and fully reliable regenerative braking will be typified by a need to address the considerably increased demand on braking applications. This is possible with the development of completely new and improved materials. The friction industry is subjected to enormous pressure when optimizing/lowering costs of brake systems, including brake materials. New and improved friction materials will be needed for the automotive technology in the not-so-distant future. It seems to be impossible, however, to achieve these demanding goals with the given group of inexpensive ingredients, and this paradigm will probably change in the near future. Application of science and knowledge about advanced materials, as well as the most recent technologies, could be expected when developing brake systems for the not-so-distant future.

7. SUMMARY

Numerous studies confirmed braking of automobiles to be a potential source of airborne particulate emissions because each application of a friction brake is accompanied with wear of brake materials and release of wear particles. Materials typically used for manufacturing of friction brake pads are multicomponent composites exposed to high-energy conditions during braking, which leads to newly formed matter released in form of wear particles having significantly different chemistry and structure compared with the originally used constituents. Driving conditions such as speed and applied pressure, formulation of brake materials (e.g., content of metals), and surrounding environment (e.g., deicing agents, humidity) have decisive impact on the extent of wear and also character of these emissions.

Numerous studies used different experimental setup to generate, measure, and characterize airborne wear particles and therefore these data are hardly comparable. Some detected emissions of microsized wear particles only, whereas others also observed particles below 100 nm in diameter. Studies using only mass distribution for quantification of wear particles most likely underestimated the contribution of the finest particles ($<1\ \mu m$) to pollution, because the studies also using number distribution confirmed that the finest particles are dominant.

Airborne brake wear emissions mostly consist of various metal compounds with Fe as the most dominant metal present in form of its oxides, followed by elemental carbon. The presence of organic compounds due to the degradation of phenolic resin binder has to be taken into account as well.

As there is still no recommended practice by governmental agencies for estimation and also control of brake wear emissions, there are plenty of issues that need to be addressed in future, such as methods for generation of wear particles, including testing devices and testing procedures, best available techniques for quantification of these emissions, and also methods for their characterization including not only elemental but also phase and organic analysis. Obviously, it is necessary to better estimate and understand the impact of brake wear emissions on the environment and health.

Acknowledgments

The chapter was created in the Faculty of Metallurgy and Materials Engineering in the Project No. LO1203 "Regional Materials Science and Technology Centre—Feasibility Program" funded by Ministry of Education, Youth and Sports of the Czech Republic.

References

Abu-Allaban, M., Gillies, J. A., Gertler, A. W., Clayton, R., & Proffitt, D. (2003). Tailpipe, resuspended road dust, and brake wear emission factors from on-road vehicles. *Atmospheric Environment, 37*, 5283—5293. Elsevier.

Alemani, M., Nosko, O., Metinoz, I., & Olofsson, U. (2016). A study on emission of airborne wear particles from car brake friction pairs. *SAE International Journal of Materials and Manufacturing, 1*, 147—157. https://doi.org/10.4271/2015-01-2665. SAE International.

Aranganathan, N., & Bijwe, J. (2016). Development of copper-free eco-friendly brake friction material using novel ingredients. *Wear, 352—353*, 79—91. Elsevier.

Beddows, D. C. S., DallÓsto, M., Olatunbosun, O. A., & Harrison, R. M. (2016). Detection of brake wear aerosols by aerosol time-of-flight mass spectrometry. *Atmospheric Environment, 129*, 167—175. Elsevier.

Chan, D., & Stachowiak, G. W. (2004). Review of automotive brake friction materials. *Proceedings of the Institution of Mechanical Engineers Part D: Journal of Automobile Engineering, 218*, 953—966. SAGE.

Day, A. (2014). *Braking of road vehicles*. Elsevier, ISBN 978-0-12-397314-6.

Filip, P. (2013). Friction brakes for automotive and aircraft. In *Encyclopedia of tribology* (pp. 1296—1304). US: Springer, ISBN 978-0-387-92897-5. https://doi.org/10.1007/978-0-387-92897-5_172.

Filip, P., Kovarik, L., & Wright, M. (1997). *Automotive brake lining characterization*. SAE technical paper 973024. Society of Automotive Engineers. https://doi.org/10.4271/973024.

Filip, P., Weiss, Z., & Rafaja, D. (2002). On friction layer formation in polymer matrix composite materials for brake applications. *Wear, 252*, 189—198. Elsevier.

Garg, B. D., Cadle, S. H., Mulawa, P. A., & Groblicki, P. J. (2000). Brake wear particulate matter. *Environmental Science & Technology, 34*(21), 4463—4469. American Chemical Society.

Gietl, J. K., Lawrence, R., Thorpe, A. J., & Harrison, R. M. (2010). Identification of brake wear particles and derivation of a quantitative tracer for brake dust at a major road. *Atmospheric Environment, 44*, 141—146. Elsevier.

Grigoratos, T., & Martini, G. (2015). Brake wear particle emissions: A review. *Environmental Science and Pollution Research, 22*, 2491—2504. Springer.

Guevara, M. (2016). Emissions of primary particulate matter. In *Airborne particulate Matter: Sources, atmospheric processes and health* (pp. 1—34). Royal Society of Chemistry, ISBN 978-1-78262-491-2. https://doi.org/10.1039/9781782626589-00001.

Hagino, H., Oyama, M., & Sasaki, S. (2015). Airborne brake wear particle emission due to braking and accelerating. *Wear, 334—335*, 44—48. Elsevier.

Hagino, H., Oyama, M., & Sasaki, S. (2016). Laboratory testing of airborne brake wear particle emissions using a dynamometer system under urban city driving cycles. *Atmospheric Environment, 131*, 269—278. Elsevier.

Hulskotte, J. H. J., Roskam, G. D., & Denier van der Gon, H. A. C. (2014). Elemental composition of current automotive braking materials and derived air emission factors. *Atmospheric Environment, 99*, 436—445. Elsevier.

Iijima, A., Sato, K., Yano, K., Tago, H., Kato, M., Kimura, H., et al. (2007). Particle size and composition distribution analysis of automotive brake abrasion dusts for the evaluation of antimony sources of airborne particulate matter. *Atmospheric Environment, 41*, 4908—4919. Elsevier.

Kazimirová, A., Peikertová, P., Barancoková, M., Staruchová, M., Tulinská, J., Vaculík, M., et al. (2016). Automotive airborne brake wear debris nanoparticles and cytokinesis-block micronucleus assay in peripheral blood lymphocytes: A pilot study. *Environmental Research, 148*, 443—449. Elsevier.

Kukutschová, J., Moravec, P., Tomášek, V., Matějka, V., Smolík, J., Schwarz, J., et al. (2011). On airborne nano/microsized wear particles released from low-metallic automotive brakes. *Environmental Pollution, 159*(4), 998—1006. Elsevier.

Kukutschová, J., Roubíček, V., Malachová, K., Pavlíčková, Z., Holuša, R., Kubačková, J., et al. (2009). Wear mechanism in automotive brake materials, wear debris and its potential environmental impact. *Wear, 267*(5—8), 807—817. Elsevier.

Kukutschová, J., Roubíček, V., Mašláň, M., Jančík, D., Slovák, V., Malachová, K., et al. (2010). Wear performance and wear debris of semimetallic automotive brake materials. *Wear, 268*, 86—93. Elsevier.

Křístková, M., Weiss, Z., Filip, P., & Peter, R. (2004). Influence of metals on the phenol-formaldehyde resin degradation in friction composites. *Polymer Degradation and Stability, 84*, 49—60. Elsevier.

Lee, P. W., & Filip, P. (2013). Friction and wear of Cu-free and Sb-free environmentally friendly automotive brake materials. *Wear, 302*, 1404—1413. Elsevier.

Limpert, R. (2011). *Brake design and safety*. SAE International, ISBN 978-0-7680-3438-7.

Malachová, K., Kukutschová, J., Rybková, Z., Sezimová, H., Plachá, D., Čabanová, K., et al. (2016). Toxicity and muta-genicity of low-metallic automotive brake pad materials. *Ecotoxicology and Environmental Safety, 131*, 37−44. Elsevier.

Matějka, V., Lu, Y., Matějková, P., Smetana, B., Kukutschová, J., Vaculík, M., et al. (2011). Possible stibnite transfor-mation at the friction surface of the semi-metallic friction composites designed for car brake linings. *Applied Surface Science, 258*, 1862−1868. Elsevier.

Megido, L., Negral, L., Castrillón, L., Maranón, E., & Fernandez-Nava, Y. (2016). Traffic tracers in a suburban location in northern Spain: Relationship between carbonaceous fraction and metals. *Environmental Science and Pollution Research, 23*, 8669−8678. Springer.

Mosleh, M., Blau, P. J., & Dumitrescu, D. (2004). Characteristics and morphology of wear particles from laboratory testing of disk brake materials. *Wear, 256*, 1128−1134. Elsevier.

Neis, P. D., Ferreira, N. F., Fekete, G., Matozo, L. T., & Masotti, D. (2017). Towards a better understanding of the structures existing on the surface of brake pads. *Tribology International, 105*, 135−147. Elsevier.

Nosko, O., & Olofsson, U. (2017). Quantification of ultrafine airborne particulate matter generated by the wear of car brake materials. *Wear, 374−375*, 92−96. Elsevier.

Österle, W., Deutsch, C., Gradt, T., Orts-Gil, G., Schneider, T., & Dmitriev, A. I. (2014). Tribological screening tests for the selection of raw materials for automtive brake pad formulations. *Tribology International, 73*, 148−155. Elsevier.

Österle, W., Prietzel, C., Kloß, H., & Dmitriev, A. I. (2010). On the role of copper in brake friction materials. *Tribology International, 43*, 2317−2326. Elsevier.

Peikertová, P., & Filip, P. (2016). Influence of the automotive brake wear debris on the environment - a review of recent research. *SAE International Journal of Materials and Manufacturing, 9*(1), 133−146. https://doi.org/10.4271/2015-01-2663. SAE International.

Perricone, G., Alemani, M., Metinöz, I., Matějka, V., Wahlström, J., & Olofsson, U. (2016). Towards the ranking of airborne particle emissions from car brakes - a system approach. *Proceedings of the Institution of Mechanical Engi-neers: Journal of Automobile Engineering, Part D*, 1−17. https://doi.org/10.1177/0954407016662800. SAGE.

Plachá, D., Peikertová, P., Kukutschová, J., Lee, P. W., Čabanová, K., Karas, J., et al. (2016). Identification of organic compounds released from low-metallic automotive model brake pad and its non-airborne wear particles. *SAE In-ternational Journal of Materials and Manufacturing, 9*(1), 123−132. https://doi.org/10.4271/2015-01-2662. SAE International.

Plachá, D., Vaculík, M., Mikeska, M., Dutko, O., Peikertová, P., Kukutschová, J., et al. (2017). Release of volatile organic compounds by oxidative wear of automotive friction materials. *Wear, 376−377*, 705−716. Elsevier.

Qi, S., Fu, Z., Yun, R., Jiang, S., Zheng, X., Lu, Y., et al. (2014). Effects of walnut shells on friction and wear perfor-mance of eco-friendly brake friction composites. *Proceedings of the Institution of Mechanical Engineers Part J; Journal of Engineering Tribology, 228*(5), 511−520. SAGE.

Roubíček, V., Raclavská, H., Juchelková, D., & Filip, P. (2008). Wear and environmental aspects of composite mate-rials for automtive braking industry. *Wear, 265*, 167−175. Elsevier.

Sanders, P. G., Xu, N., Dalka, T. M., & Maricq, M. M. (2003). Airborne brake wear debris: Size distributions, compo-sition, and a comparison of dynamometer and vehicle tests. *Environmental Science and Technology, 37*, 4060−4069. American Chemical Society.

Sanderson, P., Su, S. S., Chang, I. T. H., Delgado Saborit, J. M., Kepaptsoglou, D. M., Weber, R. J. M., et al. (2016). Characterization of iron-rich atmospheric submicrometre particles in the roadside environment. *Atmospheric Envi-ronment, 140*, 167−175. Elsevier.

Shupert, L. A., Ebbs, S. D., Lawrence, J., Gibson, D. J., & Filip, P. (2013). Dissolution of copper and iron from auto-motive brake pad wear debris enhances growth and accumulation by the invasive macrophyte *Salvinia molesta* Mitchell. *Chemosphere, 92*(1), 45−51.

Sousanis, J. (2011). *World vehicle population tops 1 billion units*. Available online http://wardsauto.com/news-analysis/world-vehicle-population-tops-1-billion-units, request on 5th January 2017.

Straffelini, G., Ciudin, R., Ciotti, A., & Gialanella, S. (2015). Present knowledge and perspectives on the role of copper in brake materials and related environmental issues: A critical assessment. *Environmental Pollution, 207*, 211−219. Elsevier.

Thorpe, A., & Harrison, R. M. (2008). Sources and properties of non-exhausst particulate matter from road traffic: A review. *Science of the Total Environment, 400*, 270−282. Elsevier.

Verma, P. C., Alemani, M., Gialanella, S., Lutterotti, L., Olofsson, U., & Straffelini, G. (2016). Wear debris from brake system materials: A multi-analytical characterization approach. *Tribology International, 94*, 249–259. Elsevier.

Verma, P. C., Menapace, L., Bonfanti, A., Ciudin, R., Gialanella, S., & Straffelini, G. (2015). Braking pad-disc system: Wear mechanism and formation of wear fragments. *Wear, 322–323*, 251–258. Elsevier.

Wahlström, J. (2015). A comparison of measured and simulated friction, wear, and particle emission of disc brakes. *Tribology International, 92*, 503–511. Elsevier.

Wahlström, J. (2016). A factorial design to numerically study the effects of brake pad properties on friction and wear emissions. *Advances in Tribology, 2016*, 8181260. https://doi.org/10.1155/2016/8181260. Hindawi.

Wahlström, J., Olander, L., & Olofsson, U. (2010). Size, shape, and elemental composition of airborne wear particles from disc brake materials. *Tribology Letters, 38*, 15–24. Springer.

Wahlström, J., & Olofsson, U. (2015). A field study of airborne particle emissions from automotive disc brakes. *Proceedings of the Institution of Mechanical Engineers: Journal of Automobile Engineering, Part D, 229*(6), 747–757. SAGE.

Wahlström, J., Söderberg, A., Olander, L., Jansson, A., & Olofsson, U. (2010). A pin-on-disc simulation of airborne wear particles from disc brakes. *Wear, 268*, 763–769. Elsevier.

Wahlström, J., Söderberg, A., Olander, L., Olofsson, U., & Jansson, A. (2010). Airborne wear particles from passenger car disc brakes: A comparison of measurements from field tests, a disc brake assembly test stand, and a pin-on-disc machine. *Proceedings of the Institution of Mechanical Engineers: Journal of Engineering Tribiology, J, 224*(2), 179–188. London; SAGE.

Yun, R., Filip, P., & Lu, Y. (2010). Performance and evaluation of eco-friendly brake friction materials. *Tribology International, 43*, 2010–2019. Elsevier.

Review of Tire Wear Emissions
A Review of Tire Emission Measurement Studies: Identification of Gaps and Future Needs

Julie Panko, Marisa Kreider, Kenneth Unice

Cardno ChemRisk, LLC, Pittsburgh, PA, United States

O U T L I N E

1. Introduction	147	3.1 Tire Wear Particle Air Concentrations	153
2. Characteristics of Tire Wear Particles	148	3.2 Tire Wear Emission Rates	154
2.1 Tire Wear Particle Morphology and Composition	149	4. Data Gaps and Needs	156
2.2 Volume Size Distribution	152	5. Conclusions	156
3. Tire Wear Studies	153	References	157

1. INTRODUCTION

As tire tread wears from the use of the tires on a vehicle, particles are formed as a result of the tread abrasion caused by the road surface. Tire wear itself is a complex physicochemical process, which is driven by the frictional energy developed at the interface between the tread and the road pavement (Veith, 1995). The amount of wear that occurs during a tire's lifetime varies greatly depending on its type and how it is used. The factors affecting tire wear include:

- *Tire characteristics*
  - Size (radius/width/depth)
  - Tread depth

- Construction
  - Tire pressure and temperature
  - Contact patch area
  - Chemical composition
  - Accumulated mileage
  - Alignment
- *Road surface characteristics*
  - Material: bitumen/concrete
  - Texture pattern
  - Texture wavelength—micro/macro/mega
  - Porosity
  - Condition, including rutting camber
  - Road surface wetness
  - Silt loading of road surface
  - Surface dressing
- *Vehicle operation*
  - Speed
  - Linear acceleration
  - Radial acceleration
  - Frequency and extent of braking and cornering
- *Vehicle characteristics*
  - Vehicle weight and distribution of load
  - Location of driving wheels
  - Engine power
  - Power/unassisted steering
  - Electronic braking systems
  - Suspension type and condition

The particles formed from the pavement interaction, known as tire wear particles, are elongated, with a "sausage-like" shape (Adachi & Tainosho, 2004; Cadle & Williams, 1979; Dannis, 1974; Kreider, Panko, McAtee, Sweet, & Finley, 2010; Padovan, Prasad, Gerrard, Park, & Lindsley, 1999; Williams & Cadle, 1978), and consist of a complex mixture of tread rubber and pavement. Tire wear particles are released directly into the environment both on the road surface and suspended in the air. These particles are included in the category of non-exhaust vehicle particulate emissions along with other vehicle-related wear particles, such as brake and clutch wear as well as pavement wear and road dust suspension. Subsequent to release, the tire wear particles can be transported to the soil and surface water via roadway runoff and air deposition. The ultimate fate of the particles in the environment is governed by their physical and chemical characteristics.

2. CHARACTERISTICS OF TIRE WEAR PARTICLES

The manufacturing of a tire is a complex process, requiring the use of a wide variety of chemicals, fillers, and polymers. Although the specific recipes used in tire manufacturing are dependent on the desired characteristics of the tire, the primary components of tire tread remain consistent. Both unreactive (polymers, fillers, oils, waxes, resins, processing aids, and antioxidants) and reactive chemicals (sulfur compounds, accelerators and retarders, adhesives, and activators) are used in the compounding of tire tread. Most reactive chemicals are consumed during tire manufacturing, primarily in the vulcanization and curing process (Lawrence Livermore National Laboratory, 1996; U.S. EPA, 1997, p. 70). The physical and chemical characteristics of tire wear particles are distinct from tread rubber and are related to the wear processes exerted on the tread during driving.

2.1 Tire Wear Particle Morphology and Composition

As a tire is used, interaction with the road is thought to alter the chemical composition and characteristics of the particles generated during contact with paved surfaces when compared with original tread through both heat and friction or via incorporation of material from the road surface (Adachi & Tainosho, 2004; Williams & Cadle, 1978). Several studies have characterized particles from tires using rolling resistance machines or other methods (Cadle & Williams, 1979; Camatini et al., 2001; Davis, Shokouhian, & Ni, 2001; Ozaki, Watanabe, & Kuno, 2004; Rogge, Hildemann, Mazurek, Cass, & Simoneit, 1993; Sadiq, Alam, El-Mubarek, & Al-Mohdhar, 1989; Williams & Cadle, 1978). However, in an on-road scenario, as the tire passes over the road, particles from the tread combine with existing road dust and minerals from the pavement to produce a unique entity. Adachi and Tainosho (2004) demonstrated that rubber-containing particles in road dust are incorporated with metals from alternative contributing traffic-related sources (i.e., paint and brake dust).

Tire wear particles created using a laboratory road simulator utilizing actual pavement surfaces have been analyzed and confirm the differences in morphology and chemical composition of tire wear particles versus tread particles (Kreider et al., 2010). Fig. 7.1 shows pictures of roadway particles (including tire wear particles) and tire wear particles (generated using a road simulator). From the images it is clear that tire wear particles contain encrustations of minerals from the pavement. This observation was also confirmed in studies using aerosol time-of-flight mass spectrometer where incorporation of the pavement into tire wear particles as they exist in the environment was observed from the test results (Dall'Osto et al., 2014).

Chemical characterizations of tread particles have primarily addressed the metallic content, although studies of tire particle toxicity in aquatic species and human cell lines have alluded to organic constituents of tires (Adachi & Tainosho, 2004; Davis et al., 2001; Gualtieri, Andrioletti, Mantecca, Vismara, & Camatini, 2005; Ozaki et al., 2004; Rogge et al., 1993; Sadiq et al., 1989; Wik, 2007; Zheng, Cass, Schauer, & Edgerton, 2002). In addition, the contribution of tires to polycyclic aromatic hydrocarbons (PAHs) in road dust has also been discussed in the literature, with contrasting results (KEMI, 2003; Macias-Zamora, Mendoza-Vega, & Villaescusa-Celaya, 2002; Rogge et al., 1993; Takada, Onda, & Ogura, 1990; Zakaria et al., 2002).

The study by Kreider et al. (2010) provides a compositional analysis of tire wear particles, and a comparison of their chemical composition with that of tread particles derived from the tires used in the study is provided in Table 7.1. Significant differences between tire wear particles and tread particles were confirmed, in particular with respect to the mineral components. This supports the microscopic evaluations regarding encrustation of pavement wear into the tread particles. The mass concentration of Zn in tire wear particles is significantly less than that in tread particles because of dilution from the pavement. Additionally,

FIGURE 7.1   Pictures of tire wear particles (TWP). Scanning electron images of roadway particles (A, B) and TWP (C, D). Scales are located below the photographs. Mineral incrustations are evident in the photographs of greater magnification (B, D). *From Kreider et al. (2010).*

differences in some of the other metals detected show a clear contribution from pavement including Al, Ca, Cu, Fe, Mg, Mn, K, Ni, Na, Ti, and V. Differences in PAH concentrations were also observed where five substances, including phenanthrene, pyrene, benzo(a)pyrene, benzo(g,h,i)perylene, and indeno-1,2,3(c,d)pyrene, were detected at higher concentrations in the tire wear particles than in the tread, indicating pavement as a contributor.

TABLE 7.1 Chemical Constituents of Tire Wear Particles and Tread Particles (TPs)

Chemical Family (%w/w)	Tire Wear Particles	Tread Particles
Plasticizers and oils	10	19
Polymers	16	46
Carbon blacks	13	19
Minerals	61	16

Metals (ppm)	Tire Wear Particles	Tread Particles
Aluminum	28,200	470
Antimony	130	76.5
Arsenic	N.D.	N.D.
Beryllium	N.D.	N.D.
Bismuth	N.D.	86.8
Boron	N.D.	N.D.
Cadmium	N.D.	N.D.
Calcium	65,300	1010
Chromium	N.D.	N.D.
Cobalt	N.D.	N.D.
Copper	634	21.5
Iron	27,700	224
Lead	N.D.	N.D.
Magnesium	14,500	65.8
Manganese	607	N.D.
Nickel	52.6	N.D.
Potassium	5810	242
Selenium	N.D.	N.D.
Silicon	87,000	54,000
Silver	N.D.	N.D.
Sodium	4750	N.D.
Sulfur	5000	12,000
Titanium	1390	29.9

(*Continued*)

TABLE 7.1   Chemical Constituents of Tire Wear Particles and Tread
Particles (TPs)—cont'd

Metals (ppm)	Tire Wear Particles	Tread Particles
Vanadium	49.6	N.D.
Zinc	3000	9000

PAHs (ppm)	Tire Wear Particles	Tread Particles
Acenaphthene	0.04	0.13
Naphthalene	0.2	1.18
Phenanthrene	1.66	1.21
Pyrene	4.77	0.06
Acenaphthalene	0.15	1.24
Anthracene	0.1	0.11
Benzo(a)anthracene	0.18	2.87
Benzo(a)pyrene	0.28	N.D.
Benzo(b)fluoranthene	0.37	0.92
Benzo(g,h,i)perylene	3.22	1.77
Benzo(k)fluoranthene	0.02	0.92
Chrysene	0.36	2.95
Dibenzo(a,h)anthracene	0.1	0.87
Fluoranthene	0.98	1.62
Fluorene	0.07	0.25
Indeno-1,2,3(c,d)pyrene	0.21	N.D.
Total	12.71	16.1

2.2 Volume Size Distribution

Historically, researchers who have studied tire wear and tread loss have reported wide size distributions in the particles ranging from 5 to more than 300 μm (Cadle & Williams, 1979; Dannis, 1974). More recently, Kreider et al. (2010) reported a wide unimodal size distribution, as measured by laser diffraction, from 5 to 220 μm, with a mode centered at ~75 μm. Additional measurements of the distribution using transmission optical microscopy were similar to the laser diffraction method and ranged from 4 to 350 μm, with a mode at 100 μm. Regardless of the method, Kreider et al. (2010) reported less than 1% by volume of the particles were less than 10 μm.

In addition to the overall size distribution of the bulk tire wear particles, studies have been conducted to understand the size distribution of the airborne fraction of tire wear particles, and the results indicate that these distributions vary with the tire type (nonstudded or studded) and pavement type (Gustafsson et al., 2008). Unpublished research wherein PM_{10} samples were collected while running passenger car tires on a roadway simulator showed that more than 60% of tire wear particles (by mass) were generally present in the coarse fraction (i.e., $2.5-10\,\mu m$) (Panko et al., 2009). In this study, the generation of PM_{10} was quantified with an aerodynamic particle sizer (APS; 0.5 and $10\,\mu m$) and the nano-sized fraction was evaluated using a scanning mobility particle sizer (SMPS; $8-311\,nm$). Elemental analysis (particle induced X-ray emission, PIXE) and particle morphology (scanning electron microscopy) were also examined to apportion the particles between pavement and tires. APS and SMPS data show that the concentration of particles generated was low, seldom greater than $10-20\,\mu g/m^3$. APS data indicated a bimodal distribution of the mass of particles, with peaks around $1\,\mu m$ and between 5 and $8\,\mu m$. The SMPS data identified a peak in the number of particles generated between 10 and 100 nm, but the overall concentration was similar to that of background nanoparticles in the laboratory. In other studies, the use of studded tires has been demonstrated to produce particles in the nanometer range; however, the sources of these particles are the stud and pavement, not the tread rubber (Gustafsson et al., 2009). When non-studded tires are used, nanoparticle concentrations do not exceed background (VTI 2009).

3. TIRE WEAR STUDIES

A variety of studies have been conducted to understand the contribution of tire wear to airborne particulate, especially PM_{10}. The studies can be grouped into two categories: air concentration measurements and emission rate estimates.

3.1 Tire Wear Particle Air Concentrations

Very few reliable estimates of tire wear particles in the ambient air are available, primarily because of limitations in the chemical markers used to quantify this type of PM. Furthermore, many measurements of tire wear have been conducted at locations where higher than average levels would be expected, such as near or in tunnels, and thus are not representative of typical human exposures (Almeida-Silva, Canha, Freitas, Dung, & Dionísio, 2011; Fauser, Tjell, Mosbaek, & Pilegaard, 1999; Gualtieri et al., 2005; Hüglin & Gehrig, 2000, p. 14; Salma & Maenhaut, 2006; Schauer, Fraser, Cass, & Simoneit, 2002).

Because tire wear particles are not a single substance, a chemical marker is necessary to detect and quantify them in the environment. Historically, most researchers had relied on Zn as an indicator of tire wear particles. After Pierson and Brachaczek (1974) established a correlation between traffic activity and elemental Zn concentrations in airborne PM as well as in roadside soil samples, Zn was proposed as a marker for tire wear. However, other confounding sources of Zn from exhaust and non-exhaust emissions such as corrosion of crash barriers and brake wear were not considered in the studies. Because the use of elemental Zn lacks specificity to tire wear particles, Fauser et al. (1999) proposed to use extractable organic zinc (zinc-mercaptobenzothiazole) as a marker because tires are, with the exception of engine oil, the only significant contributors to extractable organic zinc in airborne particulates. Nevertheless, given the interference from additives used in engine oils (zinc-dithiophosphates), which were widely used in Europe, significant uncertainties remained for determinations of tire wear particles. Building off of initial research by Harada, Shbata, Panko, and Unice (2009) and Kitamura, Kuroiwa, Harada, and Kato (2007), Unice, Kreider, and Panko (2012) published a marker for tire wear that uses a pyrolysis GC/MS method wherein the presence of the particles are qualitatively confirmed using butadiene, isoprene, and styrene and quantified using the dimeric markers vinylcyclohexene and dipentene, which have good specificity for tread rubber polymer with no other appreciable environmental sources.

Panko, Chu, Kreider, and Unice (2013) presented the results of ambient air sampling conducted at roadside locations throughout parts of France, the United States, and Japan, wherein tire wear particles were quantified using the pyrolysis marker. The results indicated that the tire wear particle concentrations in the PM_{10} fraction were low with averages ranging from 0.05 to 0.70 $\mu g/m^3$, representing an average contribution to total PM_{10} of 0.84%. When concentrations are converted to a tread basis (i.e., divide the mass of tire wear particles by 2 to account for an approximate twofold difference between polymer content in tread [$\sim$50%] and tire wear particles [$\sim$25%]), the results from this study are consistent with the low end of the range of historically reported values (Table 7.2). The difference between results reported by Panko et al. (2013) and the upper portion of the range of historical estimates of tread concentration in PM_{10} could be a result of differences in collection or chemical markers, which have included benzothiazole compounds, metal concentration, and the analysis of pyrolysis products with a nonspecific flame ionization or flame photometric detectors (Kumata et al., 2011; Miguel, Cass, Weiss, & Glovsky, 1996; Sakamoto, Hirota, Nezu, & Okuyama, 1999; Toyosawa, Umezawa, Ikoma, & Kameyama, 1977). Kumata et al. (2011) reported that tread contributed 0.68% of the total PM_{10} in roadside samples collected in Japan based on the analysis of dihydroresin acid markers (Kumata et al., 2011). This estimate is also in close proximity to, albeit higher than, the percent contribution of tread to PM_{10} (0.42%) predicted for all samples collected in Japan by Panko et al. (2013). Additional factors that may affect the contribution of tire wear to total PM_{10} include differences in road types and vehicles, local driving behavior, background concentrations and wind (direction and strength) between the sampling locations, type of sampling site (background, urban, etc.) and distance of sampling sites from the road, and meteorological conditions and climate season.

3.2 Tire Wear Emission Rates

Over time, many estimates of emission rates for airborne tire wear particles have been made and were recently summarized by Grigoratos & Martini (2014, p. 53). For PM_{10} the most recent emission rate estimates range from 2.4 to 13 mg/vkm with an average of 6.3 mg/vkm (Table 7.3). The differences in the reported rates are likely explained by the varied methods used to calculate the emission rates such as derivation from emission inventories, receptor modeling, and statistical models using source profiles, as well as direct measurements in road simulator laboratory experiments and during roadside air sampling campaigns. Furthermore, although the PM_{10} emission rates listed in Table 7.3 are the most recent available, many originate from studies conducted 10–20 years ago and should be used with caution, as tire wear rates have improved with advanced tire technologies. Emission rates for tire wear particles in the $PM_{2.5}$ fraction are sparse and generally estimated as a percent of PM_{10}; their reliability is uncertain.

TABLE 7.2 Historical Estimates of Tread Contribution to Airborne Particulate Matter (PM) (%)

Location	TSP	SPM	PM$_{10}$	PM$_{2.5}$	Reference
United States	1.5–9.2		–	–	Cardina (1974), Cardina (1973)
	1–4		–	–	Pierson and Brachaczek (1974)
	1.6–2.4		–	–	Cadle and Williams (1979)
	1.6–3.5		–	–	Rogge et al. (1993)
	–		–	1	Fishman and Turner (1999)
	–		–	1.3, 3	Schauer et al. (2002)
	–	–	0.44	–	Panko et al. (2013)
Japan		5.2 ± 0.4	–	–	Toyosawa et al. (1977)
		1.3–3	–	–	Kima, Yagawaa, Inoue, Lee, and Shirai (1990)
		0.5–5.6	–	–	Yamaguchi, Yamazaki, Yamauchi, and Kakiuchi (1995)
		4.3	–	–	Sakamoto et al. (1999)
		3.6 ± 4.7			Kumata, Takada, and Ogura (1997)
		3	–	–	Miguel et al. (1996), Doki, Kunimi, and Takahasi (2002)
		3–4	2.3	1.3	JATMA (2001)
			0.68		Kumata et al. (2011)
	–	–	0.2	–	Panko et al. (2013)
Europe	–		5	–	Annema, Booij, Hesse, and van derMeulenSlooff (1996, p. 102)
	16		10	–	Israël, Pesch, and Schlums (1994)
	–		8.6	–	Rauterberg-Wulff (1998)
	–		5	–	Fauser et al. (1999)
	–		1–7.5	–	Hüglin and Gehrig (2000, p. 14)
	5		–	–	Fauser, Tjell, Mosbaek, and Pilegaard (2002)
	–		3–7	–	Gualtieri et al. (2005)
	–		6	3	BLIC (2005)
			0.1–3.9		Sjödin et al. (2010, p. 100)
			3–4	4–7	Kwak, Kim, Lee, and Lee (2013)
	–		0.62	–	Panko et al. (2013)

PM_{10}, particulate matter with aerodynamic diameter <10 μm; $PM_{2.5}$, particulate matter with aerodynamic diameter <2.5 μm; SPM, suspended particulate matter; TSP, total suspended particulate.

TABLE 7.3   Summary of PM_{10} Emission Factors for Tire Wear

Reference	PM_{10} Emission Factor (mg/vkm)	EF Method
U.S. EPA (1995)	5	Emission Inventory
Keuken, Denier van der Gon, and van der Valk (2010)	5	Emission Inventory
Rauterberg-Wulff (1998)	6.1	Receptor Modeling
Hüglin and Gehrig (2000)	13	Receptor Modeling
Luekewille et al. (2001)	6.5	Receptor Modeling
CEPMEIP (2002)	4.5	Emissions Inventory
Luhana, Sokhi, Warner, Mao, and Boulter (2004)	7.4	Receptor Modeling
Kupiainen et al. (2005)	9	Road Simulation Study
Ten Broeke, Hulskotte, and Denier van der Gon (2008)	8	Emissions Inventory
Sjödin et al. (2010)	3.8	Road Simulation Study
CEPMEIP (2012)	3.5—4.5	Emissions Inventory
NAEI (2012)	7	Emissions Inventory
Panko et al. (2013) (central tendency)	2.4	Roadside Study
Panko et al. (2013) (95th percentile)	7	Roadside Study

PM_{10}, particulate matter with aerodynamic diameter <10 μm.

4. DATA GAPS AND NEEDS

As many researchers have noted, the percent contribution of non-exhaust vehicle emissions to traffic-related PM levels will continue to increase as exhaust emissions decrease. As such, an accurate picture of the sources in the non-exhaust vehicle emission category is important for making appropriate risk management decisions. Much of the existing data for tire-associated emissions are confounded by the lack of a consistent marker for measuring tire wear particles. Recently, the pyrolysis GC/MS method proposed by Unice et al. (2012) was accepted as an ISO Technical Specification (TS-20593). Use of this marker in conducting more measurements of tire wear particles in the ambient air would significantly strengthen the aging data set with more modern values. Additionally, very few data are available to characterize tire wear particle contributions to $PM_{2.5}$; therefore measurements in this fraction would help complete the tire wear emission profile and provide more reliable emission factors for tire wear particles in the $PM_{2.5}$ fraction.

5. CONCLUSIONS

Emphasis on minimizing human exposures to respirable particulate in the ambient air is a global effort. Regulations governing the allowable PM in the air have been established

TABLE 7.4 International Standards for $PM_{2.5}$ and PM_{10}

Pollutants	Australia	Canada	China	South Korea	EU	United States	India	WHO Guidelines
$PM_{2.5}$ annual ($\mu g/m^3$)	8	10	15	25	25	12	40	10
$PM_{2.5}$ for 24-h ($\mu g/m^3$)	25	28	35	50	—	35	60	25
PM_{10} annual ($\mu g/m^3$)	—	—	40	50	40	—	60	20
PM_{10} for 24-h ($\mu g/m^3$)	50	—	50	100	50	150	100	50

internationally, and programs are in place to measure progress toward achieving the PM goals (Table 7.4). Data obtained from established monitoring networks have indeed shown a measurable decline. For example, in the United States, from 2000 to 2016, the average nationwide levels of PM_{10} and $PM_{2.5}$ have decreased 40% and 42%, respectively, with the vast majority of the measurements below the national standards (U.S. EPA, 2017). Similarly, in Europe significant decreasing trends in the PM_{10} annual mean were observed between 2000 and 2014, and for $PM_{2.5}$ concentrations, a decreasing trend was observed between 2006 and 2014 (EEA, 2016). Nevertheless, some regional PM air concentrations, particularly in urban environments, still exceed air quality standards and are often attributable to vehicle traffic.

Available data indicate that the contribution of tire wear particles to the non-exhaust emission category is important, although the contribution of tire wear particles to the total ambient PM_{10} and $PM_{2.5}$ is low. Nevertheless, as vehicle exhaust emissions are curbed through various control measures, the contribution of tire wear particles may increase. To date, there are no known regulations that specifically govern tire wear particle emissions and no single approach or technology to control the emissions. Given the variety of factors that influence the generation of tire wear particles, any future efforts at reducing this source would need to consider not only the characteristics of the tire, but also the vehicle to which it is mounted, the pavement on which the vehicle is driven, and the design of roadway systems that affect the manner in which the vehicle is operated.

References

Adachi, K., & Tainosho, Y. (2004). Characterization of heavy metal particles embedded in tire dust. *Environment International, 30*(8), 1009–1017.

Almeida-Silva, M., Canha, N., Freitas, M. C., Dung, H. M., & Dionísio, I. (2011). Air pollution at an urban traffic tunnel in Lisbon, Portugal: An INAA study. *Applied Radiation and Isotopes: Including Data, Instrumentation and Methods for Use in Agriculture, Industry and Medicine, 69*(11), 1586–1591.

Annema, J., Booij, H., Hesse, J. M., van der Meulen, A., & Slooff, W. (1996). *Integrated criteria document fine particulate matter*. Bilthoven, the Netherlands: RIVM.

BLIC. (2005). *BLIC annual report 2004–2005*. Brussels.

Cadle, S. H., & Williams, R. L. (1979). Gas and particle emissions from automobile tires in laboratory and field studies. *Rubber Chemistry and Technology, 52*(1), 146–158.

Camatini, M., Crosta, G. F., Dolukhanyan, T., Sung, C., Giuliani, G., Corbetta, G. M., et al. (2001). Microcharacterization and identification of tire debris in heterogeneous laboratory and environmental specimens. *Materials Characterization, 46*(4), 271–283.

Cardina, J. A. (1973). The determination of rubber in atmospheric dusts. *Rubber Chemistry and Technology, 46*, 232–241.

Cardina, J. A. (1974). Particle size determination of tire tread rubber in atmospheric dusts. *Rubber Chemistry and Technology, 47*, 271–283.

CEPMEIP. (2002). *CEPMEIP database*. Available from http://www.air.sk/tno/cepmeip/.

CEPMEIP. (2012). *CEPMEIP database*. Available from http://www.air.sk/tno/cepmeip/.

Dall'Osto, M., Beddows, D. C. S., Gietl, J. K., Olatunbosun, O. A., Yang, X., & Harrison, R. M. (2014). Characteristics of tyre dust in polluted air: Studies by single particle mass spectrometry (ATOFMS). *Atmospheric Environment, 94*, 224–230.

Dannis, M. (1974). Rubber dust from the normal wear of tires. *Rubber Chemistry and Technology, 47*, 1011–1037.

Davis, A. P., Shokouhian, M., & Ni, S. (2001). Loading estimates of lead, copper, cadmium, and zinc in urban runoff from specific sources. *Chemosphere, 44*(5), 997–1009.

Doki, S., Kunimi, H., & Takahasi, K. (2002). Estimation of tire emission factors by roadside observation – Japan Clean Air Program (JCAP II). In *43rd annual meeting of Japan Society for atmospheric environment* (Japan).

EEA. (2016). *Air quality in Europe – 2016 report*. Luxembourg: Publications Office of the European Union.

Fauser, P., Tjell, J. C., Mosbaek, H., & Pilegaard, K. (1999). Quantification of tire-tread particles using extractable organic zinc as tracer. *Rubber Chemistry and Technology, 72*(5), 969–977.

Fauser, P., Tjell, J. C., Mosbaek, H., & Pilegaard, K. (2002). Tire-tread and bitumen particle concentrations in aerosol and soil samples. *Petroleum Science and Technology, 20*(1&2), 127–141.

Fishman, R., & Turner, J. (1999). *Tyre wear contributions to ambient particulate matter*. St. Louis: Washington University.

Grigoratos, T., & Martini, G. (2014). *Non-exhaust traffic related emissions. Brake and tyre wear PM*. European Commission.

Gualtieri, M., Andrioletti, M., Mantecca, P., Vismara, C., & Camatini, M. (2005). Impact of tire debris on in vitro and in vivo systems. *Particle and Fibre Toxicology, 2*(1), 1.

Gustafsson, M., Blomqvist, G., Gudmundsson, A., Dahl, A., Swietlicki, E., Bohgard, M., et al. (2008). Properties and toxicological effects of particles from the interaction between tyres, road pavement and winter traction material. *The Science of the Total Environment, 393*(2–3), 226–240.

Gustafsson, M., Blomqvist, G., Lunden, E. B., Dahl, A., Gudundsson, A., Hjort, M., et al. (2009). *NanoWear: Nanoparticles from the abrasion of tires and pavement*. SE-581 95 Linköping Sweden: VTI (Swedish National Pavement and Transport Research Institute).

Harada, M., Shbata, T., Panko, J. M., & Unice, K. M. (2009). Analysis methodology of rubber fraction in fine particles. In *Fall 176th technical meeting of the rubber division of the American Chemical Society, Inc.* (Pittsburgh, PA).

Hüglin, C., & Gehrig, R. (2000). *Contributions of road traffic to ambient PM_{10} and $PM_{2.5}$ concentrations - chemical Speciation of fine particulates and source attribution with a receptor model*. Dübendorf: Swiss Federal Laboratories for Materials Testing and Research (EMPA).

Israël, G. W., Pesch, M., & Schlums, C. (1994). Bedeutung des Reifenabriebs für die Rußemission des Kfz-Verkehrs. *Staub - Reinhaltung der Luft, 54*, 423–430. Springer-Verlag.

JATMA. (2001). Japan Automobile Tire Manufacturers Association.

KEMI. (2003). *HA oils in automotive tyres – prospects of a national ban*. Stockholm, Sweden: Swedish National Chemicals Inspectorate.

Keuken, M., Denier van der Gon, H., & van der Valk, K. (2010). Non-exhaust emissions of PM and the efficiency of emission reduction by road sweeping and washing in The Netherlands. *Science of the Total Environment, 408*(20), 4591–4599.

Kima, M. G., Yagawaa, K., Inoue, H., Lee, Y. K., & Shirai, T. (1990). Measurement of tyre tread in urban air by pyrolysis-gas chromatography with flame photometric detection. *Atmospheric Environment, 24A*, 1417–1422.

Kitamura, Y., Kuroiwa, C., Harada, M., & Kato, N. (2007). *Analysis of rubber fraction in suspended particulate matter (SPM) on road*.

Kreider, M. L., Panko, J. M., McAtee, B. L., Sweet, L. I., & Finley, B. L. (2010). Physical and chemical characterization of tire-related particles: Comparison of particles generated using different methodologies. *Science of the Total Environment, 408*(3), 652–659.

Kumata, H., Mori, M., Takahashi, S., Takamiya, S., Tsuzuki, M., Uchida, T., et al. (2011). Evaluation of hydrogenated resin acids as molecular markers for tire-wear debris in urban environments. *Environmental Science & Technology, 45*(23), 9990–9997.

Kumata, H., Takada, H., & Ogura, N. (1997). 2-(4-Morpholinyl)benzothiazole as an indicator of tire-wear particles and road dust. In R. P. Eganhouse (Ed.), *Molecular markers in environmental geochemistry* (pp. 291–305). Washington D.C: American Chemical Society.

Kupiainen, K. J., Tervahattu, H., Räisänen, M., Mäkelä, T., Aurela, M., & Hillamo, R. (2005). Size and composition of airborne particles from pavement wear, tires, and traction sanding. *Environmental Science & Technology, 39*(3), 699–706.

Kwak, J. H., Kim, H., Lee, J., & Lee, S. (2013). Characterization of non-exhaust coarse and fine particles from on-road driving and laboratory measurements. *The Science of the Total Environment, 458–460*, 273–282.

Lawrence Livermore National Laboratory. (1996). *Effect of waste tires, waste tire facilities and waste tire projects on the environment.* Department of Energy.

Luekewille, A., Bertok, I., Amann, M., Cofala, J., Gyarfas, F., Heyes, C., et al. (2001). A Framework to estimate the potential and costs for the control of fine particulate emissions in Europe. In *Interim report.* Laxenburg, Austria: IIASA.

Luhana, L., Sokhi, R., Warner, L., Mao, H., & Boulter, P. (2004). Measurement of non-exhaust particulate matter. In *Deliverable 8 from EU Project particulates - characterisation of exhaust particulate emissions from road vehicles* (p. 96). European Commission 5th Framework Programme.

Macias-Zamora, J. V., Mendoza-Vega, E., & Villaescusa-Celaya, J. A. (2002). PAHs composition of surface marine sediments: A comparison to potential local sources in Todos Santos Bay, B.C., Mexico. *Chemosphere, 46*(3), 459–468.

Miguel, A. G., Cass, G. R., Weiss, J., & Glovsky, M. M. (1996). Latex allergens in tire dust and airborne particles. *Environmental Health Perspectives, 104*(11), 1180–1186.

NAEI. (2012). *Road transport emission factors from 2010 NAEI.* Available at http://naei.defra.gov.uk/datawarehouse/3_9_323_136259_roadtransportefs_naei10_v2.xls.

Ozaki, H., Watanabe, I., & Kuno, K. (2004). Investigation of the heavy metal sources in relation to automobiles. *Water, Air, & Soil Pollution, 157*, 209–223.

Padovan, J., Prasad, N., Gerrard, D., Park, S. W., & Lindsley, N. (1999). Topology of wear particles. *Rubber Chemistry and Technology, 72.*

Panko, J., McAtee, B., Gustafsson, M., Blomqvist, G., Gudmundsson, A., Sweet, L., et al. (2009). Physio-chemical analysis of airborne tire wear particles. In *46th congress of the European Societies of Toxicology, Eurotox* (Dresden, Germany).

Panko, J. M., Chu, J., Kreider, M. L., & Unice, K. M. (2013). Measurement of airborne concentrations of tire and road wear particles in urban and rural areas of France, Japan, and the United States. *Atmospheric Environment, 72*, 192–199.

Pierson, W., & Brachaczek, W. (1974). Airborne particulate debris from rubber tyres. *Rubber Chemistry and Technology, 47*, 1275–1299.

Rauterberg-Wulff, A. (1998). Beitrag des Reifen- und Bremsenabriebs für Rußemissionen an Straßen. In *Dissertation am Fachgebiet Luftreinhaltung der Technischen Universität Berlin.*

Rogge, W. F., Hildemann, L. M., Mazurek, M. A., Cass, G. R., & Simoneit, B. R. T. (1993). Sources of fine organic aerosol. 3. Road dust, tire debris, and organometallic brake lining dust: Roads as sources and sinks. *Environmental Science & Technology, 27*, 1892–1904.

Sadiq, M., Alam, I., El-Mubarek, A., & Al-Mohdhar, H. M. (1989). Preliminary evaluation of metal pollution from wear of auto tires. *Bulletin of Environmental Contamination and Toxicology, 42*(5), 743–748.

Sakamoto, K., Hirota, Y., Nezu, T., & Okuyama, M. (1999). Contribution of rubber dust generated from tire-dust by travelling of heavy-duty and passenger vehicles on the road to atmospheric particulate matter. *Journal of Aerosol Research, 14*(3), 242–247.

Salma, I., & Maenhaut, W. (2006). Changes in elemental composition and mass of atmospheric aerosol pollution between 1996 and 2002 in a Central European city. *Environmental Pollution, 143*(3), 479–488.

Schauer, J. J., Fraser, M. P., Cass, G. R., & Simoneit, B. R. T. (2002). Source reconciliation of atmospheric gas-phase and particle-phase pollutants during a severe photochemical smog episode. *Environmental Science & Technology, 36*(17), 3806–3814.

Sjödin, Å., Ferm, M., Björk, A., Rahmberg, M., Gudmundsson, A., Swietlicki, E., et al. (2010). *Wear particles from road traffic - a field, laboratory and modelling study.* Göteborg: IVL Swedish Environmental Research Institute Ltd.

Takada, H., Onda, T., & Ogura, N. (1990). Determination of polycyclic aromatic hydrocarbons in urban street dusts and their source materials by capillary gas chromatography. *Environmental Science & Technology, 24*(8), 1179–1186.

Ten Broeke, H., Hulskotte, J., & Denier van der Gon, H. (2008). *Road traffic tyre wear. Emission estimates for diffuse sources, Netherlands Emission Inventory, Netherlands national water board — water unit.*

Toyosawa, S., Umezawa, Y., Ikoma, Y., & Kameyama, Y. (1977). Analysis of tire tread rubber in airborne particulate matter by pyrolysis-gas chromatography. *Analytical Chemistry of Japan (Bunseki Kagaku), 26*(38).

U.S. EPA. (1995). *Draft user's guide to part 5: A program for calculating particle emissions from motor vehicles.* Ann Arbor, MI: United States Environmental Protection Agency.

U.S. EPA. (1997). *U.S. EPA.AP-42 section 4.12, Manufacture of rubber products.* United States Environmental Protection Agency.

U.S. EPA. (2017). *Air trends. Particulate matter air quality (PM_{10} and $PM_{2.5}$).* [cited 2017 June 28, 2017]; Available from https://www.epa.gov/air-trends/particulate-matter-pm10-trends.

Unice, K. M., Kreider, M. L., & Panko, J. M. (2012). Use of a deuterated internal standard with pyrolysis-GC/MS dimeric marker analysis to quantify tire tread particles in the environment. *International Journal of Environmental Research and Public Health, 9*, 4033—4055.

Veith, A. G. (1995). Tyre tread wear — the joint influence of compound properties and environmental factors. *Tire Science and Technology, 23*, 212—237.

Wik, A. (2007). Toxic components leaching from tire rubber. *Bulletin of Environmental Contamination and Toxicology, 79*(1), 114—119.

Williams, R. L., & Cadle, S. H. (1978). Characterization of tire emissions using an indoor test facility. *Rubber Chemistry and Technology, 51*(1), 7—25.

Yamaguchi, T., Yamazaki, H., Yamauchi, A., & Kakiuchi, Y. (1995). Analysis of tire tread rubber particles and rubber additives in airborne-particulates at a roadside. *Journal of Toxicology and Environmental Health (Japan), 41*(2), 155—162.

Zakaria, M. P., Takada, H., Tsutsumi, S., Ohno, K., Yamada, J., Kouno, E., et al. (2002). Distribution of polycyclic aromatic hydrocarbons (PAHs) in rivers and estuaries in Malaysia: A widespread input of petrogenic PAHs. *Environmental Science & Technology, 36*(9), 1907—1918.

Zheng, M., Cass, G. R., Schauer, J. J., & Edgerton, E. S. (2002). Source apportionment of $PM_{2.5}$ in the southeastern United States using solvent-extractable organic compounds as tracers. *Environmental Science & Technology, 36*(11), 2361—2371.

Review of Road Wear Emissions
A Review of Road Emission Measurement Studies: Identification of Gaps and Future Needs

Mats Gustafsson

Swedish National Road and Transport Research Institute, Linköping, Sweden

O U T L I N E

1. Background	161		3.2.1 Asphalt Pavements	167
2. Road Wear Particle Emissions From Nonstudded Tires	162		3.2.2 Influence of Alternative Pavement Materials and Constructions	170
2.1 Source Apportionment Studies	162		3.3 Mobile Measurement Studies	172
2.1.1 Source Apportionment of Road Dust	163		4. Road Wear Emission Factors	172
2.1.2 Source Apportionment of PM$_{10}$ in Ambient Air	163		5. Road Wear Particle Properties	173
2.2 Controlled Studies	165		6. Discussion and Identification of Gaps and Future Needs	177
3. Road Wear Particle Emissions From Studded Tires	167		References	179
3.1 Source Apportionment Studies	167			
3.2 Controlled Studies	167			

1. BACKGROUND

Road wear, and its rate, is crucial for emissions of airborne particles from road traffic. Road wear is the fragmentation and breakdown of road pavements' surfaces when interacting with vehicle tires. Pavements normally have a rock ballast, resulting in wear composed of minerals dominated by elements such as Si, Ca, K, Fe, and Al (Lindgren, 1996). On pavements, dust particles from many sources are deposited, where several are mineral rich. Dust from building activities, winter time traction sanding, adjacent gravel roads, or just windblown dust from surrounding bare soils deposits on the road surfaces and is suspended by vehicles together with the wear dust from the pavements themselves. These processes make road wear contribution to particle emission from the interaction between tires and pavements problematic, as it is difficult to discern the wear contribution in a mineral dust mix.

The interest for road wear dust contribution to particle emissions has been highest in regions where road wear is high, which coincides with countries where studded tires are used during winter (Denier van der Gon et al., 2012). Studded tires are used to increase traction and road safety during icy and snowy conditions and are allowed from late autumn to spring in the Nordic countries, some states in the United States and Canada, as well as in Russia. Studded tires were introduced in the 1970s, and since then, pavements have been adapted to withstand studded tires, employing aggregate-rich and coarse pavements with abrasion-resistant aggregates on many heavily trafficked roads and streets in the Nordic countries. Nevertheless, studded tire use in Sweden causes approximately 100,000 tons of worn pavement each winter season (Gustafsson et al., 2006). Only a small fraction of this is directly contributing to particulate air pollution, but coarser fractions are present in the road and street environments and can be ground to finer fractions through the influence of traffic. In the Nordic countries, the pavement wear is a main contributor to high PM_{10} concentrations during winter and early spring, which is why efforts are made to understand the influence of pavement parameters as well as efforts to reduce the use of studded tires (Norman & Johansson, 2006). A main question regarding the emission of PM_{10} from road wear is how it is affected by various pavement parameters that govern the total abrasion of an asphalt pavement, such as construction design, largest aggregate size, and the abrasion resistance of the aggregate material. A complementary question has been how alternatives to standard asphalt pavements can help to reduce road wear particle emissions.

In countries where no studded tires are used, the situation is different. Wear resistance of pavements have a lower priority than in the Nordic countries, which means that weaker rocks and less wear-resistant constructions are used. Therefore, road wear from nonstudded tire contribution to particle emissions is potentially higher and one of the important fractions of non-exhaust particle pollution.

The aim of this chapter is to summarize the current knowledge on the contribution to PM_{10} from road surface wear, excluding the suspension processes dealt with in Chapter 10 of this issue. In this area, there is a massive Nordic dominance of available literature and a lot is published in reports in the native languages. Available English literature is referred to, but also some Nordic reports are included. Most of these have at least English summaries and might therefore be valuable to the reader.

2. ROAD WEAR PARTICLE EMISSIONS FROM NONSTUDDED TIRES

2.1 Source Apportionment Studies

As pointed out in previous reviews (Amato, Cassee et al., 2014; Denier van der Gon et al., 2013; Thorpe & Harrison, 2008), separating particle emission from direct road wear and road dust suspension is a difficult task owing to the problems with separating similar mineral-based sources emanating from the same contact between the tire and pavement. Therefore, the mineral components are often combined as road dust or a crustal source or as road wear/suspension (Keuken et al., 2013; Kholdebarin, Biati, Moattar, & Shariat, 2015; Mooibroek et al., 2016; Pant et al., 2015; Vu, Delgado-Saborit, & Harrison, 2015). In source apportionment studies, detailed source profiles are needed, and because the road surface may change in composition over very short distances, as well as other sources to mineral dust suspended from the road surface, this is truly a problematic issue. Bitumen, being the other main component of asphalt pavements containing organic components, might be used as tracers for road wear particles (Thorpe & Harrison, 2008).

2.1.1 Source Apportionment of Road Dust

A few studies aim to separate the road wear component, e.g., Amato et al. (2016). The authors still state that the road wear component might be influenced by dust from unpaved areas and urban works though. In this study, positive matrix factorization(PMF) modeling was used on Paris road dust data from a vacuum cleaner—based road dust sampler (Amato et al., 2009) sampling RD10 (road dust smaller than 10 μm), and the road chemical profile was characterized by a crustal composition including Al, Ca, K, Ti, Fe, and Mg (Fig. 8.1).

The study concluded that the three main sources identified (road wear, brake wear, and carbonaceous) together contributed in similar extents to 96% of RD10 (Fig. 8.2). The highest dust load (10.3 mg/m^2) was found on a city street with cobble stones, whereas a ring road asphalt site had higher dust amounts than those in an urban one, attributed to the poor state of the asphalt at this site. The authors concluded that the dust load showed agreement with emission factors at both the inner roads of Paris and the ring road site.

In a similar study in Spain, Amato, Alastuey et al. (2014) used PMF modeling and found a higher relative contribution of road wear/mineral and carbonaceous (now named tire wear/bitumen) sources to road dust mass loadings (Fig. 8.3). The road wear factor in the PMF could not be separated from other possible sources from unpaved roads and urban work, but was, in average, contributing to about 20% of road dust mass load (RD10). The variation was high (0.3—5.8 mg/m^2) but could reach as high as 10 mg/m^2 in areas close to unpaved areas, urban work, or in pavements in poor condition.

2.1.2 Source Apportionment of PM$_{10}$ in Ambient Air

Using principal component analysis on data from tunnel measurements in England, Lawrence et al. (2013) found a component dominated by Mg with moderate loadings of Ba and Na that was interpreted as road wear. In the source apportionment, the contribution from road surface wear was calculated to be 11% of PM$_{10}$ (Fig. 8.4). In Lawrence, Sokhi, and Ravindra (2016), the source apportionment was used to estimate an emission factor of 3.9—4.5 mg/vkm for road wear PM$_{10}$.

Fauser, Tjell, Mosbaek, and Pilegaard (2000) suggested and used a method to identify bitumen in aerosols. They found that asphalt is the only source contributing to organic molecules with a molecular weight more than 2000 g/mol. These were separated and identified. Using this analysis technique in combination with size-segregated aerosol sampling, Fauser, Tjell, Mosbaek, and Pilegaard, (2002) found that bitumen constitutes about 5 wt.% of roadside total suspended particles and that bitumen particles have a mean aerodynamic size of about 1 μm.

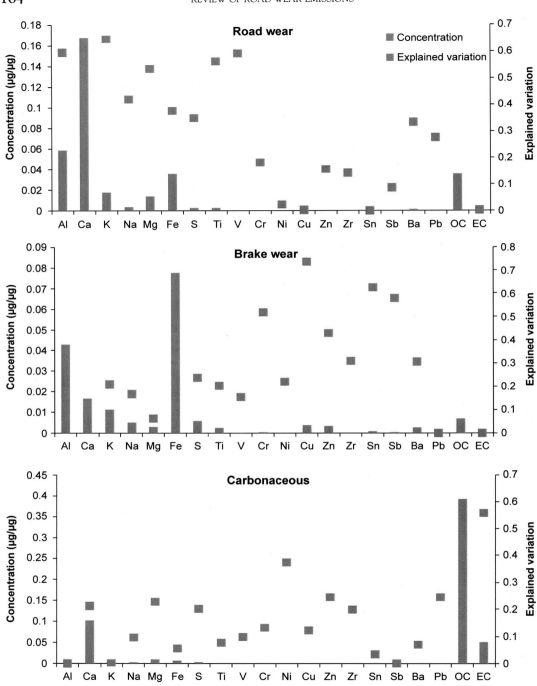

FIGURE 8.1   Factor profiles for RD10 used by Amato et al. (2016).

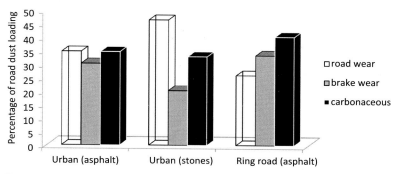

FIGURE 8.2 Percentage of RD10 separating source contributions for different road categories (Amato et al., 2016).

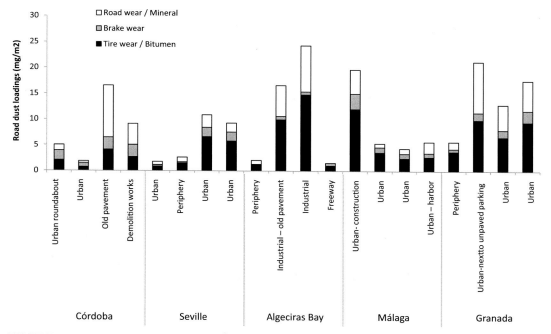

FIGURE 8.3 Absolute contribution (mg/m²) of road dust sources to road dust mass loadings in different areas in Spain. *From Amato, Alastuey et al. (2014).*

2.2 Controlled Studies

To avoid the source apportionment problems related to real-world studies, controlled experiments can be of help. An experimental approach was used by Gehrig et al. (2010), where mobile load simulators were used to study the emission from road surfaces. By separating the initial suspension during measurements from the following stable PM concentration interpreted as the "clean" road wear contribution, they could calculate emission factors for both

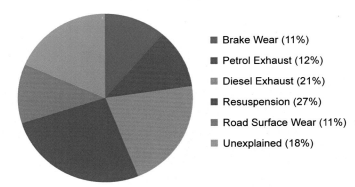

- ■ Brake Wear (11%)
- ■ Petrol Exhaust (12%)
- ■ Diesel Exhaust (21%)
- ■ Resuspension (27%)
- ■ Road Surface Wear (11%)
- ■ Unexplained (18%)

FIGURE 8.4   Source apportionment of the Hatfield tunnel PM_{10} (Lawrence et al., 2013).

fresh particle production and suspension (Fig. 8.5). They concluded that the emission factor for fresh particle abrasion of a standard asphalt concrete in good condition was around 3 mg PM_{10}/vkm, which is about one magnitude lower than exhaust PM emission factors from light-duty vehicles (LDV) used in the emission model HBEFA (http://www.hbefa.net) and also much lower than the derived suspended dust emission factor of 76 mg PM_{10}/vkm. For tests with heavy-duty vehicle (HDV) tires, the figures were 7 and 110 mg PM_{10}/vkm, respectively, for direct wear emission and suspension. It should be noted that these emission factors are calculated from devices with a tire speed of only 9 km/h (LDV) and 25 km/h[1] (HDV), so real-world emission factors are likely to be higher. Interesting observations on the same study were that a new porous asphalt pavement tended to give lower emissions than a concrete asphalt pavement in poor condition (cracks and damages), which emitted much higher amounts of both direct wear and suspension.

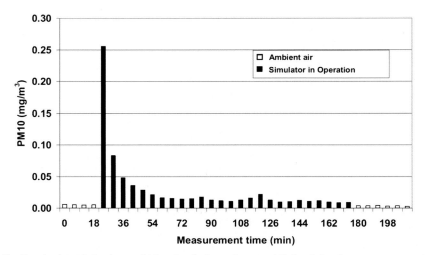

FIGURE 8.5   Results from light-duty vehicles simulation using a mobile load simulator on a new asphalt concrete pavement (Gehrig et al., 2010).

3. ROAD WEAR PARTICLE EMISSIONS FROM STUDDED TIRES

3.1 Source Apportionment Studies

Using roadside measurements and the mobile measurement laboratory Sniffer (Pirjola, Kupiainen, Perhoniemi, Tervahattu, & Vesala, 2009), Kupiainen et al. (2016) investigated the contribution to urban PM_{10} from road pavements and winter traction sand. They used the same source apportionment method as suggested by Räisänen, Kupiainen, and Tervahattu (2003) (see below), choosing a sand material with different mineralogy than the pavement aggregates. The results confirmed the laboratory experiments and showed that, even though large amounts of traction sand were used, the main contributor (about 50%) to dust and PM_{10} was pavement wear, whereas traction sand contributed to about 25%. The results are not only due to the use of studded tires but also due to the previously mentioned "sandpaper effect" contributing with a few percent to the pavement wear (Fig. 8.6).

3.2 Controlled Studies

3.2.1 Asphalt Pavements

In Finland and Sweden, research has been conducted in similar laboratories regarding the influence of pavement properties on particle emissions, with a strong focus on PM_{10} and wear resistant pavement constructions, such as stone mastic asphalt (SMA). Early Finnish studies (Kupiainen, Tervahattu, & Räisänen, 2003; Kupiainen et al., 2005; Räisänen et al., 2003; Räisänen, Kupiainen, & Tervahattu, 2005) used a road simulator to study the contribution from pavement and traction sand to emitted PM_{10}. The analysis technique was based on

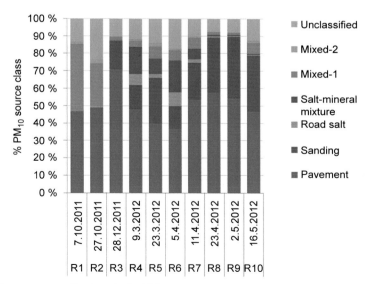

FIGURE 8.6   The source contributions in the PM_{10} suspension emission samples as percentages. *From Kupiainen et al. (2016).*

choosing mineralogically different rock aggregates for pavement and traction sand to be able to separate the sources in PM_{10}. Tests were made using both studded and friction tires. Their results show that the contribution from the pavement not only increases with studded tire use but is also enhanced when traction sand is used. This effect was called the "sandpaper effect" (Kupiainen et al., 2003). Even though most data are on how properties of traction sand influences PM_{10} emission, Räisänen et al., (2005) concluded that a pavement made with a granite with higher resistance to abrasive wear resulted in lower PM_{10} emissions than a pavement made with a mafic volcanic rock when tested with the same traction sand (Fig. 8.7).

Similar studies, but with a stronger focus on the influence on PM_{10} emission of pavement properties, have also been performed using a road simulator at the Swedish National Road and Transport Research Institute. Early tests showed a marked difference in PM_{10} production resulting from studded tire wear of different pavement constructions with different rock aggregates (Gustafsson, Gudmundsson et al., 2009). Most studies have been conducted on the most abrasion-resistant and usual pavement designs on heavily trafficked roads in the Nordic countries, namely SMA, even though a number of other designs have also been tested. Many studies are reported in Swedish reports, but tests until 2012 have been summarized in an English report, Gustafsson and Johansson (2012). For SMA pavements, rock aggregate properties and maximum aggregate size (D_{max}) have been studied. Furthermore, alternative designs and materials, such as rubber mixed asphalt, furnace slag asphalt, porous asphalt, and cement concrete, have been tested for PM_{10} emission. A brief summary of the results produced using this facility is given as follows.

It is well known from previous wear studies that the D_{max} in the pavements (Fig. 8.8) influences the total wear inasmuch as coarser material results in lesser wear (Jacobson & Wågberg, 2007). This also holds for the PM_{10} studies made in the road simulator (Fig. 8.9).

As shown in Fig. 8.10, the total mean wear is also related to the PM_{10} concentrations in the tests.

The most important data, reflecting wear properties, for the aggregate material are the Nordic abrasion test, which is a ball mill value and the Los Angeles value. The Nordic abrasion value is a measure of the resistance to studded tire wear, whereas the fragmentation capacity of the material is determined in the Los Angeles test. Both these tests comprise certain tumbling fractions of the material in drums together with steel balls. The mass percentage of the fraction below a certain size after tumbling, in relation to the mass of the total

FIGURE 8.7 PM_{10} concentration of two asphalts tested using traction sand of diabase (DB) and mafic volcanic rock (MF). F denotes friction tyres and S denotes studded tyres. The asphalt in "stage I and II" was a mafic volcanic rock, and in stage III a granite (Räisänen et al., 2005).

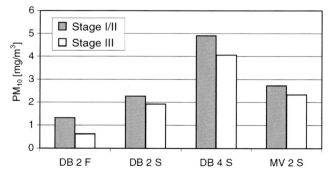

FIGURE 8.8   Three stone mastic asphalt pavements with largest aggregate sizes of 8, 11, and 16 mm.

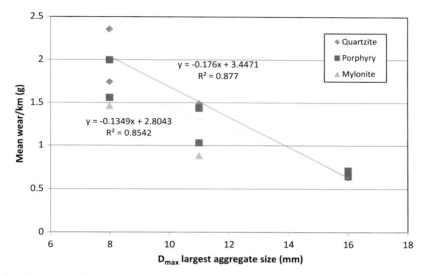

FIGURE 8.9   The relation between maximum aggregate size (D_{max}) and mean wear for three stone mastic asphalt 11 pavements with different rock aggregates (Gustafsson & Johansson, 2012).

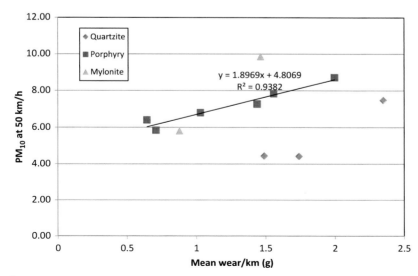

FIGURE 8.10 PM_{10} at 50 km/h as a function of mean wear of stone mastic asphalt 11 pavements tested in the road simulator. The linear regression line is based on the porphyry pavement's data (Gustafsson & Johansson, 2012).

sample, gives the Nordic abrasion and Los Angeles values. A high value thus indicates less resistant materials.

It has not been considered self-evident that a material of high abrasion resistance also generates less PM_{10}, even though this was actually the main result. Döse and Åkesson (2011) studied the amount of PM_{10} produced in the Nordic ball mill test and showed that some materials have higher ball mill values (less resistant) but still do not produce higher PM_{10} amounts than far more resistant rock aggregates (Fig. 8.11). For rocks with Nordic ball mill values below 10, the authors suggested that for each unit lower value, each ton of rock will produce 4 kg less PM_{10}.

Fig. 8.12 shows the Nordic ball mill abrasion values and Los Angeles values for all the SMA 11 pavements tested in the road simulator (Gustafsson & Johansson, 2012). It is seen that both the Nordic abrasion and Los Angeles values correlate with PM_{10}, supporting the results from Döse and Åkesson (2011), at least within the range of the Nordic ball mill values tested. The Nordic abrasion value has a slightly higher correlation, but the analysis is weakened slightly by the fact that many of the aggregate materials in the tested pavements have similar Nordic abrasion values.

The results shown here are based on a single experimental design. More robust conclusions could be drawn if other experimental methods were used, but this has yet to be carried out.

3.2.2 Influence of Alternative Pavement Materials and Constructions

Alternative materials for use in asphalt pavements as well as alternative constructions are often considered not only to improve pavement duration and properties but also to find ways to use or reuse residual or waste materials. The use of furnace slag as aggregates in pavements and mixing ground tire rubber into the bitumen phase of the asphalt are examples of the latter. These methods might also improve asphalt properties, which, of course, is beneficial. As studded tire use and the resulting pavement wear is an important aspect in the

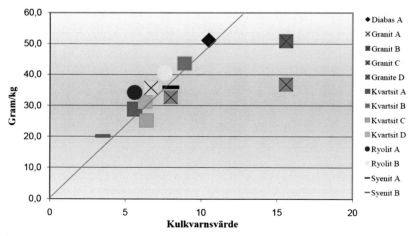

FIGURE 8.11   Produced amount of PM_{10} (in g/kg) in relation to the Nordic ball mill value ("Kulkvarnsvärde" in the figure) for a number of different rock aggregates (diabas = diabase, granit = granite, kvartsit = quartzite, ryolit = rhyolite, syenit = syenite). *From Döse and Åkesson (2011).*

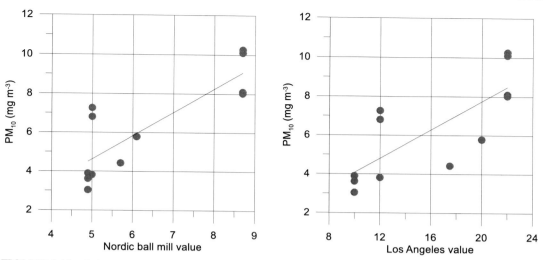

FIGURE 8.12   Relations between the Nordic abrasion value (left) and Los Angeles value (right) and PM_{10} at 50 km/h, measured with tapered element oscillating microbalance for all stone mastic asphalt 11 (Gustafsson & Johansson, 2012).

Nordic countries, projects testing these kinds of pavements sometimes also include particle emission tests.

Rubber mixed pavements were tested in Gustafsson, Blomqvist, and Bennet (2011) as well as Gustafsson, Blomqvist, Jonsson, and Gudmundsson (2009) and summarized in Gustafsson and Johansson (2012). Rubber asphalt with a gap grading was shown to slightly reduce wear PM_{10} production, whereas open-graded rubber asphalt did not differ from a reference SMA pavement. Furnace slag pavements were tested in Viman and Gustafsson (2015). PM_{10} production from SMA 8 and SMA 11 slag asphalts were at similar levels to most asphalt wearing courses made from natural aggregates.

An alternative construction mainly used to reduce tire—pavement noise and to increase water drainage from the surface is porous pavement. These pavements have been reported to decrease suspension of road dust (e.g., Gehrig et al. (2010)) through reducing the amount of dust available for suspension on the road surface. The direct wear emission does not seem to be affected by the construction though. Stationary and mobile measurements in Stockholm showed that the emission was of the same order as for other pavements when studded tires were used (Gustafsson & Johansson, 2012; Johansson, 2006). The difference in PM_{10} emissions between the pavements (porous and reference) was less than c.15%.

In cement concrete, bitumen as a binder is substituted by cement. Cement concrete pavements are often durable and wear resistant, but more expensive to build than asphalt pavements (Wiman, Carlsson, Viman, & Hultqvist, 2009). As they have advantages from a fire perspective, they are often considered for use in road tunnels (Bonati, Merusi, Polacco, Filippi, & Giuliani, 2012). Cement concrete has been tested for PM_{10} emission, and the results show that they emit more PM_{10} than a reference asphalt with the same rock aggregates, even if the total wear is lower (Gustafsson, Kraft, & Silfwerbrand, 2015). An obvious increase in calcium content of PM_{10} worn from cement concrete compared with a reference asphalt implies that the explanation lies in the calcium rich cement binder, giving rise to "extra" PM_{10} emissions, that are not associated with bitumen.

3.3 Mobile Measurement Studies

In Gustafsson and Johansson (2012), field work performed by Stockholm University with a mobile measurement system, called Emma (Hussein et al., 2008) is also reported. Emma measures the PM_{10} concentration behind both front wheels in relation to background in front of the car in line with the TRAKER system (Etyemezian et al., 2003). The measurements relate results to pavement properties and constructions and indicate that emissions are higher from pavements with smaller maximum aggregate size (Table 8.1). They also suggest that aggregate size might be more important than construction type, but more data are needed to support this. In contrast, similar Finnish studies using the Sniffer (Pirjola et al., 2009) show the opposite; pavements with smaller maximum aggregate size seem to generate lower PM_{10} emissions than reference pavements with larger sizes. On-road measurements on real roads suffer from the complexity of separating direct wear from suspension of road dust, which makes the interpretation complex.

4. ROAD WEAR EMISSION FACTORS

As stated earlier, few studies have tried to separate the road wear emissions from general road dust emission (including suspension of dust accumulated on the road surface). The latest version of the EMEP EEA air pollutant emission inventory guidebook (Ntziachristos & Boulter, 2016) uses an approach originally used by Klimont et al. (2002), based on a

TABLE 8.1 Particle Measurements Behind the Studded Tire on Pavements With Different Aggregate Sizes

Pavement	Traffic Flow (Veh/day)	Nordic Abrasion Value	Studded Tire	Summer Tire	Ratio Studded/ Summer
ROAD 859					
Quiet: AC11	4600	<7	43.5 ± 2.6	10.7 ± 4.0	20
Ref: TSK 16			*29.0 ± 4.2*	*6.9 ± 1.3*	*4.8*
ROAD 268					
Quiet: AC11	12,800	<7	30.7 ± 2.3	19.8 ± 7.7	14
Ref: SMA 11			*33.5 ± 6.3*	*2.9 ± 2.1*	*24*
ROAD 260					
Quiet: AC11	10,200	<7	32.4 ± 2.4	7.6 ± 1.9	9.7
Ref: SMA16 and SMA11			*22.2 ± 5.8*	*3.2 ± 4.2*	*17*

The values relate to concentrations behind each tire ($\mu g/m^3$) and the ratios between the studded and summer tire.
TSK, thin layer asphalt.
After Gustafsson and Johansson (2012).

compilation of basically four previous studies (a couple of studies from tunnel measurements were discarded). Since 2002, estimates of specific road wear emission factors have been made by Luhana et al. (2004) and Gehrig et al. (2010) for general nonstudded tire traffic and by Sjödin et al. (2010) for Nordic conditions including studded tire wear (Table 8.2).

5. ROAD WEAR PARTICLE PROPERTIES

Studies on road wear PM properties, where road dust suspension contribution can be neglected, are rare. Gehrig et al. (2010), using mobile load simulators, could investigate the size distribution of the wear particles produced and found a mass size distribution with a maximum at 6–7 μm and with no particles below 0.5 μm (Fig. 8.13). Even though at lower concentrations, this is a very similar distribution to tests made in the VTI road simulator using studded tires (Gustafsson & Johansson, 2012), indicating that road wear PM is controlled by rock disintegration properties rather than the tire type wearing the pavement.

Wearing pavements with studded tires in the VTI road simulator results in PM_{10} totally dominated by rock aggregate fragments, as shown in Fig. 8.14 (Gustafsson et al., 2011). The particle morphology is typical for freshly worn mineral grains, with sharp edges and conchoidal fractures. No morphological signs of bitumen or tire particles are normally found in these samples, probably because these components wear in larger (or even smaller) size fractions.

In a street environment, the particles produced in the interaction between tires and pavements, called road particles (RP) are commonly forming aggregates, where mineral particles from pavement wear stick to rubber and bitumen particles (Kreider, Panko, McAtee, Sweet, & Finley, 2010). This process is likely to reduce the "clean" pavement wear ending up as PM_{10} in the air.

TABLE 8.2 Road Wear PM_{10} Emission Factors

Source	Method	Vehicle Type	EF_{PM10} (mg/vkm)
Ntziachristos and Boulter (2016) based on Klimont et al. (2002)	Compilation of previous data	LDV	7.5
		HDV	38
Luhana et al. (2004)	PCA and MLR analysis	LDV	3.1
		HDV	29
Gehrig et al. (2010)	Mobile load simulator	LDV	3 (new AC), 0–2 (new PA)
		HDV	7 (good AC), 80 (poor AC)
Sjödin et al. (2010) Nordic conditions with studded tire use	Hybrid model (COPREM)	Mean value for Hornsgatan (Stockholm)	226.9

AC, asphalt concrete; *EF,* emission factor; *HDV,* heavy-duty vehicles; *LDV,* light-duty vehicles; *MLR,* multiple linear regression; *PA,* porous asphalt; *PCA,* principal component analysis.

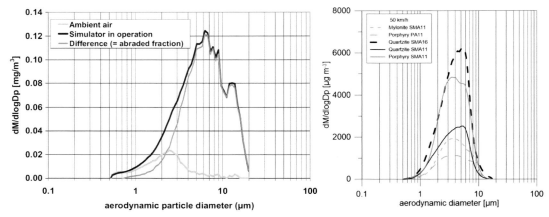

FIGURE 8.13 Typical particle size distribution of PM_{10} (10 μm cutoff) obtained from an experiment with a mobile load simulator (Gehrig et al., 2010) (left) and from a road simulator using studded tires (right) (Gustafsson & Johansson, 2012). *PA*, porous asphalt; *SMA*, stone mastic asphalt.

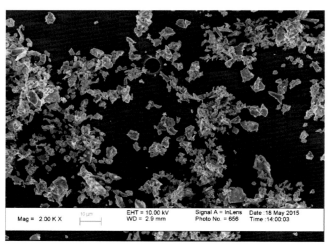

FIGURE 8.14 Typical morphology of PM_{10} from pavement wear caused by studded tires. *SEM-photo by Mats Gustafsson, VTI.*

The dominating mineral content of road wear PM_{10} is reflected also in size-segregated analyses of the elemental composition of PM_{10}. Using granites, quartzites, or similar rocks, the coarser fractions are composed mainly of mineral-related elements, such as Si, Ca, K, and Fe. An example of the composition of PM_{10} worn from granite, diabase, and syenite pavements of SMA type is given in Fig. 8.15 (Gustafsson & Johansson, 2012). Below about 1 μm, the relative contribution from a source of sulfur (S) is strong, which might indicate a contribution from the tire or bitumen. Tungsten is related to the wear of the studs themselves, the tips of which are made of tungsten carbide. Zinc is usually associated with tire wear and is found sporadically in quite small quantities in varying fractions.

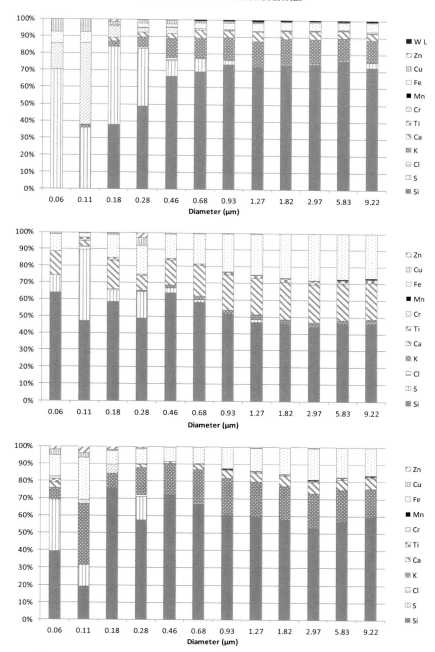

FIGURE 8.15 Relative elemental composition in different particle sizes for PM_{10} from stone mastic asphalt pavements with main aggregates consisting of granite (upper), diabase (middle), and syenite (lower), worn using studded tires (Gustafsson & Johansson, 2012).

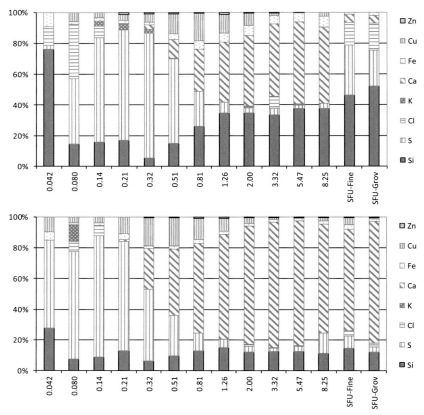

FIGURE 8.16 Relative elemental composition in different particle sizes for PM$_{10}$ worn from an AC with diorite (upper) and an AC with limestone (lower) using summer tires (Ravnikar Turk, 2009). The rightmost columns are a composition of the total fine (<2.5 μm) and coarse (2.5–10 μm) fractions.

Even though studded tires enhance the wear and PM$_{10}$ production strongly, the pavement rock composition is reflected in PM$_{10}$ also when using summer tires, as shown in data produced in the SPENS project using the VTI road simulator (Ravnikar Turk, 2009). Fig. 8.16 shows how the size segregated elemental composition varies when testing asphalt concrete pavements with diorite or limestone rock aggregates. As is the case when using studded tires, the coarser fractions are dominated by rock-related elements. In particular, the calcium content of the limestone is evident.

Introducing alternative materials in asphalt naturally affects the composition of the produced particles. For example, Viman and Gustafsson (2015) showed that PM$_{10}$ from slag asphalts contained more metal components, such as Fe, Mn, and Cr, compared with asphalts with standard rock aggregates (Fig. 8.17), whereas cement concrete produced PM$_{10}$ with a high Ca content owing to a contribution from the cement binder (Gustafsson, Blomqvist, & Hultqvist, 2013). This contribution resulted in higher PM$_{10}$ production than the reference asphalt, despite the fact that the total wear of the cement concrete was lower.

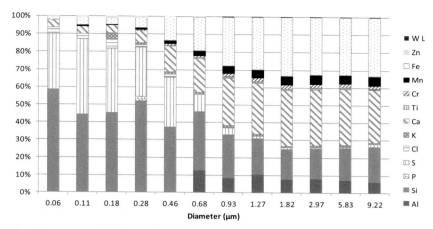

FIGURE 8.17   Size-segregated relative element composition of PM_{10} produced during wear of a slag asphalt using studded tires (Viman & Gustafsson, 2015).

When comparing the composition of RP, tire wear particles (TWP), and tire particles (TP), Kreider et al. (2010) found that the TWP, with only tire and pavement as possible sources, were enriched compared with TP in Al, Ca, Fe, Mg, Mn, K, Na, Ti, and V, indicating that these elements originate in the pavement. The laboratory-produced TWP were also similar in composition to the RP collected from behind a vehicle wheel when sampling on asphalt roads, confirming that road pavement is an important particle source even when studded tires are not present.

6. DISCUSSION AND IDENTIFICATION OF GAPS AND FUTURE NEEDS

Generally, the knowledge on the contribution of pavement wear to airborne particles is, as this chapter shows, scarce. Studies in Scandinavia, where road wear is considered a main source to PM_{10}, have pointed out the importance of pavements and aggregate construction and properties. Most data are laboratory based because of the difficulties in separating road aggregate rocks from other crustal sources and to separate the direct wear contribution from suspension of dust deposited on the road surface. In the world outside the studded tire countries, most data and information are still lacking.

As the wear resistance of rocks used for road construction, as well as the road constructions themselves, varies significantly geographically, it is likely that the contribution from road wear to PM_{10} also varies a lot. Even though no studded tires are used, a pavement using limestone in a nondurable construction is likely to emit higher wear emissions than a porphyry-based SMA. Durable bedrock might be rare and its use might not be economically defendable in areas where wear is of low priority compared with deformation or other pavement degrading processes. However, from an air quality point of view, the road wear contribution might also be of importance, which is why data on PM emissions from road constructions would be valuable.

Today, there are standardized methods to test wear resistance of rock aggregates and pavement samples. None of these methods are adapted to give any information on PM emissions. Total wear has been shown to be related to PM production for some tested rock samples, but laboratory tests also indicate exceptions. Full road simulator tests, as used in some studies, are close to reality, but expensive and can only give information about one pavement type at a time. A simpler test, based on standardized wear tests or, for that matter, completely new methods, could offer a possibility to scan the PM emission properties of pavement samples. A new test method could help to understand which pavement construction types and which pavement and rock aggregate properties are most important for PM emission. This could facilitate the choice of pavements in environments where pavement wear is an important PM source. The Prall and the Tröger test methods might be useful methods as have been shown by Snilsberg (2008), but need further development for the specific purpose of particle emission investigations.

To be able to identify pavement wear as a source for airborne particles, there is a need for better source apportionment possibilities. These must be based on knowledge and data of the pavement used at the site where sampling is made. It is rarely seen that air pollution studies have put any effort into characterizing the pavement or even gathered available information from the road owner of rock aggregates and other components used in the pavement construction. Moreover, the road keeper will hopefully have information on what kind of traction material is used during winter. A petrographic and/or elemental analysis of pavement and traction material used at a specific site is a good start, but it might not be sufficient because the materials have to be discernible from each other and, in best case, from any additional dust source accumulating on and suspending from the road surface. If a tracer or a combination of tracer minerals and/or elements can be found, this will of course be of great value.

Road wear is one of the many sources of road dust that will vary greatly in strength in both time and space. The wear properties of pavements are parts of many factors important for the understanding of the road dust system, but pavement surface properties, such as surface texture and state, might be even more important in areas where the pavement wear is low, but other dust sources are strong. The surface texture has been shown to affect particle suspension. For example, China and James (2011) as well as Blomqvist, Gustafsson, and Lundberg (2013) could show that for a given road dust amount, a coarser texture will emit lower amounts of PM_{10} because of dust not being available for suspension as it accumulates further down in the deeper texture. On the other hand, if road dust sources are strong and the supply is high, a coarser texture will also have the ability to hold more dust available for suspension, as indicated by Gustafsson, Blomqvist, Johansson, and Norman (2012).

Other factors, such as road operation, meteorology, surface humidity, and traffic properties, are all affecting what is finally emitted from the road surface and all need to be managed for full understanding of the road dust system. These issues will be further discussed in the following chapter. An ambitious initiative to understand and describe the road dust system is made in the development of the NORTRIP model (Denby et al., 2016, 2013a, 2013b; Norman et al., 2016).

References

Amato, F., Alastuey, A., de la Rosa, J., Gonzalez Castanedo, Y., Sánchez de la Campa, A. M., Pandolfi, M., & Querol, X. (2014). Trends of road dust emissions contributions on ambient air particulate levels at rural, urban and industrial sites in southern Spain. *Atmospheric Chemistry and Physics, 14*(7), 3533—3544. https://doi.org/10.5194/acp-14-3533-2014.

Amato, F., Cassee, F. R., Denier van der Gon, H. A. C., Gehrig, R., Gustafsson, M., Hafner, W., & Querol, X. (2014). Urban air quality: The challenge of traffic non-exhaust emissions. *Journal of Hazardous Materials, 275*, 31—36.

Amato, F., Favez, O., Pandolfi, M., Alastuey, A., Querol, X., Moukhtar, S., & Sciare, J. (2016). Traffic induced particle resuspension in Paris: Emission factors and source contributions. *Atmospheric Environment, 129*, 114—124. https://doi.org/10.1016/j.atmosenv.2016.01.022.

Amato, F., Pandolfi, M., Viana, M., Querol, X., Alastuey, A., & Moreno, T. (2009). Spatial and chemical patterns of PM10 in road dust deposited in urban environment. *Atmospheric Environment, 43*(9), 1650—1659.

Blomqvist, G., Gustafsson, M., & Lundberg, T. (2013). Road surface dust load is dependent on road surface macro texture. In *Paper presented at the European aerosol conference, Prague, Czech Republic.*

Bonati, A., Merusi, F., Polacco, G., Filippi, S., & Giuliani, F. (2012). Ignitability and thermal stability of asphalt binders and mastics for flexible pavements in highway tunnels. *Construction and Building Materials, 37*, 660—668. https://doi.org/10.1016/j.conbuildmat.2012.07.096.

China, S., & James, D. E. (2011). Comparison of laser-based and sand patch measurements of pavement surface mac- rotexture. *Journal of Transportation Engineering, 138*(2), 176—181.

Denby, B. R., Ketzel, M., Ellermann, T., Stojiljkovic, A., Kupiainen, K., Niemi, J. V., & Sundvor, I. (2016). Road salt emissions: A comparison of measurements and modelling using the NORTRIP road dust emission model. *Atmospheric Environment, 141*, 508—522. https://doi.org/10.1016/j.atmosenv.2016.07.027.

Denby, B. R., Sundvor, I., Johansson, C., Pirjola, L., Ketzel, M., Norman, M., & Omstedt, G. (2013a). A coupled road dust and surface moisture model to predict non-exhaust road traffic induced particle emissions (NORTRIP). Part 1: Road dust loading and suspension modelling. *Atmospheric Environment, 77*, 283—300. https://doi.org/10.1016/j.atmosenv.2013.04.069.

Denby, B. R., Sundvor, I., Johansson, C., Pirjola, L., Ketzel, M., Norman, M., & Omstedt, G. (2013b). A coupled road dust and surface moisture model to predict non-exhaust road traffic induced particle emissions (NORTRIP). Part 2: Surface moisture and salt impact modelling. *Atmospheric Environment, 81*, 485—503. https://doi.org/10.1016/j.atmosenv.2013.09.003.

Denier van der Gon, H., Gerlofs-Nijland, M. E., Gehrig, R., Gustafsson, M., Janssen, N., Harrison, R. M., & Cassee, F. R. (2012). The policy relevance of wear emissions from road transport, now and in the future - an international workshop report and consensus statement. *Journal of the Air & Waste Management Association.* https://doi.org/10.1080/10962247.2012.741055.

Denier van der Gon, H. A. C., Gerlofs-Nijland, M. E., Gehrig, R., Gustafsson, M., Janssen, N., Harrison, R. M., & Cassee, F. R. (2013). The policy relevance of wear emissions from road transport, Nowand in the future-an inter- national workshop report and consensus statement. *Journal of the Air and Waste Management Association, 63*(2), 136—149.

Döse, M., & Åkesson, U. (2011). *Förhållandet mellan nötning och polering samt deras förutsättningar att producera PM_{10}.*

Etyemezian, V., Kuhns, H., Gillies, J., Chow, J., Hendrickson, K., McGown, M., et al. (2003). Vehicle-based road dust emission measurement (III): Effect of speed, traffic volume, location, and season on PM_{10} road dust emissions in the Treasure Valley, ID. *Atmospheric Environment, 37*(32), 4583—4593.

Fauser, P., Tjell, J. C., Mosbaek, H., & Pilegaard, K. (2000). Quantification of bitumen particles in aerosol and soil sam- ples using HP-GPC. *Petroleum Science and Technology, 18*(9—10), 989—1007.

Fauser, P., Tjell, J. C., Mosbaek, H., & Pilegaard, K. (2002). Tire-tread and bitumen particle concentrations in aerosol and soil samples. *Petroleum Science and Technology, 20*(1—2), 127—141.

Gehrig, R., Zeyer, K., Bukowiecki, N., Lienemann, P., Poulikakos, L. D., Furger, M., et al. (2010). Mobile load simu- lators - a tool to distinguish between the emissions due to abrasion and resuspension of PM_{10} from road surfaces. *Atmospheric Environment, 44*(38), 4937—4943.

Gustafsson, M., Berglund, C. M., Forsberg, B., Forsberg, I., Forward, S., Grudemo, S., & Öberg, G. (2006). *Effekter av vinterdäck, en kunskapsöversikt.* VTI Rapport 543. Available at http://vti.diva-portal.org/smash/get/diva2: 675283/FULLTEXT01.pdf.

Gustafsson, M., Blomqvist, G., & Bennet, C. (2011). *Wear particles from road pavements with rubbermixed bitumen — Comparison with reference pavement.* VTI Notat 8-2011. Available at http://vti.diva-portal.org/smash/get/diva2: 1084428/FULLTEXT01.pdf.

Gustafsson, M., Blomqvist, G., Gudmundsson, A., Dahl, A., Jonsson, P., & Swietlicki, E. (2009). Factors influencing PM_{10} emissions from road pavement wear. *Atmospheric Environment, 43*(31), 4699—4702. https://doi.org/ 10.1016/j.atmosenv.2008.04.028.

Gustafsson, M., Blomqvist, G., & Hultqvist, B.-Å. (2013). *Slitage av och partikelemission från betongbeläggning.* VTI Rapport 780. Avaiable at http://vti.diva-portal.org/smash/get/diva2:670649/FULLTEXT01.pdf.

Gustafsson, M., Blomqvist, G., Johansson, C., & Norman, M. (2012). *Driftåtgärder mot PM10 på Hornsgatan och Sveavägen i Stockholm - utvärdering av vintersäsongen 2011-2012.* VTI Rapport 767. Available at http://vti.diva-portal.org/smash/get/diva2:670638/FULLTEXT01.pdf.

Gustafsson, M., Blomqvist, G., Jonsson, P., & Gudmundsson, A. (2009). *Slitagepartiklar från vägbeläggning med gummiinblandad bitumen - jämförelser med referensbeläggning* (VTI Notat 10—2009). Available at http://vti.diva-portal.org/smash/get/diva2:670405/FULLTEXT02.pdf.

Gustafsson, M., & Johansson, C. (2012). *Road pavements and PM10. Summary of the results of research funded by the Swedish Transport Administration on how the properties of road pavements influence emissions and the properties of wear particles.* Trafikverket, Report 2012:241, ISBN 978-91-7467-432-3. Available at http://vti.diva-portal.org/ smash/get/diva2:674206/FULLTEXT01.pdf.

Gustafsson, M., Kraft, L., & Silfwerbrand, J. (2015). Wear and particle generation of three pavement alternatives, a reference concrete, an experimental photocatalytic concrete, and a standard asphalt pavement. In *Paper presented at the 3rd international conference on best practices for concrete pavements, Bonito, Brazil.*

Hussein, T., et al. (2008). Factors affecting non-tailpipe aerosol particle emissions from paved roads: On-road measurements in Stockholm, Sweden. *Atmospheric Environment, 42*(4), 688—702.

Jacobson, T., & Wågberg, L. (2007). *Utveckling och uppgradering av prognosmodell för beläggningsslitage från dubbade däck samt en kunskapsöversikt över inverkande faktorer.* Sweden. Available at http://vti.diva-portal.org/smash/get/ diva2:670344/FULLTEXT01.pdf.

Johansson, C. (2006). *Betydelsen av bullerreducerande beläggning för partikelhalterna.* SLB Rapport 3:2006. Available at http://slb.nu/slb/rapporter/pdf8/slb2006_003.pdf.

Keuken, M. P., Moerman, M., Voogt, M., Blom, M., Weijers, E. P., Röckmann, T., et al. (2013). Source contributions to $PM_{2.5}$ and PM_{10} at an urban background and a street location. *Atmospheric Environment, 71*, 26—35. https:// doi.org/10.1016/j.atmosenv.2013.01.032.

Kholdebarin, A., Biati, A., Moattar, F., & Shariat, S. M. (2015). Outdoor PM_{10} source apportionment in metropolitan cities—a case study. *Environmental Monitoring and Assessment, 187*(2). https://doi.org/10.1007/s10661-015-4294-z.

Klimont, Z., Cofala, J., Bertok, I., Amann, M., Heyes, C., & Gyarfas, F. (2002). *Modelling particulate emissions in Europe - a framework to estimate reduction potential and control costs.* Interim Report IR-02-076, 109.

Kreider, M. L., Panko, J. M., McAtee, B. L., Sweet, L. I., & Finley, B. L. (2010). Physical and chemical characterization of tire-related particles: Comparison of particles generated using different methodologies. *Science of the Total Environment, 408*(3), 652—659.

Kupiainen, K., Ritola, R., Stojiljkovic, A., Pirjola, L., Malinen, A., & Niemi, J. (2016). Contribution of mineral dust sources to street side ambient and suspension PM_{10} samples. *Atmospheric Environment, 147*, 178—189. https://doi.org/ 10.1016/j.atmosenv.2016.09.059.

Kupiainen, K., Tervahattu, H., & Räisänen, M. (2003). Experimental studies about the impact of traction sand on urban road dust composition. *Science of the Total Environment, 308*(1—3), 175—184. https://doi.org/10.1016/ S0048-9697(02)00674-5.

Kupiainen, K. J., Tervahattu, H., Räisänen, M., Mäkelä, T., Aurela, M., & Hillamo, R. (2005). Size and composition of airborne particles from pavement wear, tires, and traction sanding. *Environmental Science and Technology, 39*(3), 699—706. https://doi.org/10.1021/es035419e.

Lawrence, S., Sokhi, R., & Ravindra, K. (2016). Quantification of vehicle fleet PM_{10} particulate matter emission factors from exhaust and non-exhaust sources using tunnel measurement techniques. *Environmental Pollution, 210*, 419—428. https://doi.org/10.1016/j.envpol.2016.01.011.

Lawrence, S., Sokhi, R., Ravindra, K., Mao, H., Prain, H. D., & Bull, I. D. (2013). Source apportionment of traffic emissions of particulate matter using tunnel measurements. *Atmospheric Environment, 77,* 548−557.

Lindgren, Å. (1996). Asphalt wear and pollution transport. *The Science of the Total Environment, 189/190,* 281−286.

Luhana, L., Sokhi, R., Warner, L., Mao, H., Boulter, P., McCrae, I., & Osborn, D. (2004). *Measurement of non-exhaust particulate matter. Deliverable 8 from EU project PARTICULATES − characterisation of exhaust particulate emissions from road vehicles.*

Mooibroek, D., Staelens, J., Cordell, R., Panteliadis, P., Delaunay, T., Weijers, E., & Roekens, E. (2016). PM_{10} source apportionment in five north western European cities − outcome of the Joaquin project. *Environmental Science and Technology, 2016,* 264−292.

Norman, M., & Johansson, C. (2006). Studies of some measures to reduce road dust emissions from paved roads in Scandinavia. *Atmospheric Environment, 40*(32), 6154−6164.

Norman, M., Sundvor, I., Denby, B. R., Johansson, C., Gustafsson, M., Blomqvist, G., et al. (2016). Modelling road dust emission abatement measures using the NORTRIP model: Vehicle speed and studded tyre reduction. *Atmospheric Environment, 134,* 96−108. https://doi.org/10.1016/j.atmosenv.2016.03.035.

Ntziachristos, L., & Boulter, P. (2016). *EMEP EEA air pollutant emission inventory guidebook.* EEA Report No 21/2016.

Pant, P., Baker, S. J., Shukla, A., Maikawa, C., Godri Pollitt, K. J., & Harrison, R. M. (2015). The PM10 fraction of road dust in the UK and India: Characterization, source profiles and oxidative potential. *Science of the Total Environment, 530−531,* 445−452. https://doi.org/10.1016/j.scitotenv.2015.05.084.

Pirjola, L., Kupiainen, K. J., Perhoniemi, P., Tervahattu, H., & Vesala, H. (2009). Non-exhaust emission measurement system of the mobile laboratory SNIFFER. *Atmospheric Environment, 43*(31), 4703−4713. https://doi.org/10.1016/j.atmosenv.2008.08.024.

Räisänen, M., Kupiainen, K., & Tervahattu, H. (2003). The effect of mineralogy, texture and mechanical properties of anti-skid and asphalt aggregates on urban dust. *Bulletin of Engineering Geology and the Environment, 62*(4), 359−368. https://doi.org/10.1007/s10064-003-0200-y.

Räisänen, M., Kupiainen, K., & Tervahattu, H. (2005). The effect of mineralogy, texture and mechanical properties of anti-skid and asphalt aggregates on urban dust, stages II and III. *Bulletin of Engineering Geology and the Environment, 64*(3), 247−256.

Ravnikar Turk, M. (2009). *Sustainable pavements for European new MemberStates.* SPENS. Deliverable D20: Final Report.

Sjödin, Å., Ferm, M., Björk, A., Rahmberg, M., Gudmundsson, A., Swietlicki, E., & Blomqvist, G. (2010). *Wear particles from road traffic - a field, laboratory and modeling study, IVL Report B 1830* (IVL report B1830). Available at http://www.ivl.se/download/18.343dc99d14e8bb0f58b756f/1445517393581/B1830.pdf.

Snilsberg, B. (2008). *Pavement wear and airborne dust pollution in Norway; characterization of the physical and chemical properties of dust particles.* Doctoral thesis 2008:133. (Ph.D.), NTNU, Trondheim.

Thorpe, A., & Harrison, R. M. (2008). Sources and properties of non-exhaust particulate matter from road traffic: A review. *Science of the Total Environment, 400*(1−3), 270−282.

Viman, L., & Gustafsson, M. (2015). *Slag asphalt, interim report C: Wear and formation of respirable particles (PM_{10}).* VTI Notat 24-2015. In Swedish. Available at http://vti.diva-portal.org/smash/get/diva2:865529/FULLTEXT01.pdf.

Vu, T. V., Delgado-Saborit, J. M., & Harrison, R. M. (2015). Review: Particle number size distributions from seven major sources and implications for source apportionment studies. *Atmospheric Environment, 122,* 114−132. https://doi.org/10.1016/j.atmosenv.2015.09.027.

Wiman, L. G., Carlsson, H., Viman, L., & Hultqvist, B.-Å. (2009). *Prov med olika överbyggnadstyper, Uppföljning av observationssträckor på väg E6, Fastarp−Heberg, 1996−2006.* VTI Rapport 632. Available at http://vti.diva-portal.org/smash/get/diva2:670553/FULLTEXT01.pdf.

Review of Road Dust Emissions

Bruce R. Denby[1], Kaarle J. Kupiainen[2], Mats Gustafsson[3]

[1]The Norwegian Meteorological Institute (MET), Norway; [2]Nordic Envicon Oy, Helsinki, Finland; [3]Swedish National Road and Transport Research Institute, Linköping, Sweden

O U T L I N E

1. Introduction	184	3. Sampling of Road Dust Loading	191
2. Overview of Sources and Processes Affecting Road Dust Loading and Emissions	185	4. Measurement of Road Dust Emissions	194
2.1 Sources	185	4.1 Mobile Platform Measurements	194
2.2 Suspension	185	4.2 Downwind Ambient Air Measurements	195
2.3 Road Dust Sources Other Than Traffic-Related Wear Sources	186	4.3 Suspension Rates	195
2.3.1 Surface Abrasion (The Sandpaper Effect)	186	4.4 Emission Factors for Road Dust Suspension	196
2.3.2 Traction Sand	189	5. Process-Based Modeling of Road Dust Loading and Emissions	196
2.3.3 Fugitive Sources	189		
2.3.4 Deposition	189	6. Conclusions and Future Research Needs	199
2.3.5 Road Salt	190		
2.4 Processes Affecting Road Dust Retention/Availability	190	References	200
2.5 Processes Affecting Road Dust Distribution	190		

1. INTRODUCTION

Road dust is a generic description for any form of solid particle distributed on the road surface that can be suspended in the air through traffic or windblown suspension. This generally limits the size of the relevant particles we are concerned with to <70 μm, or even <10 μm if we are only concerned with respirable particles. However, because larger particles can be crushed to generate smaller particles, it may be relevant to include even larger sizes in a general overview.

There are a large number of sources (Section 2) that can contribute to road dust. These include traffic-related wear production, friction sanding application, deposition of exhaust, migration of particles from nearby environments, migration from construction and other "dusty" sites, deposition from the atmosphere, and the application of road salt. The vehicle wear—related sources and their direct emissions are treated in Chapters 8—10.

The processes affecting road dust loading are many (Section 2). First and foremost is the generation of road dust from the various sources that determines the road dust loading. How these contribute will depend on the state of the road surface, wet or dry, and the processes governing dust removal will also be dependent on the state of the road. Wet removal processes, such as drainage and vehicle spray, likely play an important role in controlling the dust loading budget. However, dry removal through vehicle suspension and the migration of larger particles out of the wheel tracks will also strongly impact the dust loading.

Direct measurements of road dust can be made using either dry or wet sampling techniques (Section 3). There has never been any intercomparison of these methods so their relative efficiency is unknown till date.

Road dust emissions are strongly controlled by the state of the road surface, dry or wet/frozen, as well as other factors such as surface texture, vehicle speed, and driving patterns. Measurement of road dust emissions (Section 4) can be made using mobile techniques that sample behind individual vehicles tires or can be estimated using roadside measurements. Under normal, real-world conditions, it is not possible to differentiate between direct (Chapters 6—8) and suspended particles from road dust, unless the measurement methodology is specifically designed to do so.

In addition to measurements, various models have been designed to calculate road dust emissions and, in some cases, road dust loadings (Section 5). These vary from straightforward emission factors to more complex mass balance schemes for calculating road dust loading.

If road dust is a significant contributor to particle emissions, then there is also a desire to reduce these emissions in some way. Various abatement measures have been implemented to reduce road dust emissions (as well as direct emissions) ranging from changes in traffic patterns, such as speed reductions, to road cleaning and the use of dust-binding chemicals. The impacts of these measures vary and are typically dependent on the technology used, as well as the local situation. This aspect is not discussed in this chapter as it is addressed in Chapters 8 and 11 of this book.

2. OVERVIEW OF SOURCES AND PROCESSES AFFECTING ROAD DUST LOADING AND EMISSIONS

There are a large number of processes that can impact the road dust loading, its distribution on the road surface, and its eventual emission into the ambient air. Many of these processes are related to the physical impact of vehicles, but a number of these are also related to meteorological conditions. Some processes, such as removal by vehicle spray, are related to both. Road maintenance activities may also contribute to production, redistribution, and loss of road dust, including salt.

2.1 Sources

Sources of road dust have been listed in various reviews (Amato, Cassee et al., 2014; Boulter, 2005; Denier van de Gon, 2013; US EPA, 2006) and include vehicle wear contributions, fugitive contributions from construction and roadsides, road maintenance activities, atmospheric deposition, plant materials, etc. The dominant contributor to road dust varies significantly depending on the environment, so individual studies in one environment are not likely to be representative for another (Amato et al., 2011). In Nordic regions, road dust is dominated by road wear from studded tires (e.g., Norman & Johansson, 2006), in arid regions it may be dominated by roadside dust entrainment, and in congested urban environments it may be dominated by tire and brake wear (e.g., Amato, Karanasiou et al., 2012; Bukowiecki et al., 2010). Under many conditions, there may be no dominant source at all, but the sum of many smaller sources.

Road dust from vehicle wear is mainly accumulated on the road surface during wet or frozen periods, but an unknown fraction will also be deposited during dry conditions. Some of this may be immediately suspended and some may accumulate between wheel tracks, on curb sides, or within the pores of the road surface texture. It is difficult to measure the difference between "direct" wear emissions and "suspended" wear emissions and so some confusion may exist when describing "road dust" separately to "vehicle wear" sources. In addition, "crustal material,", identified in chemical speciation of road dust, may come from either road wear or deposited soil dust sources. These ambiguities must be considered when trying to specify emission factors for the different sources of road dust emissions.

2.2 Suspension

The process of traffic-induced suspension, leading to emissions of road dust, can be conceptually divided into the process of mechanical suspension from the passage of vehicle tires, induced by physical contact (or air pressure gradients) from the tire, and turbulent suspension caused by the vehicles. Tire induced suspension has been shown to be fairly efficient and the wheel tracks of roads are quickly cleaned of dust after just a few vehicle passages (Langston et al., 2008; Nicholson & Branson, 1990). This means that longer term suspension of road dust is controlled by turbulent suspension and the redistribution of dust into the tire tracks. Traffic meandering due to lane changing, parking, or variability in lateral driving position across the road will access dust outside the normal wheel tracks. Most experiments where dust is applied to the road surface to determine emissions and suspension rates (Langston et al., 2008; Patra et al., 2008) are addressing only the initial phase of suspension induced by the vehicle tires and not the "real-world" longer term suspension rates after the wheel tracks have been cleaned.

An additional ambiguity arises in regard to the production and loss of road dust. When the main production terms of road dust are related to "direct" non-exhaust emissions (road, tire, and brake wear), some of these will accumulate on the road surface. As has been discussed in Berger and Denby (2011), Denby et al. (2013b), and Amato, Schaap et al. (2012), under long-term dry conditions, the amount of dust on the road surface will be a balance between production and loss terms. Under such conditions, the direct wear accumulated on the road surface will be suspended at the same rate as it is produced and the total non-exhaust emissions will simply be the total direct wear contributions. Road dust contributions do not need to be included as an additional emission source under such conditions, but if the sources of road dust are deposition or migration, or equilibrium is not achieved through external dust inputs and/or variable road surface conditions, then road dust suspension needs to be included as an additional source. The time taken to reach this equilibrium is dependent mainly on the suspension rate and the traffic volume. Equilibrium is likely reached quickly in the wheel tracks, but it may take much longer, many weeks or months, for such an equilibrium to be reached for the rest of the road surface. This trend toward equilibrium is disrupted when the surface becomes wet, as then, a large portion of the wear emissions are accumulated on the road surface.

Although suspended emissions have been shown to vary with speed when sampled using mobile platforms (Langston et al., 2008; Pirjola et al., 2010), long-term suspension rates are more likely connected to driving patterns than speed, as the tire tracks are quickly emptied of the available dust loading.

In Table 9.1, processes governing road dust loading are listed, briefly described, and references to studies addressing these are given (when applicable).

2.3 Road Dust Sources Other Than Traffic-Related Wear Sources

2.3.1 Surface Abrasion (The Sandpaper Effect)

Early Finnish studies (Kupiainen, Tervahattu, & Räisänen, 2003; Kupiainen et al., 2005; Räisänen, Kupiainen, & Tervahattu, 2003, 2005) used a road simulator to study the contribution from pavement and traction sand to emitted PM_{10}. The analysis technique was based on choosing mineralogically different rock aggregates for pavement and traction sand to be able to separate the sources in PM_{10}. Tests were made using both studded and friction tires. Their results show that the contribution from the pavement not only increases with studded tire use, but is also enhanced when traction sand is used. This effect was called the "sandpaper effect" (Kupiainen et al., 2003). Even though most data are on how properties of traction sand influences PM_{10} emission, Räisänen et al. (2005) concluded that a pavement made with a granite with higher resistance to abrasive wear than a pavement made with a mafic volcanic rock also resulted in lower PM_{10} emissions when tested with the same traction sand.

TABLE 9.1 Summary of the Sources and Processes Related to Road Lust Loading and Suspension

Process	Description	Type of Process[a]	Relevant Dependencies
PRODUCTION PROCESSES (SOURCES)			
Road wear	Wear of roads through tire contact. Dependent on tire and road characteristics. Maximum retention during wet road conditions	P	Tire type (studded or not), road pavement properties, vehicle weight and speed, surface wetness
Tire wear	Wear of tires through road contact. Dependent on temperature and tire characteristics. Maximum retention during wet road conditions	P	Tire properties, temperature, vehicle speed, surface wetness
Brake wear	Wear of brake pads and discs during breaking. Dependent on driving characteristics. Maximum retention during very wet road conditions	P	Brake properties, driving patterns, surface wetness
Exhaust deposition	Exhaust emissions are released just above the surface for passenger vehicles. Under wet, and to a lesser extent dry, conditions a small fraction will deposit on the road surface	P	Exhaust emissions, surface wetness
Traction sanding	Addition of sand material during freezing periods	P	Size distribution of sand, sand properties
Deicing salt (NaCl)	Application of road salt during freezing conditions. Generally, only particulate matter (PM) emissions after the surface has dried	P	Mass applied, surface wetness, loss mechanisms
Atmospheric dry and wet deposition	Rainout or deposition of particles in the air, dust, or pollen episodes	P	Precipitation, column concentration of PM
Roadside and pavement entrainment	Dust sources on the roadside, road shoulder, foot paths, or bicycle tracks that migrate onto the road surface through cleaning, drainage, or other activities	P	Dust and salt availability on the roadside, meteorological conditions
Fugitive loading from construction	Migration of dust from construction sites and unpaved roads or loss from trucks and other vehicles	P	Local sources and traffic
Organic sources	Deposition of pollen, organic material such as leaves on the road surface	P	Local sources
Road abrasion ("sand paper effect")	Existing road dust or sand scraping the road surface to produce additional dust. Likely higher during wet periods	P	Sand loading, dust loading

(Continued)

TABLE 9.1　Summary of the Sources and Processes Related to Road Lust Loading and Suspension—cont'd

Process	Description	Type of Process[a]	Relevant Dependencies
Disintegration of the road surface	Road pavements may, particularly as a result of freezing and thawing or salt crystallization, disintegrate over time releasing large particles that can be crushed	P	Road pavement characteristics

DRY LOSS AND REDISTRIBUTION PROCESSES

Process	Description	Type of Process[a]	Relevant Dependencies
Vehicle-induced suspension	Suspension due to mechanical contact of tires and vehicle turbulence	E, DL	Vehicle speed, tire type, surface texture, road surface conditions
Windblown suspension	Wind-induced suspension of road dust	E, DL	Wind, surface texture, road surface conditions
Dry cleaning	Dry cleaning processes using vacuum cleaning systems and sweeping	DL, R	Cleaning method

WET LOSS AND REDISTRIBUTION PROCESSES

Process	Description	Type of Process[a]	Relevant Dependencies
Drainage	Significant amounts of dust may be lost during precipitation events	WL, R	Road slope, surface texture, precipitation
Vehicle splash and spray	Significant amounts of water, along with dust and salt, can be removed from the road surface as a result of vehicle-induced spray	WL, R	Driving speed, vehicle and tire type, surface texture
Snow removal	Under snowy conditions, snow removal will also remove road dust bound in the snow layer	WL, R	Frequency and efficiency of snow removal
Wet cleaning	Flushing or combined flushing, sweeping, and vacuum systems	WL, R	Cleaning method

PROCESSES AFFECTING SIZE DISTRIBUTION

Process	Description	Type of Process[a]	Relevant Dependencies
Crushing	Sand or larger particles are crushed into smaller particle sizes	SR	Available road dust loading and type, driving characteristics
Aggregation	Particle, such as tire and other wear particles, can aggregate either by physical contact or through freezing	SR	Little is known about such processes

REDISTRIBUTION PROCESSES

Process	Description	Type of Process[a]	Relevant Dependencies
Crossroad migration	Larger particles not suspended in the air are lifted and redistributed resulting in redistribution to the side of the road	R	Traffic volume, lateral variability, driving speed, road surface conditions
Along road migration	Movement of road dust along the road. Typically high migration from construction sites that effectively generate road dust on paved roads	P, R	Dependent on the source and on meteorological conditions

[a]*Type of process divided into production (P), redistribution (R), wet loss (WL), dry loss (DL), emission (E), and size redistribution (SD).*

2.3.2 Traction Sand

The addition of sand to the road surface, to increase surface friction, during icy conditions is common in many colder countries around the world. Sand may be applied to the road surface or to the curbside. The size distribution of the sand and its resistance to crushing varies significantly. Using roadside measurements and the mobile measurement laboratory Sniffer (Pirjola, Kupiainen, Perhoniemi, Tervahattu, & Vesala, 2009), Kupiainen et al. (2016) investigated the contribution to urban PM_{10} from the road pavement and winter traction sand. They used the same source apportionment method as suggested by Räisänen et al. (2003), choosing a sand material with different mineralogy than the pavement aggregates. The results confirmed the laboratory experiments and showed that, even though large amounts of traction sand were used, the main contributor (about 50%) to dust and PM_{10} was pavement wear, whereas traction sand contributed to about 25%. The results are not only due to the use of studded tires but also due to the previously mentioned "sandpaper effect."

2.3.3 Fugitive Sources

"Fugitive sources" covers a range of possible source contributions to road dust. These sources are important only if there is a significant and continual replenishment of the dust loading. The migration of dust on vehicle tires from construction sites or unpaved roads is one possible source, but these will be limited spatially. In countries where sanding and salting is used on pavements in winter, migration of this from pavements to the road surface may also contribute to the dust loading on the road surface. This migration can occur through drainage, pavement cleaning, or general use of the side pavements by pedestrians. In more arid countries where dust is available along the side of roads, meandering traffic or wind can migrate dust to the road surface. These dust sources may be coarse-grained but can be crushed and suspended by traffic to produce PM_{10} emissions.

2.3.4 Deposition

The atmospheric deposition of dust on the road surface is a possible source. Given a dry deposition rate of 2 cm/s and a "high urban" average PM_{10} concentration of 50 $\mu g/m^3$, the total dry deposition would be around 100 $mg/m^2/day$. This can be a significant contributor to the road dust loading, especially during wet and frozen winter periods where the deposited material can be accumulated on the surface. However, during dry periods when an equilibrium between production and resuspension may exist, the rate of deposition, and thus resuspension, will be directly dependent on the atmospheric concentration (Amato, Schaap et al., 2012). In this situation, the contribution of resuspended dust (the portion of road dust emission due only to atmospheric deposition) to the atmospheric concentration will be small, compared with the atmospheric concentration causing the deposition. Amato, Karanasiou et al. (2012) estimated the dry deposition of PM_{10} during a low-intensity Saharan dust event to be 1 mg/m^2, which resulted in a 30% increase in the road surface dust loading. However, the road dust loading quickly returned to its pre-Saharan event levels. Atmospheric deposition from other sources, such as pollen, or wet deposition may, at different periods, lead to more deposition, but because of the sporadic nature of these events, their impact is expected to be small.

2.3.5 Road Salt

Various measurements combined with receptor modeling (Ducret-Stich et al., 2013; Gianini et al., 2012; Hagen, Larssen, & Schaug, 2005; Richard et al., 2011; Wåhlin, Berkowicz, & Palmgren, 2006) have indicated that the contribution of road salt to PM_{10} concentrations can be significant if dry residual salt is left on the road after deicing is carried out. Contributions of salt vary, but during winter months this can range from 10% to 35% of the average PM_{10} concentration. In some cases (Hagen et al., 2005; Wåhlin et al., 2006), daily mean salt concentrations measured at roadside can be larger than $40\,\mu g/m^3$. The study carried out by Denby et al. (2016) looked at both measured and modeled road salt loading and emissions at a number of sites in Nordic countries. They concluded that the contribution of salt to curbside winter mean PM_{10} concentration was around $4.1 \pm 3.4\,\mu g/m^3$ for every kg/m^2 of salt applied on the road surface during the winter season. On roads where salt is extensively used during winter, this means that road salt emission contributions to PM_{10} are as large as exhaust emissions.

2.4 Processes Affecting Road Dust Retention/Availability

The retention of dust on the road is chiefly due to the moisture level of the road surface. Wet or frozen roads retain the road dust produced by wear or from other fugitive sources and prohibit its suspension. Norman and Johansson (2006) showed a strong correlation between PM_{10} concentrations and road surface wetness on streets in Stockholm. Amato, Schaap et al. (2012) showed that road dust mobility (ability of dry vacuum sampling to collect road dust particles) was dependent on time since the last rainfall. Denby et al. (2013a) showed that correlation between modeled and observed PM_{10} concentrations was significantly better when using observed road wetness levels.

Water-related processes, such as vehicle spray and drainage, that remove water from the surface will also remove dust from the surface. Measurements of water spray along roads have generally been limited to salt content, rather than dust content, but modeling studies, such as Denby et al. (2016), have shown that vehicle spray can be a significant removal mechanism for both salt and dust on high-speed roads.

Road surface texture plays an important role in the retention and availability of road dust. Blomqvist, Gustafsson, and Lundberg (2013) and China and James (2012) have shown, under controlled experiments, that coarse surface textures limit the availability of road dust for suspension. Measurements made on streets in Stockholm (Gustafsson, Blomqvist, Janhäll, Johansson, & Norman, 2014, 2015; Gustafsson, Blomqvist, Johansson, & Norman, 2013) have shown generally higher levels of retained dust loading on streets with coarser surface texture. Surface texture also impacts vehicle spray. Spray removal is less efficient with coarser surface texture.

2.5 Processes Affecting Road Dust Distribution

Patra et al. (2008) studied the rate of traffic-induced movement/dispersion of particulate matter in a busy public road with a unidirectional flow of vehicles from south to north at the center of London. They applied $20\,g/m^2$ of road surface gritting salt to a section of the road to act as a tracer to detect dust movement at the site. They measured particulate matter (PM) concentrations on both sides of the roads in two locations to estimate the fluxes. The material moved fast along the road right after the gritting. However, the movement rate slowed down with time as the material available in the road structure decreased. Part of the grit moved also across the street. Suspension accounted for 40% of the total material removed from a road segment and 70% of the material was removed either along or across the road. On average, a single vehicle pass removed 0.08% of material present on a road segment. Measurements on roads in Stockholm (Gustafsson et al., 2015, 2017) have shown a large variation in road dust loading laterally across the road, but have not quantified the mechanisms causing this.

3. SAMPLING OF ROAD DUST LOADING

Road dust samplings have been carried out in a variety of ways. The earliest road dust samplings were carried out using street cleaners, sweeping, and vacuum to collect samples (e.g., Hildemann, Markowski, & Cass, 1991; Rogge, Hildemann, Mazurek, & Cass, 1993). To determine "silt loading" (<75 μm) for the AP42 model (US EPA, 2006), sweeping and vacuum collection methods were also used. The combined use of sweeping and dry brushing was used by Vaze and Chiew (2002) to determine road dust levels before and after precipitation events. Amato et al. (2009) and Amato, Schaap et al. (2012) have used a dry vacuum method to sample road dust with sizes <10 μm (Fig. 9.1B and Fig. 9.2B). This method has been used in several European cities (Paris, Zurich, Madrid, Barcelona, Athens, Porto, among

(A)   **(B)**

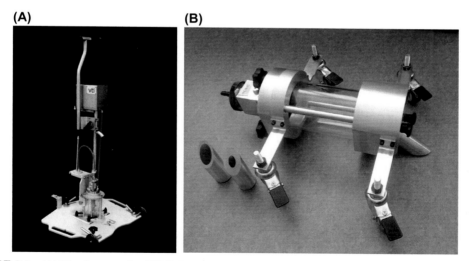

FIGURE 9.1   (A) Wet dust sampler (WDS) III used in Stockholm (Gustafsson et al., 2017). Photo: Mats Gustafsson, VTI (B) Dry road dust sampler described in Amato et al. (2009). *Photo: Fulvio Amato, IDAEA-CSIC*

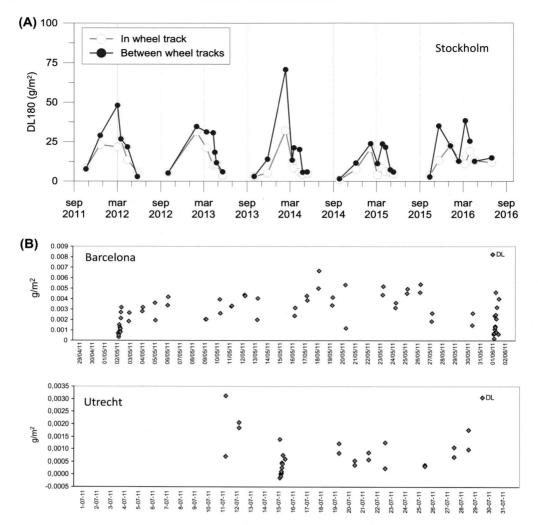

FIGURE 9.2 (A) Time series between 2011 and 2016 of sampled road dust load expressed as DL180 (road dust load smaller than 180 μm) sampled with a wet dust sampler on Sveavägen, a central city street in Stockholm (Gustafsson et al., 2017). (B) Time series over a 1-month period in 2011 of sampled road dust load DL10 (road dust less than 10 μm) using a dry dust sampler in Barcelona and Utrecht (Amato, Schaap et al., 2012).

others) and at industrial and airport areas. Road dust sampling during controlled experiments (Snilsberg, Myran, & Uthus, 2008) is also carried out using vacuum methods. Most road dust measurement campaigns are short-term studies, but long-term measurements of road dust sampling has been carried out in Stockholm (Gustafsson et al., 2014, 2015, 2013) using a wet sampler (Jonsson, Blomqvist, & Gustafsson, 2008), which uses a high-pressure water jet and sampling of the water in containers (Fig. 9.1A). Five years of data are available for a number of roads in Sweden using this method, Fig. 9.2A.

The levels of road dust reported worldwide vary significantly depending on the particle size sampled, traffic conditions, tire types, road types, road maintenance practices, and climate. In countries using studded tires and traction sanding, road dust levels can be as high as 200 g/m^2 (<200 μm) of which 5%–40% is in the size range <10 μm, when collected with a wet dust sampler (Gustafsson et al., 2014, 2015, 2013). Sampling using a dry vacuum technique in Barcelona, where no studded tires or traction sanding are used, indicates road dust levels two to three orders of magnitude less (Amato et al., 2009; Amato, Schaap et al., 2012) for particles <10 μm. The wet dust sampling technique is effective at removing a large amount of dust from the road surface under both wet and dry conditions, whereas dry vacuum methods may not completely sample the road dust, dependent on the moisture retention of the surface and on the strength of the vacuum being used. In this regard, a dry dust sampler may be more representative of the dust load that is available for suspension, whereas the wet dust sampler may also sample dust that is not currently available. To date, no intercomparison between the different methods has been undertaken, no assessment of their efficiency has been made, and no standards exist as to the size distribution that should be sampled.

A number of studies have been carried out concerning chemical analysis and source apportionment of ambient air samples to characterize traffic-related emissions. These studies are largely discussed in Chapters 2 and 1. There are also specific chemical speciation source and apportionment studies made directly on road dust samples. These include analyses carried out by Hildemann et al. (1991), Amato et al. (2009, 2011, 2016), Amato, Alastuey et al. (2014), and Gustafsson et al. (2013, 2014, 2015). The size distribution of road dust samples has also been addressed in a number of studies (Gustafsson et al. (2013, 2014, 2015), Janhäll, Gustafsson, Andersson, Järlskog, and Lindström (2016); Fig. 9.3). It should be noted that the sampled size distribution will be dependent on the sampling method.

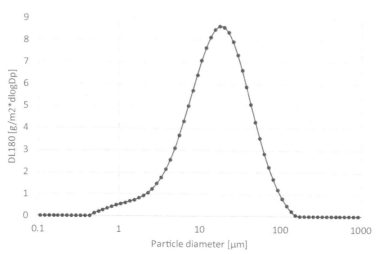

FIGURE 9.3   Particle mass size distribution of DL180 sampled on street in Trondheim, Norway. *Extracted from Janhäll et al. (2016).*

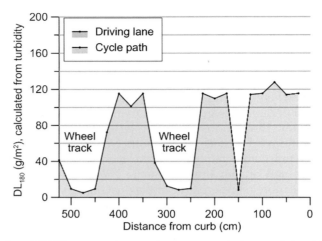

FIGURE 9.4   Road dust (DL180) transect from curb across a cycle lane and the right lane on Hornsgatan (Gustafsson et al., 2017).

The local spatial variability in road dust loading has also been investigated in Sweden. Gustafsson, Blomqvist, Janhäll, Johansson, and Norman (2014, 2015) have measured both along and across roads. Significant variability has been shown across the road with typically low levels in the wheel tracks, medium levels between tracks, and highest levels at the curbside (Fig. 9.4). Such local spatial variability makes quantifying road dust loadings uncertain without a large number of samples.

4. MEASUREMENT OF ROAD DUST EMISSIONS

4.1 Mobile Platform Measurements

Mobile measurement platforms have been developed to measure road dust emission levels in on-road conditions. Fitz, Bumiller, Etyemezian, Kuhns, and Nikolich (2005) and Mathissen, Scheer, Kirchner, Vogt, and Benter (2012) determined PM_{10} emission factors from roadways using a trailer with sensors mounted in front and behind the vehicle in the well-mixed wake (SCAMPER: System of Continuous Aerosol Monitoring of Particulate Emissions from Roadways). In the on-road measurement system TRAKER (Testing Reentrained Aerosol Kinetic Emissions from Roads), the sampling system is mounted into a vehicle and the road dust emissions are measured from behind the rotating tires (Etyemezian et al. 2003; Kuhns et al. 2001). The Swedish measurement vehicle "Emma" used a similar design as the TRAKER (Hussein, Johansson, Karlsson, & Hansson, 2008), whereas the Finnish "Sniffer" vehicle (Pirjola et al., 2009) uses a different inlet design behind the tire to collect road dust.

Mobile systems have been used for monitoring variation of road dust levels along the road and street networks, studying emissions of winter tires, as well as studying the influence of driving speed, traction sanding, and cleaning on road dust emissions (Hussein et al., 2008; Kupiainen & Pirjola, 2011; Pirjola et al., 2009, 2010). They are also able to detect key features of the emission dynamics, for example, emission seasonality in the Nordic countries, including the effect of winter maintenance measures and cleaning, hot spots such as construction works and quarries, as well as the effects of tire studs. The differences in the characteristics of the sampling and platform vehicle or trailer make direct comparisons of the results obtained with the different systems difficult (Pirjola et al., 2010).

4.2 Downwind Ambient Air Measurements

Upwind–downwind measurements of PM can also be used to determine emission factors, for both exhaust and non-exhaust particles, as described by Gertler et al. (2006) and applied in Pirjola et al. (2012). The idea is to establish a pair of measurement positions on different sides of the road or the vehicle path so that wind direction is approximately perpendicular to the road. In such situations, the upwind signal represents the background and the downwind signal represents the background plus the vehicle's plume of emissions. If the exhaust emissions are known and the background deducted, the rest of the plume is non-exhaust emissions. This method has been used for determining the relationship between mobile measurement systems and vehicle emission factors, but in principle it can also be used to determine the emission of individual vehicles in the traffic flow.

The use of NO_x emissions from vehicles as a tracer to determine PM emission factors is widespread (e.g., Bukowiecki et al. 2010; Denby et al. 2013a; 2013b; Ferm and Sjöberg, 2015; Gehrig et al., 2004; Ketzel et al., 2007). Other tracers, such as CO_2, have also been used (Patra et al., 2008). The method requires simultaneous measurements of NO_x and PM at a background and a traffic station. To differentiate between road dust, or other non-exhaust, emissions, the fine fraction ($PM_{2.5}$ or PM_1) is subtracted from PM_{10}. With knowledge of traffic, emission factors can be determined. The method is robust but is dependent on the uncertainty of the tracer emission factor.

4.3 Suspension Rates

A commonly used concept when dealing with road dust suspension is the suspension rate. This describes the fraction of surface road dust loading that is suspended with the passage of a vehicle. Experiments are undertaken by distributing dust on the road surface and then measuring its emission or its reduction in loading after a number of vehicle passages have been used to determine these rates. Experiments undertaken by Langston et al. (2008) indicate that applied deposited dust is quickly removed from the surface, with suspension rates of the order of $10^{-2}-10^{-3}$ veh^{-1}, i.e., an e-folding time of 100–1000 vehicles. Patra et al. (2008) estimated this rate to be 3×10^{-4} veh^{-1} based on distributions of rock salt on a road in London. Kupiainen and Pirjola (2011) found that traction sanding, added to the surface under dry conditions, increased the suspended emissions by a factor of 15 but that the PM_{10} emissions reduced quickly, over a matter of hours. Optimizing suspension rates using the NORTRIP model on roads in Stockholm (Denby et al., 2013b) gave suspension rates of $3-8 \times 10^{-6}$ veh^{-1}. This variation between suspension rates reflects the difference between the immediate tire-induced suspension, as seen in Langston et al. (2008), and the long-term "whole road" suspension seen in Denby et al. (2013b). The suspension rates determined by Patra et al. (2008) likely reflect a combination of both.

4.4 Emission Factors for Road Dust Suspension

Estimates of road dust suspension emission factors are difficult to make because most measurements will include both the direct and the suspended emissions, as well as the exhaust emissions. Mobile measurements behind tires will avoid exhaust emissions but may still include direct tire and road wear. Methods most often used to derive total non-exhaust emissions is to subtract literature-based exhaust emissions from the total PM_{10} or to only address PM_{10} concentrations larger than 1 or 2.5 μm. Some studies have used receptor modeling (e.g., Amato et al., 2016; Bukowiecki et al., 2010) to identify road wear specifically.

In Table 9.2 a number of results are listed, but the reader is also referred to Schaap et al. (2009) and Boulter (2005) for a further review. The study by Thorpe, Harrison, Boulter, and McCrae (2007), the mobile measurements from Gertler et al. (2006) and Kauhaniemi et al. (2014), and calculations using road dust emission models (Denby et al., 2013b; US EPA, 2006) actually try to estimate the suspended road dust emissions, rather than the total non-exhaust emissions. Whenever light-duty (LDV) and heavy-duty vehicles (HDV) are reported separately, the HDV contribution is estimated to be from 5 to 10 times larger than the LDV emission factors. A comparison of modeled and observed emission factors, based on mobile measurements, was made by Kauhaniemi et al. (2014), showing road dust emission factors ranging from 20 to 1800 mg/veh/km that were also reproduced by the models. The averaging times and methods also vary significantly. Some report average emission factors only during dry periods, either by choice or owing to the methodology used, and others also include wet periods in the average. During the year, emission factors may also vary significantly, especially in Nordic or Alpine regions, where dust loading increases during the winter.

In principle, the road dust suspension emission factor is dependent on the product of the road dust available for suspension and the suspension rate (Section 4.3). The suspension rate is also a function of vehicle speed, vehicle type, road surface roughness, and road surface conditions. These are the concepts used in modeling road dust suspension (e.g., Denby et al. 2013b; US EPA, 2006) and a significant range of emission factors can then be expected dependent on these factors. Road dust loading of PM_{10} may differ by three orders of magnitude (Section 3), and a similar range is found in the emission factors reported in Table 9.2. As such, it is unlikely that the use of a constant emission factor for road dust suspension is a viable methodology. More information is required, e.g., the use of emission models, to provide useful emission factors for road dust suspension.

5. PROCESS-BASED MODELING OF ROAD DUST LOADING AND EMISSIONS

Although emission factors for non-exhaust emissions are available in the literature (e.g., Boulter, 2005; EMEP/EEA, 2009), they relate, in principle, to the individual sources and often do not refer to road dust suspension as a separate source. Determining the road dust suspension emissions needs to take into account the more complex processes described in Section 2 and should not include emissions from sources that are already included. As a result, a number of models have been developed to better describe road dust emissions.

TABLE 9.2 Examples of Road Dust Emission Factors Provided in the Literature

EF PM$_{10}$ (mg/veh/km)	City, Country	Period	Source Identification	Method Used	References
5.4–17.0	Paris, France	1-year period excluding wet days on four roads	Road dust identified by receptor modeling	Based on deposition measurements and assumed equilibrium between deposited and suspended particles	Amato et al. (2016)
12.0–47.1	Barcelona, Spain	2-month period in summer on a freeway	Road dust identified by receptor modeling	Based on deposition measurements and assumed equilibrium between deposited and suspended particles	Amato, Karanasiou et al. (2012)
97	Barcelona, Spain	5-week period in winter. City center	Road dust identified as subtraction of exhaust	Ambient air measurements combined with NOx concentrations and emission factors as a tracer for dispersion	Amato et al. (2010)
LDV: 24–50 HDV: 498–288	Zurich, Switzerland	6-week winter period on two roads	Road dust identified by receptor modeling	Ambient air measurements combined with NOx concentrations and emission factors as a tracer for dispersion	Bukowiecki et al. (2010)
LDV: 0.6–2.8 HDV: 146–176	London, United Kingdom	4-year period	Suspended road dust identified as PM$_{10}$ > 2.5 μm with subtraction of direct road wear emission factors based on literature	Ambient air measurements combined with NOx concentrations and emission factors as a tracer for dispersion	Thorpe et al. (2007)[a]
53–74	Stockholm, Sweden	5-year average including wet periods	Based on model calculated road dust loading and suspension	Calculated suspension of wear particles (road, brake, tire) deposited on the road using the NORTRIP road dust emission model	Denby et al. (2013b)[a]
20–1800	Helsinki, Finland	Range over a 2-year period excluding wet days	PM$_{10}$ collected from behind back wheel	Mobile measurement using SNIFFER converted to emission factors	Kauhaniemi et al. (2014)[a]
LDV: 40–780 HDV: 230–7800	Reno and Durham, USA	Range over a 2-week period	Road dust identified by receptor modeling	Ambient measurements and use of a multilag regression approach	Abu-Allaban, Gillies, Gertler, Clayton, and Proffitt (2003)
76–660	Lake Tahoe, USA	Range over a 4-month period	PM$_{10}$ collected from behind vehicle	Mobile measurement using TRAKER converted to emission factors using vertical profile measurements	Gertler et al. (2006)[a]

When emission factors for light and heavy duty vehicles (LDV, HDV) are determined separately, then these are included.

[a]*Methods that differentiate between suspended road dust and direct wear emissions.*

The first of such models is the US EPA model AP42 (US EPA, 2006). This model is a semi-empirical model that relates "silt loading," vehicle weight, and time from the last rainfall to road dust emissions. It is based on a set of measurements carried out in different conditions. The model requires the user to specify the silt loading and as such only transfers the uncertainty in the emissions one step back, from emission factor to silt loading. Its dependence on rainfall is also highly parameterized, dependent only on hours since last rainfall.

The first real process-based model for road dust emissions was developed by Omstedt, Andersson, Gidhagen, and Robertson (2011) and Omstedt, Bringfelt, and Johansson (2005). This model was applied in Sweden where studded tires generate large road dust loadings and emissions during the winter and spring months. The model generates its own road dust loading, includes the application of traction sanding, and uses a simplified but physically based surface water mass balance model to account for road wetness and its impact on emissions. However, even in this model the road dust loading was not directly quantified and emission factors were related to a maximum observed emission factor. The model has been successfully applied in Sweden for many years, but is only applicable in regions where studded tires and traction sanding dominate the production of road dust.

Owing to the reliance of the Omstedt model on measurement data to quantify the emission factors, a more general approach to road dust emission modeling was undertaken by Berger and Denby (2011). In that model, road dust was generated based on road wear information and the road dust load was quantified, providing for the first time estimates of road dust emissions generated by road wear by studded tires without the use of empirical fitting. That model, however, also lacked a realistic description of the road surface conditions (dry/wet/frozen) and did not adequately describe its impact on the road dust emissions.

Currently the most advanced process-based model of road dust emissions is the NORTRIP model (Denby et al. 2013a, 2013b). This model bases road wear due to studded tires on an existing road wear model (Jacobson & Wågberg, 2007), follows the mass balance of road and other wear sources, including road salt, and uses an energy and water balance model to predict road surface wetness and road dust suspension/retention. The model has been successfully applied in a number of studies in Scandinavia (Denby et al. 2013a, 2013b; Kauhaniemi et al., 2014; Norman et al., 2016) and has also been used to quantify the emissions of salt (Denby et al., 2016). An example of such a model calculation is shown in Fig. 9.5.

Several emission models used for air quality modeling use alternative methods for road dust emission calculations. The method described by Denier van der Gon, Jozwicka, Hendriks, Gondwe, and Schaap (2010), based on soil moisture levels and an on/off precipitation method, has been implemented in both the EMEP (Simpson et al., 2012) and the LOTOS-EUROS (Schaap et al., 2009) regional scale models. Pay, Jimenez-Guerrero, and Baldasano (2011) applies the parameterization from the US EPA (2006) to control road dust emissions after precipitation events. The parameterization from Amato et al. (2009) has been implemented in local scale modeling by Amato et al. (2016). Both these parameterizations reduce the road dust suspension emission factors based on time since the last precipitation event. Generally, however, road dust emissions, if included at all in air quality models, are specified by a predefined emission factor.

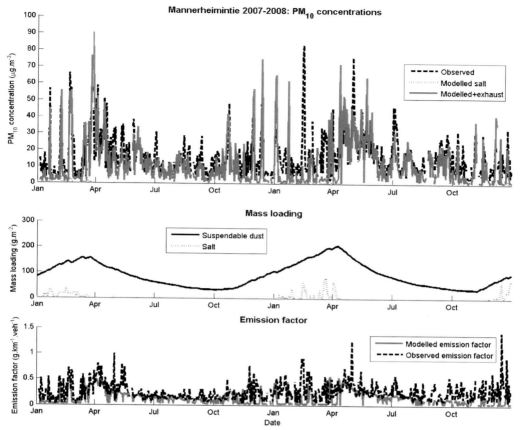

FIGURE 9.5 Modeled PM_{10} concentrations, road dust (<200 μm) and salt loading, and effective emissions factor for PM_{10} over a 2-year period on a road in Helsinki, Finland. Measured concentrations and emission factors are also shown (Denby et al., 2013b).

6. CONCLUSIONS AND FUTURE RESEARCH NEEDS

There are large differences between road dust loadings in different environments due to the different sources, meteorological conditions, and traffic patterns. It is therefore difficult to use a study in one area and apply it to another, unless these conditions are shown to be similar and relevant. There are a number of studies available, looking at source apportionment of road dust and ambient non-exhaust contributions. These indicate wide ranges in contributions, but rarely provide quantifiable links to the local traffic and environmental conditions. Such links are required if the processes leading to road dust generation and depletion are to be used in modeling studies.

To better quantify road dust and its contribution, it is necessary to relate the road dust loading to the ambient air or emission measurements. For this reason, future measurement campaigns should include road dust loading measurements, mobile vehicle measurements, and ambient air measurements. In this way, the three major measurement techniques can be linked and a clearer understanding of direct emission contribution, suspension rates, and road dust contributions can be achieved.

Measurement methods for road dust sampling have been developed on an ad hoc basis. To date, there has been no intercomparison of the different methods and no assessment of the residual left after sampling. The efficiency of the sampling methods, in regard to size distribution, has not yet been assessed. Wet dust sampling methods with high pressure water remove virtually all of the surface material. However, in doing so, they also remove road dust, deep in the road texture pores, that may otherwise not have been available for suspension (Gustafsson et al., 2015). It is recommended that an intercomparison of dry and wet dust sampling methods be undertaken.

Many of the processes described in Table 9.1 are poorly quantified. More information on the loss processes, particularly spray and drainage, is necessary to calculate the mass balance of the road dust loading system. It would also be beneficial for the practical implementation of measures to enhance knowledge in regard to the efficiency of cleaning methods and effective methods for dust binding.

References

Abu-Allaban, M., Gillies, J. A., Gertler, A. W., Clayton, R., & Proffitt, D. (2003). Tailpipe, resuspended road dust, and brake-wear emission factors from on-road vehicles. *Atmospheric Environment, 37*, 5283–5293.

Amato, F., Alastuey, A., de la Rosa, J., Gonzalez Castanedo, Y., Sánchez de la Campa, A. M., Pandolfi, M., et al. (2014). Trends of road dust emissions contributions on ambient air particulate levels at rural, urban and industrial sites in southern Spain. *Atmospheric Chemistry and Physics, 14*, 3533–3544.

Amato, F., Cassee, F. R., Denier van der Gon, H. A. C., Gehrig, R., Gustafsson, M., Hafner, W., et al. (2014). Urban air quality: The challenge of traffic non-exhaust emissions. *Journal of Hazardous Materials, 275*, 31–36.

Amato, F., Favez, O., Pandolfi, M., Alastuey, A., Querol, X., Moukhtar, S., et al. (2016). Traffic induced particle resuspension in Paris: Emission factors and source contributions. *Atmospheric Environment, 129*, 114–124.

Amato, F., Karanasiou, A., Moreno, T., Alastuey, A., Orza, J. A. G., Lumbreras, J., et al. (2012). Emission factors from road dust resuspension in a Mediterranean freeway. *Atmospheric Environment, 61*, 580–587. http://dx.doi.org/10.1016/j.atmosenv.2012.07.065. ISSN 1352-2310.

Amato, F., Nava, S., Lucarelli, F., Querol, X., Alastuey, A., Baldasano, J. M., et al. (2010). A comprehensive assessment of PM emissions from paved roads: Real-world emission factors and intense street cleaning trials. *Science of the Total Environment, 408*(20), 4309–4318.

Amato, F., Pandolfi, M., Moreno, T., Furger, M., Pey, J., Alastuey, A., et al. (2011). Sources and variability of inhalable road dust particles in three European cities. *Atmospheric Environment, 45*(37), 6777–6787.

Amato, F., Pandolfi, M., Viana, M., Querol, X., Alastuey, A., & Moreno, T. (2009). Spatial and chemical patterns of PM_{10} in road dust deposited in urban environment. *Atmospheric Environment, 43*(9), 1650–1659.

Amato, F., Schaap, M., Denier van der Gon, H. A. C., Pandolfi, M., Alastuey, A., Keuken, M., et al. (December 2012). Effect of rain events on the mobility of road dust load in two Dutch and Spanish roads. *Atmospheric Environment, 62*, 352–358. http://dx.doi.org/10.1016/j.atmosenv.2012.08.042. ISSN 1352-2310.

Berger, J., & Denby, B. (2011). A generalised model for traffic induced road dust emissions. Model description and evaluation. *Atmospheric Environment, 45*, 3692–3703.

Blomqvist, G., Gustafsson, M., & Lundberg, T. (2013). Road surface dust load is dependent on road surface macro texture. In *Extended abstract presented at the 2013 European Aerosol Conference, September 1–6, 2013 in Prague Czech Republic.*

Boulter, P. G. (2005). *A review of emission factors and models for road vehicle non-exhaust particulate matter*. Wokingham, UK: TRL Limited (TRL Report PPR065). http://uk-air.defra.gov.uk/reports/cat15/0706061624_Report1__Review_of_Emission_Factors.PDF.

Bukowiecki, N., Lienemann, P., Hill, M., Furger, M., Richard, A., Amato, F., et al. (2010). PM_{10} emission factors for non-exhaust particles generated by road traffic in an urban street canyon and along a freeway in Switzerland. *Atmospheric Environment, 44*, 2330−2340.

China, S., & James, D. E. (2012). Influence of pavement macrotexture on PM_{10} emissions from paved roads: A controlled study. *Atmospheric Environment, 2012*(63), 313−326.

Denby, B. R., Ketzel, M., Ellermann, T., Stojiljkovic, A., Kupiainen, K., Niemi, J. V., et al. (2016). Road salt emissions: A comparison of measurements and modelling using the NORTRIP road dust emission model. *Atmospheric Environment, 141*, 508−522.

Denby, B. R., Sundvor, I., Johansson, C., Pirjola, L., Ketzel, M., Norman, M., et al. (2013a). A coupled road dust and surface moisture model to predict non-exhaust road traffic induced particle emissions (NORTRIP). Part 2: Surface moisture and salt impact modelling. *Atmospheric Environment, 81*, 485−503.

Denby, B. R., Sundvor, I., Johansson, C., Pirjola, L., Ketzel, M., Norman, M., et al. (2013b). A coupled road dust and surface moisture model to predict non-exhaust road traffic induced particle emissions (NORTRIP). Part 1: Road dust loading and suspension modelling. *Atmospheric Environment, 77*, 283−300.

Denier van der Gon, H. A. C., Gerlofs-Nijland, M. E., Gehrig, R., Gustafsson, M., Janssen, N., Harrison, R. M., et al. (2013). The policy relevance of wear emissions from road transport, now and in the future-an international workshop report and consensus statement. *Journal of the Air & Waste Management Association, 63*, 136−149.

Denier van der Gon, H., Jozwicka, M., Hendriks, E., Gondwe, M., & Schaap, M. (2010). *Mineral dust as a component of particulate matter, Tno, bop - wp2-report, report 500099003*. The Netherlands: TNO Delft. ISSN:1875−2322 (print) ISSN: 1875-2314 (online) www.pbl.nl.

Ducret-Stich, R. E., Tsai, M. Y., Thimmaiah, D., Kunzli, N., Hopke, P. K., & Phuleria, H. C. (2013). PM_{10} source apportionment in a Swiss Alpine valley impacted by highway traffic. *Environmental Science & Pollution Research, 20*(9), 6496−6508.

EMEP/EEA. (2009). *EMEP/EEA air pollutant emission inventory guidebook*. EEA Technical report No 9/2009. http://dx.doi.org/10.2800/23924 http://www.eea.europa.eu/publications/emep-eea-emission-inventory-guidebook-2009.

Etyemezian, V., Kuhns, H., Gillies, J., Green, M., Pitchford, M., & Watson, J. (2003). Vehicle-based road dust emission measurement: I. Methods and calibration. *Atmospheric Environment, 37*, 4559−4571.

Ferm, M., & Sjöberg, K. (October 2015). Concentrations and emission factors for $PM_{2.5}$ and PM_{10} from road traffic in Sweden. *Atmospheric Environment, 119*, 211−219. http://dx.doi.org/10.1016/j.atmosenv.2015.08.037. ISSN 1352-2310.

Fitz, D. R., Bumiller, K., Etyemezian, V., Kuhns, H., & Nikolich, G. (2005). Measurement of PM_{10} emission rate from roadways in Las Vegas, Nevada using a SCAMPER mobile platform with real-time sensors. In *Presented at the 14th international emission inventory conference, April 12−14, 2005, Las Vegas, NV*.

Gehrig, R., Hill, M., Buchmann, B., Imhof, D., Weingartner, E., & Baltensperger, U. (2004). Separate determination of PM_{10} emission factors of road traffic for tailpipe emissions and emissions from abrasion and resuspension processes. *International Journal of Environment and Pollution, 22*, 312−325.

Gertler, A., Kuhns, H., Abu-Allaban, M., Damm, C., Gillies, J., Etyemezian, V., et al. (October 2006). A case study of the impact of winter road sand/salt and street sweeping on road dust re-entrainment. *Atmospheric Environment, 40*(31), 5976−5985. http://dx.doi.org/10.1016/j.atmosenv.2005.12.047. ISSN 1352-2310.

Gianini, M. F. D., Fischer, A., Gehrig, R., Ulrich, A., Wichser, A., Piot, C., et al. (2012). Comparative source apportionment of PM_{10} in Switzerland for 2008/2009 and 1998/1999 by positive matrix factorisation. *Atmospheric Environment, 54*, 149−158. http://dx.doi.org/10.1016/j.atmosenv.2012.02.036.

Gustafsson, M., Blomqvist, G., Janhäll, S., Johansson, C., & Norman, M. (2014). *Operational measures against PM_{10} pollution in Stockholm − evaluation of winter season 2012−2013*. VTI report 802 (In Swedish) http://www.vti.se/sv/publikationer/driftatgarder-mot-pm10-i-stockholm--utvardering-av-vintersasongen-20122013/.

Gustafsson, M., Blomqvist, G., Janhäll, S., Johansson, C., & Norman, M. (2015). *Operational measures against PM_{10} pollution in Stockholm − evaluation of winter season 2013−2014*. VTI report 847 (In Swedish) http://www.vti.se/en/publications/operational-measures-against-pm10-pollution-in-stockholm-evaluation-of-winter-season-20132014/.

Gustafsson, M., Blomqvist, G., Janhäll, S., Johansson, C., Norman, M., & Silvergren, S. (2017). *Driftåtgärder mot PM$_{10}$ i Stockholm, Utvärdering av vintersäsongen 2015−2016.* VTI report (In Swedish), (in press).

Gustafsson, M., Blomqvist, G., Johansson, C., & Norman, M. (2013). *Operational measures against PM$_{10}$ pollution in Stockholm − evaluation of winter season 2011−2012.* VTI report 767 (In Swedish) http://www.vti.se/sv/publikationer/driftatgarder-mot-pm10-pa-hornsgatan-och-sveavagen-i-stockholm–utvardering-av-vintersasongen-20112012/.

Hagen, L. O., Larssen, S., & Schaug, J. (2005). *Environmental speed limit in Oslo. Effects on air quality of reduced speed limit on RV4.* NILU, Kjeller (OR 41/2005) (In Norwegian).

Hildemann, L. M., Markowski, G. R., & Cass, G. R. (1991). Chemical composition of emissions from urban sources of fine organic aerosol. *Environmental Science & Technology, 25*(4), 744−759.

Hussein, T., Johansson, C., Karlsson, H., & Hansson, H.-C. (2008). Factors affecting non-tailpipe aerosol particle emissions from paved roads: On-road measurements in Stockholm, Sweden. *Atmospheric Environment, 42,* 688−702.

Jacobson, T., & Wågberg, L. G. (2007). *Developing and upgrading of a prediction model of wear caused by studded tyres and an overview of the knowledge of the factors influencing the wear − Version 3.2.03.* Linköping: The Swedish National Road and Transport Research Institute (VTI notat 7−2007) (in Swedish) http://www.vti.se/en/publications/developing-and-upgrading-of-a-prediction-model-of-wear-caused-by-studded-tyres-and-an-overview-of-the-knowledge-of-the-factors-influencing-the-wear–version-3203/.

Janhäll, S., Gustafsson, M., Andersson, K., Järlskog, I., & Lindström, T. (2016). *Utvärdering av städmaskiners förmåga att reducera vägdammsförrådet i gatu- och tunnelmiljöer i Trondheim.* VTI rapport 883.

Jonsson, P., Blomqvist, G., & Gustafsson, M. (2008). Wet dust sampler: Technological innovation for sampling particles and salt on road surface. In *Seventh international symposium on snow removal and ice control technology, transportation research circular E-C126: 102−111.*

Kauhaniemi, M., Stojiljkovic, A., Pirjola, L., Karppinen, A., Härkönen, J., Kupiainen, K., et al. (2014). Comparison of the predictions of two road dust emission models with the measurements of a mobile van. *Atmospheric Chemistry and Physics, 14,* 9155−9169. http://dx.doi.org/10.5194/acp-14-9155-2014.

Ketzel, M., Omstedt, G., Johansson, C., Düring, I., Pohjola, M., Oettl, D., et al. (2007). Estimation and validation of PM$_{2.5}$/PM$_{10}$ exhaust and non-exhaust emission factors for practical street pollution modelling. *Atmospheric Environment, 41,* 9370−9385.

Kuhns, H., Etyemezian, V., Landwehr, D., MacDougall, C., Pitchford, M., & Green, M. (2001). Testing re-entrained aerosol kinetic emissions from roads (TRAKER): A new approach to infer silt loading on roadways. *Atmospheric Environment, 35,* 2815−2825.

Kupiainen, K., & Pirjola, L. (2011). Vehicle non-exhaust emissions from the tyre-road interface - effect of stud properties, traction sanding and resuspension. *Atmospheric Environment, 45,* 4141−4146.

Kupiainen, K., Ritola, R., Stojiljkovic, A., Pirjola, L., Malinen, A., & Niemi, J. (2016). Contribution of mineral dust sources to street side ambient and suspension PM$_{10}$ samples. *Atmospheric Environment, 147,* 178−189.

Kupiainen, K., Tervahattu, H., & Räisänen, M. (2003). Experimental studies about the impact of traction sand on urban road dust composition. *Science of the Total Environment, 308,* 175−184.

Kupiainen, K. J., Tervahattu, H., Räisänen, M., Mäkelä, T., Aurela, M., & Hillamo, R. (2005). Size and composition of airborne particles from pavement wear, tires, and traction sanding. *Environmental Science & Technology, 39,* 699−706.

Langston, R., Merle, R. S., Jr., Etyemezian, V., Kuhns, H., Gillies, J., Zhu, D., et al. (2008). *Clark County (Nevada) paved road dust emission studies in support of mobile monitoring technologies.* Las Vegas, NV: Clark County Department of Air Quality and Environmental Management. http://www.epa.gov/ttnchie1/ap42/ch13/related/Final_Test_Report.pdf.

Mathissen, M., Scheer, V., Kirchner, U., Vogt, R., & Benter, T. (November 2012). Non-exhaust PM emission measurements of a light duty vehicle with a mobile trailer. *Atmospheric Environment, 59,* 232−242. http://dx.doi.org/10.1016/j.atmosenv.2012.05.020. ISSN 1352-2310.

Nicholson, K. W., & Branson, J. R. (1990). Factors affecting resuspension by road traffic. *The Science of the Total Environment, 93,* 349−358. Elsevier Science.

Norman, M., & Johansson, C. (2006). Studies of some measures to reduce road dust emissions from paved roads in Scandinavia. *Atmospheric Environment, 40,* 6154−6164.

Norman, M., Sundvor, I., Denby, B. R., Johansson, C., Gustafsson, M., Blomqvist, G., et al. (2016). Modelling road dust emission abatement measures using the NORTRIP model: Vehicle speed and studded tyre reduction. *Atmospheric Environment, 134,* 96–108.

Omstedt, G., Andersson, S., Gidhagen, L., & Robertson, L. (2011). Evaluation of new model tools for meeting the targets of the eu air quality directive: A case study on the studded tyre use in Sweden. *International Journal of Environment and Pollution, 47,* 79–96.

Omstedt, G., Bringfelt, B., & Johansson, C. (2005). A model for vehicle-induced non-tailpipe emissions of particles along Swedish roads. *Atmospheric Environment, 39,* 6088–6097.

Patra, A., Colvile, R., Arnold, S., Bowen, E., Shallcross, D., Martin, D., et al. (2008). On street observations of particulate matter movement and dispersion due to traffic on an urban road. *Atmospheric Environment, 42,* 3911–3926.

Pay, M. T., Jimenez-Guerrero, P., & Baldasano, J. M. (2011). Implementation of resuspension from paved roads for the improvement of CALIOPE air quality system in Spain. *Atmospheric Environment, 45,* 802–807.

Pirjola, L., Johansson, C., Kupiainen, K. J., Stojiljkovic, A., Karlsson, H., & Hussein, T. (2010). Road dust emissions from paved roads measured using different mobile systems. *Journal of the Air & Waste Management Association, 60,* 1422–1433.

Pirjola, L., Kupiainen, K. J., Perhoniemi, P., Tervahattu, H., & Vesala, H. (2009). Non-exhaust emission measurement system of the mobile laboratory SNIFFER. *Atmospheric Environment, 43,* 4703–4713.

Pirjola, L., Kupiainen, K. J., Ritola, R., Malinen, M., Niemi, J. V., Julkunen, A., et al. (2012). Non-exhaust PM_{10} emissions factors. In *8th international conference, air quality - science and application, Athens, 19–23 March 2012.*

Räisänen, M., Kupiainen, K., & Tervahattu, H. (2003). The effect of mineralogy, texture and mechanical properties of anti-skid and asphalt aggregates on urban dust. *Bulletin of Engineering Geology and the Environment, 62,* 359–368.

Räisänen, M., Kupiainen, K., & Tervahattu, H. (2005). The effect of mineralogy, texture and mechanical properties of anti-skid and asphalt aggregates on urban dust, stages II and III. *Bulletin of Engineering Geology and the Environment, 64,* 247–256.

Richard, A., Gianini, M. F. D., Mohr, C., Furger, M., Bukowiecki, N., Minguillón, M. C., et al. (2011). Source apportionment of size and time resolved trace elements and organic aerosols from an urban courtyard site in Switzerland. *Atmospheric Chemistry and Physics, 11,* 8945–8963. http://dx.doi.org/10.5194/acp-11-8945-2011.

Rogge, R. F., Hildemann, L. M., Mazurek, M. A., & Cass, G. R. (1993). Sources of fine organic aerosol. 3. Road dust, tire debris, and organometallic brake lining dust: Roads as sources and sinks. *Environmental Science & Technology, 27*(9), 1892–1904.

Schaap, M., Manders, A. M. M., Hendriks, E. C. J., Cnossen, J. M., Segers, A. J. S., Denier van der Gon, H. A. C., et al. (2009). *Regional modelling of particulate matter for The Netherlands.* www.rivm.nl/bibliotheek/rapporten/500099008.pdf.

Simpson, D., Benedictow, A., Berge, H., Bergström, R., Emberson, L. D., Fagerli, H., et al. (2012). The EMEP MSC-W chemical transport model — technical description. *Atmospheric Chemistry and Physics, 12,* 7825–7865. http://dx.doi.org/10.5194/acp-12-7825-2012.

Snilsberg, B., Myran, T., & Uthus, N. (2008). The influence of driving speed and tires on road dust properties. In B. Snilsberg (Ed.), *Pavement wear and airborne dust pollution in Norway. Characterization of the physical and chemical properties of dust particles. Doctoral Thesis.* Trondheim: Norwegian University of Science and Technology (Doctoral theses at NTNU, 2008:133).

Thorpe, A. J., Harrison, R. M., Boulter, P. G., & McCrae, I. S. (2007). Estimation of particle resuspension source strength on a major London road. *Atmospheric Environment, 41*(37), 8007e8020.

US EPA. (2006). AP 42. In *Chapter 13: Miscellaneous sources* (5th ed., Vol. I) http://www.epa.gov/ttn/chief/ap42/ch13/.

Vaze, J., & Chiew, F. H. S. (2002). Experimental study of pollutant accumulation on an urban road surface. *Urban Water, 4,* 379–389.

Wåhlin, P., Berkowicz, R., & Palmgren, F. (2006). Characterisation of traffic-generated particulate matter in Copenhagen. *Atmospheric Environment, 40,* 2151–2159.

CHAPTER

10

Technological Measures for Brake Wear Emission Reduction
Possible Improvement in Compositions and Technological Remediation: Cost Efficiency

Sebastian Gramstat

Development Foundation Brake, AUDI AG, Ingolstadt, Germany

OUTLINE

1. Introduction   205

2. Reduction Measures of Emitted Brake Particles   206
   2.1 Gray Cast Iron Brake Discs   206
   2.2 Aluminum Brake Discs   210
   2.3 Ceramic Brake Discs   212
   2.4 Titanium Brake Discs   213
   2.5 Brake Linings   214

3. Collection of Emitted Brake Particles   218

4. Prevention of Brake Particle Generation   220
   4.1 Role of Regenerative Braking   221
   4.2 Role of Driver Assistance Systems   223

5. Summary   225

Glossary   225

References   226

1. INTRODUCTION

Besides the very relevant question on how to characterize brake particle emissions (BPEs), another crucial point deals with the possibilities of influencing, reducing, or even avoiding the generation and emission of brake particles. It is believed that mechanical, chemical, and thermal wear should be reduced, which helps automatically to reduce BPEs. Therefore possible reduction measures are discussed, focusing on the friction partners brake rotor

Non-Exhaust Emissions
https://doi.org/10.1016/B978-0-12-811770-5.00010-8

205

and pad. Another issue deals with the collection of emitted brake particles. During the last years some approaches and concepts were invented, developed, and discussed; they are presented in this chapter. Finally, the possibility of prevention of brake particle generation is discussed in terms of the role of regenerative braking and driver assistance systems (DAS). Besides the technical description, an evaluation of how likely market introductions are, also with regard to the cost situation, is given as well.

2. REDUCTION MEASURES OF EMITTED BRAKE PARTICLES

The discussion of reduction measures includes the tribological point of view, in other words, how brake rotor and brake pad can be optimized in the future to reduce wear and BPEs.

2.1 Gray Cast Iron Brake Discs

Brake discs made of gray cast iron represent the majority of brake rotors produced today because their manufacturing process is very well understood, they are easy to handle, and, most of all, they are very cheap in terms of the raw material and their manufacturing.

However, cast iron rotors still allow some improvements to reduce wear and particle emissions. In the early 1990s AUDI tested some brake disc concepts with varying content of titanium to influence the frictional and wear behavior (AUDI AG, 1990). An addition of 0.05 wt% of titanium to the cast iron melt helps to bind nitrogen by the generation of titanium nitride, which is distributed homogeneously. Much bigger amounts lead to surface defects and cavities. The basic structure becomes ferritic then (previously, pearlitic), and titanium carbides (TiC) are generated that leads to an increased mechanical wear.

The investigations of AUDI AG (1990) revealed that the wear can be reduced down to 45% when 0.05 wt% of titanium was added. Higher additions result in an increasing wear, even if amounts up to 0.2 wt% still mean an improvement of 30% in terms of wear reduction. Interestingly, the inventors of Sichuan Province Vanadium and Titanium Brake Ltd. (2014) have discussed to consider additions of titanium in the range of 0.1%–0.25% for high-carbon cast iron brake discs to improve the hardness and reduce wear.

Another interesting idea deals with the addition of molybdenum to increase the wear resistance of cast iron rotors. According to Piwowarsky (1942), an addition of up to 1.5 wt% to the iron cast is feasible and results in wear improvement without affecting other characteristics. Also Patterson, Standke, and Kocheisen (1967) confirm the positive effect of additional molybdenum in terms of wear.

In Shangdong Hong MA Construction Machinery Co. (2014) a new disc material composition is presented using a high-carbon, low-silicon cast iron containing niobium. According to the inventors, a high temperature strength and high wear resistance can be expected by the addition of 0.1–0.3 wt% niobium.

However, all of these recommendations have not been introduced widely to the market because the addition of titanium or molybdenum results in increased costs, which are intended to be avoided in most of the brake system applications.

Besides influencing the formulation of the cast iron melt, in Mayer and Lembach (2012) a layer concept is described to improve the wear resistance of the disc. The protective layer is made of an iron-based matrix alloy, chromium or nickel; some hard materials (carbides or ceramic oxides) are added as well. Between the cast iron base body and the protection layer, a so-called sealing layer is established to support the bonding of the friction surface layer. The same authors mentioned in Mayer and Lembach (2013) an antiwear layer on the rubbing ring of a brake rotor. A thickness of 10—200 µm is explained, 10—50 µm is preferred, the layer shall consist of a metal alloy or a ceramic material. One claim deals with an integrated wear indicator, which would become visible when the protection layer is worn out. The indicator material should be made of a colored, black or white ceramic material, or a woven or braided fabric and could also consist of mineral pigments. A wear-indicating element in combination with a wear-reducing surface coating is also presented in Dupuis (2013). In this case a raised portion between the friction ring and the surface coating forms the wear indicator, running around the rubbing ring in the form of a wavy line. These indicator ideas are very relevant for wear-optimized brake rotors because their mass loss is very low due to hardened surface coatings, and it might be difficult to estimate the moment to replace the rotor.

In Qihong (2011) a cast iron disc is presented where a surface-hardened layer on the rubbing can be found. This layer is described as a composite layer of hard ceramic particles and a nodular cast iron. Its thickness is specified by 2—5 mm, which is expected to be very thick and could lead to crack or delamination processes over lifetime. However, an enhanced service life of the rotor by the factor 4.6 is stated in the patent.

Another patent that deals with coated brake discs is introduced by Keith (2009). The authors present a dual-coated concept to enhance the wear performance of the rubbing ring as well as the corrosion prevention of the hat part of the rotor. The rubbing ring is first prepared by roughening of the surface followed by the deposit of a bond coat layer of adhesion-promoting material. This can consist of a nickel—alloy material or pure nickel material. The expected thickness is 10—100 µm, while 20—30 µm are preferred. Subsequently, the wear-reducing friction surface coating is applied. The layer can be made of pure alumina material or alumina alloy material. The finished layer thickness should be in the range of 100—400 µm, preferably between 150 and 250 µm. To complete the manufacturing process the hat part is coated as well to establish a corrosion protection.

A detailed description of a coating material is proposed in Özer and Lampke (2012). It is a spraying material to produce a thermal spray coating for wear reduction purposes. A ceramic powder (30—35 wt%) and a metallic powder mixture (65—70 wt%) are considered. The latter consists of a first powder made from zinc alloy (25—30 wt%) and a second powder made from iron alloy (70—75 wt%). In terms of the ceramic powder, reinforced particles of aluminum oxide, zirconium oxide, and titanium oxide are considered. The proposed spraying material ensures an economical preparation process and an improved service life of the cast iron base body.

The design of a wear-reducing surface coating is described by Lembach and Mayer (2012). Interestingly the thickness of the coating is varying according to the cooling channel and fin design, as it can be seen in Fig. 10.1.

Because the base body material and the coating own different thermal expansion coefficients, it can be expected that mechanical strains occur while a layer of homogenous thickness is used. This can lead to cracks in the coating or even to delamination. Also hot spots, which cause crucial comfort complaints such as brake judder, can be caused. The authors invented therefore a varying layer thickness to compensate those strains. They propose to make the coating thicker in the area of the cooling channels where the thermal expansion is smaller.

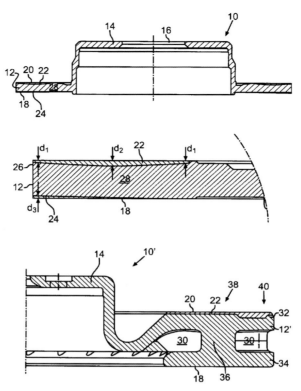

FIGURE 10.1 Illustration of the brake disc design according to Lembach and Mayer (2012).

Close to the cooling fins, where more cast iron material can be found that expands more, the antiwear coating should be thinner. This leads finally to varying thickness profiles for radial and circumferential orientation of coated surface layer.

For brake disc concepts for commercial vehicles, especially trucks, it is not unusual to reach a service life of 300,000 km and more, but an increase in wear resistance is also very beneficial for them. For this purpose it is strictly required to consider the cost situation and to follow approaches that ensure low costs. Khambekar, Baumgartner, and Pahle (2004) believe to fulfill this requirement with a metallic, nonceramic coating on a cast iron brake rotor. Additionally, they describe also the production method of such a coated disc and give some recommendations in terms of the cast iron composition of the base body as well. The metallic character of the layer means an optimized adherence to the cast iron base body compared with ceramic composites; a roughening of the friction ring surface is recommended only as a pretreatment. In the case of this invention, the coating is applied by a thermal spray coating method that leads to metallurgical reactions with the cast iron. As a result, the wear in general is not only affected positively, but also the very important high temperature wear is reduced. The application method is described in detail in terms of the coating time (3 min), the number of spray runs (5 times), the rotational workpiece speed (60 rpm), current (300 A), and voltage (30 V). A preferred composition of the coating consists of 16% of chrome and a steel alloy. In addition, the composition of the base body of the rotor should be modified in the way that a molybdenum content of 6%—8% and a chrome content of less than 5% are considered. The authors predict a doubled service life by using a coating thickness of 0.7 mm. In terms of material it means 250—450 g of coating composite per rubbing ring in the case of brake rotor with a diameter of 430 mm and a weight of 25 kg.

Summarized in terms of disc coatings it can be stated that they are very promising from a technical point of view, but the market introduction is mostly denied because of additional costs.

Besides the classical coating methods, the ferritic nitrocarburizing (FNC) method is also aimed to protect the cast iron rotor from corrosion and also to increase hardness that leads to improvement in terms of wear. The application of the FNC method is claimed, among others, in Hanna (2009); the positive effect in terms of wear reduction was confirmed by Riefe, Holly, and Learman (2016). Actually, FNC-brake rotors show a very high market infiltration for passenger cars in North America, while other markets are still dominated by conventional gray cast iron rotors.

The FNC-technology is a thermochemical treatment method, no coating in a traditional meaning. It considers enrichment with nitrogen and carbon that leads to a characteristic "metallic" diffusion layer and "ceramic" compound layer. Although the diffusion layer is dominated by iron-nitrides and shows a four to five times higher hardness than the gray cast iron basic material, the compound layer is characterized by the generation of nitrides and carbides and an increased hardness by the factor 5 (compared with cast iron). A sectional view of the different layers can be seen in Fig. 10.2, where the white layer represents the hard ceramic compound.

The brake lining concept is, of course, very relevant; FNC is working with adhesive friction mechanisms very well, which means that NAO concepts are the best friction partners and was also presented in Riefe et al. (2016). However, it must be mentioned that coefficients of friction lowered by 10%—15% are obtained Riefe et al. (2016) and AUDI AG (2016b). The use of low steel or ECE brake lining concepts would mean higher friction levels, but it removes, on the other hand, almost 80% of the compound layer after only 2000 km, which is demonstrated in Fig. 10.3.

In summary, it can be stated that cast iron rotors are still and will be, probably also in the future, the most important concepts for realizing brake discs. They are not only much cheaper

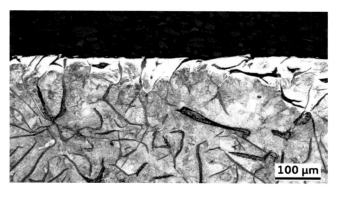

FIGURE 10.2 Sectional view of a ferritic nitrocarburizing—treated cast iron brake rotor, new condition (AUDI AG, 2016b).

FIGURE 10.3 Sectional view of ferritic nitrocarburizing—treated rotor in combination with an ECE brake pad after 2000 km (AUDI AG, 2016b).

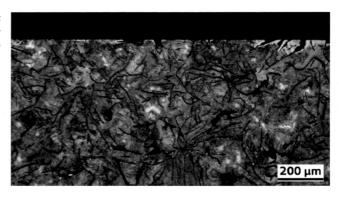

than other concepts but have also a high potential in optimization as shown previously. Coatings and treatments such as FNC seem to be the most promising approaches to improve wear and BPEs; costs on the other side is the most critical criterion and can prevent a strong market penetration.

2.2 Aluminum Brake Discs

Even if gray cast iron brake rotors have many advantages, other material concepts for the base body have become interesting for automotive applications during the last 20 years. Brake discs made of aluminum might represent one possibility for future brake systems not only to reduce weight but also to minimize the mechanical wear.

In Özer (2009) an aluminum metal matrix compound (Al-MMC) brake rotor is described for automotive applications. The presented concept considers a metal matrix made of $AlSi_2O$ with a fraction of silicon carbide of 15—25 wt%. Using a spray-forming process allows to apply additional hard material particles, which can be made of Al_2O_3, SiO_2, TiO_2, SiC, tungsten carbide, boron carbide, titanium nitride, or/and titanium diboride and showing diameters between 3 and 50 μm. In total the percentage of hard material particles is stated with 45 wt%. The same author explains the detailed production process of such a brake rotor in Lampke and Özer (2013).

The inventors Hino and Miyashita (1984) introduced a concept of an aluminum brake rotor, which is especially dedicated to an excellent abrasion resistance and to be less aggressive to the brake pads as well. This aim is realized by adding a ceramic antiwear material and/or a solid lubricant to an aluminum hypereutectic silicon alloy. This ceramic material shall consist of Al_2O_3, SiC, or SiN_4 and is added in an amount of 3—20 wt%. Additionally, a solid lubricant (such as graphite) might be added as well; not more than 5 wt% is desired. A reduced roughness of the material surface is also stated to be helpful with regard to reducing the aggressive wear behavior of the brake pad.

Another idea is introduced in Lampke and Özer (2012), where a base body made of an aluminum material in combination with a wear protection layer is mentioned. This layer is made of steel, prefabricated with the dimensions of the friction ring and tumbled to this, which allows improved wear resistance, corrosion resistance, and thermal load carrying ability.

In addition, Smolen (1991) introduced an aluminum rotor with a protection layer. The author presented a ceramic coating adhered to the braking surface with a thickness of 0.015–0.020 inch. The coating consists of aluminum oxide, aluminum titanium, and magnesium zirconate, which means that it is has a ceramic character. In terms of the manufacturing method, plasma spraying is preferred. Because the coating acts as a heat reflector, it prevents the rotor of excessive temperatures that could cause cracks or disintegrate the disc. Besides the thermal improvements, the coating is also expected to decrease wear. However, the brake pad as the other friction partner has to be adapted to this new brake disc surface characteristics..

The importance of the influence of the brake lining is investigated in Nakanishia, Kakiharaa, Nakayamaa, and Murayamab (2002). The authors used a brake dynamometer and a test cycle in conformity with JASO C406-82 to carry out a comparison of a cast iron rotor and an aluminum composite material with embedded ceramic particles (Al-MMC rotor) for improved wear resistance. In addition, the lining material, a cast iron rotor–based lining composition, was adapted by adding some hard particles. The first tests revealed the generation of a third body layer, which helps to improve rotor wear in such a way that the rotor thickness was even increasing.

Besides the material transfer from the pad to the disc, the hard particles in the disc increase the pad wear significantly; they act like a whetstone and cause abrasive wear. This experience leads to the development of an improved pad material. Because the pad also consists of hard particles, the ratio of particle diameters of rotor and pad was investigated in a deeper way. It was varied and several test samples were tested. It can be summarized that an optimal ratio of the particle diameters exist in the range of 2–20 for an optimized wear behavior of both, brake rotor and brake pad. As a consequence, brake pads with different hard particles were prepared to realize the optimal particle diameter ratio and tested subsequently on the brake dynamometer. The results reveal a wear behavior of the pads that is comparable to that of experiments with cast iron discs, valid for temperatures between 50 and 300°C.

Additionally, it must be stated that an analysis of the friction zone was also carried out and the hard particles were worn down to a round shape, which means that the tribological film could be established much better and abrasive wear is reduced significantly. In conclusion, it can be stated that the overall wear of this tribological system (Al-MMC rotor with adapted brake pad) is lower than for traditional cast iron rotors when the right hard particles are selected, especially in terms of the ratio of particle diameters of rotor and pad.

Another study revealed the advantages of Al-MMC brake rotors for a different use-case; brake discs of railway vehicles are discussed in Zeuner, Ruppert, and Engels (1998). A comparison between cast iron brake rotors (nodular graphite, GGG) and different aluminum rotors is presented. In terms of the aluminum discs, the first investigated rotor (called "Al 1") represents an SiC-reinforced Al-MMC while the second disc (called "Al 2") is made of LM 30 alloy. Standard railway brake lining materials were used for the full-scale dynamometer tests; the brake system design and the used test procedure are derived from a high speed train application. The analysis of the disc wear was carried out by the determination of the thinning of the friction ring for several measuring points in radial direction, starting at the outer diameter. It reveals that the Al-MMC shows the lowest material loss and best wear performance, respectively. In addition, the LM30 alloy performed better than the standard cast iron rotor, even if its wear is by the factor 2–3 higher than that for the Al-MMC. Additionally, the brake pad wear was also investigated in terms of the thinning of the brake pads for an increasing mileage.

The result is similar to that of the rotor comparison. In combination with the SiC-reinforced Al-MMC brake rotor the wear is the lowest, while the LM 30 alloy shows a higher material loss. The brake linings in combination with the cast iron rotor show the highest wear. It can be stated that using SiC-reinforced Al-MMC brake rotors leads to a reduction in the pad wear by 50%−65%. As a conclusion, the authors state that the additional costs of the Al-MMC brake rotors are more than compensated by the improved service life.

Finally, it can be stated that rotors made of aluminum, especially reinforced Al-MMC concepts, represent a very promising approach to reduce brake wear and particle emissions. Even if many questions, such as the critical temperature limit of around 400°C and the material and manufacturing costs, are unanswered so far, Al-MMC rotor concepts are expected to become more relevant in the future. The introduction of hybrid and full-electric vehicles to the markets might help at least to overcome the drawback of low temperature limits because regenerative braking will lead automatically to much lower operating temperatures of the friction couples. Up to now only a few examples are known, such as the Lotus Elise (in the mid-1990s) or some Volvo models 10 years later, that have been commercialized. However, the introduced Al-MMC disc concepts were all replaced by conventional cast iron discs after a short period.

2.3 Ceramic Brake Discs

Ceramic composite brake discs, mostly just called ceramic brake discs, represent another possibility for reducing wear and BPEs. Originally designed for high-performance applications, this type of brake rotors has also shown a significantly better wear performance because of its high hardness. In terms of their design, they own mostly a multilayer structure consisting of carrier bodies and friction layers, both reinforced with carbon fibers. Even if many efforts were carried out to improve the manufacturing process of ceramic brake discs, their production is still very time-consuming and energy-consuming, and the level of automation is lower compared with that of cast iron discs, which leads finally to significantly higher costs.

The inventors Kienzle and Kratschmer (2007) also state a lower wear. However, they also discuss the difficulties of crack formation due to different fiber lengths in the carrier and friction layers. As a new approach, they suggest to use carbon in a spherical form to reduce the content of fibers or even to replace them completely. Therefore the inventors introduce a ceramic matrix material that consists of at least one carbide, one carbide-forming element, and carbon, which furthermore contains a dispersed phase of carbon particles with a spherical shape. These spherical carbon particles should preferably have an average diameter range of 0.8−400 µm. It can be also expected that the spherical shape helps to reduce the brake lining wear and supports the generation of a friction film.

In Johnson (2009) it is believed that machining the friction surface influences the wear behavior. It is stated that carbon fiber reinforced brake discs can be machined in a way that the friction surface has an average roughness R_a lower than 2.5 µm. The authors propose grinding as a suitable machining method, especially when a diamond-embedded grinding disc is used. They mentioned that it might be helpful to employ a wet grinding process by using a coolant solution. As an option it is also stated that the grinding process could be realized in several stages using successively finer grinding; honing could be also be carried out as

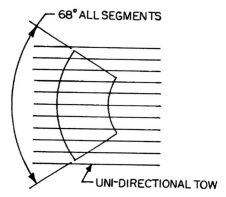

FIGURE 10.4 Proposed fiber layout according to James, Murdie, and Walker (2002).

an additional treatment. As a result of the surface smoothing, the authors obtained not only higher friction levels but also a significantly reduced level of surface wear.

In James, Murdie, and Walker (2002), an idea that is dedicated to the orientation of the used carbon fibers is introduced. The inventors present a layout where most of the continuous fibers in the fabric segments are arranged within an angle of 60 degrees from the inner to the outer diameter of the disc (radial direction), as it is shown in Fig. 10.4.

As a result of this method, the fiber pull-out is expected be minimized, which leads to reduced mechanical wear.

A carbon/carbon composite brake disc material is discussed by Matsumoto and Forsythe (2002). The composite should comprise a crystalline silicon carbide, which is uniformly distributed on the surfaces of the brake rotor in only low concentrations. In the patent, the concentration range of the silicon carbide is 0.001−1.00 wt% based on the total weight of the composite. Preparing such a brake rotor leads according to the inventors to a reduced wear with no change to the coefficient of friction of even a slight increase. They expect a wear rate of 0.101−0.13 inch/s/s $\times$ 10^{-4} and a friction coefficient in the range of 0.27−0.4.

As a conclusion it can be stated that ceramic brake rotors represent an interesting alternative to reduce BPEs and wear from the technical point of view, but their costs are still much higher and it is expected that ceramic brake discs will be found in the future only in niche applications such as super sport cars.

2.4 Titanium Brake Discs

An even more exotic material for brake rotors is titanium. However, some fundamental research activities have revealed interesting and promising results.

Muthukannan, Valarmathi, Shariff, and Padmanabham (2014) discuss the use of titanium-made brake rotors (laser nitrided Ti-6Al-4V, Ti-LSN) and carried out a comparison with commercial gray cast iron (GCI) and untreated titanium alloy (Ti). They used rotors with diameters of 80 mm and a thickness of 10 mm; Ti-LSN rotors were polished before the tribological investigations. As a frictional counterpart, a low steel NAO lining material was prepared. Because pin-on-disc studies were realized, a square pin with the dimensions 8 mm $\times$ 8 mm was used.

The authors carried out weight loss measurements to compute the wear volume loss and wear rates; the used test procedure included measurements at room temperature and at 200°C.

In Table 10.1 the results of the wear investigations are shown. The authors presented only the wear behavior of the rotors while the brake linings were not observed. The tests reveal that the untreated titanium alloy is not suitable as a brake rotor material in terms of wear; significantly higher wear rates and volume losses are stated. The authors explain them by a very poor resistance to abrasion processes and severe plastic deformations. The laser-treated material concept, on the other hand, shows a much better wear performance during all operating conditions while the frictional behavior is comparable to that of cast iron. As a remark it should be stated that no adapted lining material was used, which means that the wear-reducing potential seems to be even higher for the future.

In summary, it can be stated that titanium concepts are a promising technical approach to reduce wear. However, similar to ceramic brake discs the use of titanium material concepts seems to be very limited and is only appropriate for niche applications because of the high material and manufacturing costs.

2.5 Brake Linings

It is obvious that the second friction partner, the brake lining or pad, also represents a source for remediation measures to reduce wear and BPEs. Because the pad is always considered as the weak partner of the friction couple, most of the approaches are focused on improvements of the hardness and mechanical strength of the lining material.

Promising results are discussed in (Santamaria Ranzo & Persoon, 2016), where a new generation of reinforced mineral fibers with varying lengths and aspect ratios (L/D) is introduced. In total three different samples (A, B, and C) are investigated. Although the average diameter of the fibers is kept constant with an average value of 4.0 μm, the samples show varying lengths, which lead to different aspect ratios (A = 31, B = 76, C = 125).

The authors developed these fiber samples with the goal to reduce disc and pad wear and to improve brake dust emissions. For the investigations, an NAO lining composition was used and pad samples were prepared to carry out AK Master tests to determine friction and wear behavior.

TABLE 10.1 Results of Wear Investigations (Muthukannan et al., 2014).

Temperature (°C)	Velocity (m/s)	Volume Loss (mm^3)		
		GCI	Ti	Ti-LSN
RT	2	0.29	163	0.22
RT	3	0.50	172	0.44
200	2	0.46	190	0.40
200	3	0.55	232	0.51
Average volume loss (mm^3)		0.45	189	0.39
Wear rate (mm^3/N−m × 10^6)		3.46	1453	3.00

Even if the authors state that no general compromise in terms of the coefficient of friction is obtained (average values of approximately $\mu = 0.45$ are observed), further important results are presented. An increase in the fiber length has shown that the friction level decreases, but the friction level stability is increasing. This fact leads to a decreased disc wear and it is also shown that the decreased disc wear helps to improve the comfort behavior in terms of less vibration generation.

To complete the picture with regard to the wear behavior, the pad and disc wear are analyzed as well as the rotor roughness in terms of three different operating temperatures. It can be concluded that the longer the fibers are, the better the wear performance is. In addition, the disc roughness is decreasing with an increase in the fiber length. It is also stated that both brake disc and brake pad show a lower mass loss independently from the brake temperature, which varied from 150–300°C and finally to 500°C. This is an important fact as well because the brake system is faced with a wide range of operating conditions and requirements. The presented approach of this new fiber generation shows that it works under almost all temperature conditions properly and means an improved wear.

The nanomodification of phenolic resin is claimed in (Ning Guofei Eagle Auto Parts Co. Ltd., 2014). This invention considers a composition of steel fibers (15–20 wt%), glass fibers (5–6 wt%), barite (20–30 wt%), granular graphite (8–12 wt%), calcium carbonate (15–25 wt%), and cashew nut shell oil frictional powder (7–8 wt%). The phenolic resin is modified by nano vermiculite (10–15 wt%) to enhance the thermal stability and wear resistance. Another advantage of vermiculite is that it is a noncarcinogenic material and does not affect human health.

In (Zhao, 2013) a brake pad concept with clean composite ceramic fibers is claimed. These fibers are designed as soluble and decomposable as well as being adopted as reinforcing fibers. The inventors emphasize that the fibers do not affect human health and are environmental friendly. Moreover, it is claimed that the brake disc is protected from excessive wear because a soft ceramic fiber is used. In terms of the brake pad design, it consists of a lining matrix and a backing plate made of steel, similar to common brake pad concepts.

Another idea of a brake pad design is claimed by (Greil, Hammer, Rosenlöcher, Steinau, Travitzky, & Zipperle, 2011); the correlating drawing is shown in Fig. 10.5. This concept is dedicated only to ceramic brake discs (C/SiC). It is known that brake systems with ceramic discs normally reveal problems because of the higher operating temperatures (caused by higher coefficients of friction, lower thermal conductivity, and lower heat capacity of the ceramic disc). These high temperatures (+200 K compared with gray cast iron discs) lead to a critical wear behavior of phenolic resin–based linings; concepts that resist the heat better are needed.

Therefore the inventors introduce a brake pad (1) consisting of a functional part (2), made of a pyrolyzed preceramic polymer and comprising a friction layer (4) that is designed for a frictional interaction with a ceramic brake disc. Additionally, a damping layer (6) and an isolating layer (7) are considered. Interestingly, the backing plate is also made of a ceramic matrix consisting of a pyrolyzed preceramic polymer and can vary in composition and structure along the length and/or width to reach local varying mechanical, physical and/or chemical characteristics. It is expected to realize not only an improved mechanical strength and less mechanical wear, but also a better heat dissipation that leads to less thermos-mechanical wear.

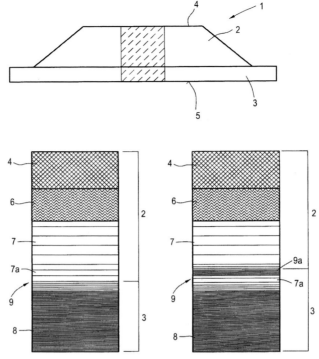

FIGURE 10.5 Schematic drawing of a brake pad concept according to (Greil, Hammer, Rosenlöcher, Steinau, Travitzky, & Zipperle, 2011).

The claim in Turani (2011) is dedicated to a brake pad that consists only of a friction portion and a mechanical support portion, which seems to be less complex than the concept that was discussed in the previous claim. The mixture of the first portion shall comprise at least one siliconic type of ceramic precursor, hard particles (abrasives), lubricants, and metal materials. Based on these materials, a preform body is obtained from a hot-pressing process carried out between 120 and 150°C. Subsequently, a pyrolysis process at 400−600°C is applied to the preformed body to realize a ceramization of the preceramic binder. As a result, a homogenous friction portion with an excellent wear resistance is obtained. Finally, the inventors state that the brake pad concept can be adapted to be suitable for combinations with composite ceramic materials as well as with gray cast iron brake discs.

In Bowei (2009), a nonmetallic ceramic brake pad for automotive applications is discussed. The inventors recommend to use a ceramic binder (10−20 wt%) with a modified aluminum sodium silicate to improve the heat resistance as well as the wear at high temperature operating conditions. However, also the use of resin and rubber (5−10 wt%) is still suggested. The idea behind is to continue the production with the already existing and used resin-based molding technologies and manufacturing equipment to reduce production costs. Additionally, ceramic and mineral fibers (15−25 wt%) as a short-fiber composite are proposed to improve high temperature stability and wear resistance. Finally, graphite (6−16 wt%) as a low-temperature lubricant and antimony sulfide (3—8 wt%), also to improve the wear behavior, are mentioned. In terms of the last mentioned ingredient, antimony sulfide, it must be stated that, at least in Europe, it is intended to avoid the use of antimony in brake linings, which means that the proposed claim should be revised and adapted in the future.

Another, natural material—based brake lining concept is claimed in Yuan (2009). Therein a reinforced material composite is introduced that is obtained by the combination of a natural fiber and crystal whisker. The natural fibers consist of bamboo and represent 4—20 wt%, while magnesium sulfate crystal whisker (15—30 wt%) is used to realize low wear rates, also under high temperature conditions. According to the inventors the bamboo fibers are intended to reinforce the phenolic resin (content of 20—30 wt%) and to lower the production costs. Additionally, steel fibers (5—15 wt%) and copper (5—12 wt%) contents are mentioned. Because copper is planned to be banned from automotive brake linings in the United States and in the European Union, it is recommended to revise the claim and exclude copper from the composition.

One possible concept of a copper-free composition is claimed by Automotive Parts Co. Ltd. Dongying Baofeng (2014). The discussed composition consists, among others, of phenolic resin (6—13 wt%), aramid fibers (1—5 wt%), iron oxide (10—22 wt%), and mineral fibers (10—22 wt%). Especially, thanks to the use of mineral fibers, a high temperature wear resistance can be obtained and environmental pollution is expected to be reduced because of the lower wear and the abstinence of copper.

The inventor Rasim (2011) is rather focused on the generation of a friction film or so-called third body layer between brake rotor and pad. A brake system with a ceramic composite brake disc and a brake lining is considered; the latter should consist at least of one metal sulfate, preferred is $BaSO_4$ with a content of 10—50 vol%. Additionally, iron can be found with 10—40 vol%, zirconia with 5—20 vol%, and at least one titanate with 5—30 vol%. With regard to a matrix-forming material, it is preferred to consider a phenolic resin (7.5—35 vol%).

Before the first brake application (I), the brake pads (3) and the disc (2) do not exhibit a friction layer, as it can be seen in Fig. 10.6. When the pads touch the disc (II), such a third body is established because of the brake pressure, relative velocity, and temperature in the friction zone and remains also when the brake is opened again (III). While the friction layer

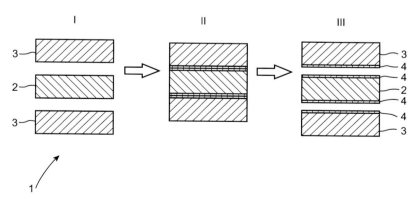

FIGURE 10.6   Development of a friction layer during a brake application according to Rasim (2011).

is very smooth and viscous during the friction process, it hardens after the brake application and remains adhered on the surfaces of the disc and pads. However, the third body layer will be viscous and softened again during the following brake application. According to the inventor, the layer is very thin, between a few nanometer and some microns, and consists of the friction products of brake disc and rotor. The inventor states that this friction layer is not only responsible for a stable friction behavior but also for an improved wear resistance.

In summary, it can be stated that brake linings reveal a huge potential of brake wear and particle emission reduction. However, during the foundation brake development, a wide range of characteristics have to be considered (friction performance, noise and vibration behavior, etc.) and the presented approaches might have a negative influence on other characteristics than wear and particle emissions. Therefore, each single discussed idea has to be considered carefully for each vehicle project that is developed—hence a general and wide market penetration is unlikely, because every vehicle project has different characteristic requirements.

3. COLLECTION OF EMITTED BRAKE PARTICLES

During the last years, some ideas occurred to collect the generated wear products by designing and installing collecting devices. However, up to now brake particle collection systems have not been introduced to the market, neither in terms of OEM (original equipment manufacturer), nor with regard to aftermarket solutions.

An interesting approach is claimed by Ono (1999). The intension is to reduce scattering of abrasion powder to prevent environmental pollution. It is discussed to install an abrasion powder adsorber on a part or the entire periphery of the friction portion of a brake pad by clamping or using an adhesive. The adsorber is using a meshlike sheet, which is made by molding and is working as a fine fiber material with the goal to adsorb the fine abrasion powder. Alternatively, a porous structure with a fine surface can be used as well. The proposed idea is able to be used for semimet as well as for NAO brake linings.

In Jessberger (2012) a more complex brake dust collecting device for motor vehicles is presented. Therein an air flow—generating device is described as a rotatable fan blade. This fan is mounted on the wheel rim and driven by this, which means that no external and additional energy is required. The inventor describes that the generated air flow is high enough to overcome the flow resistance of the used filter. Establishing this air flow allows transferring the brake particles from the friction zone to the collecting device, which is embodied as a filter element. The filtered air can then be released into the atmosphere without any concerns. In case when the filter is soiled, it can be removed or cleaned easily.

Another concept is invented and claimed in Rocca Serra (2014). Therein a nonpolluting brake assembly is described, consisting besides the brake system with brake rotor and lining also the collecting device. This autonomous suction device is expected to draw off the generated wear particles and consists of an intake opening and a collection chamber to collect the wear debris. A pipe leads the particle flow from the intake opening to the chamber. The device is powered by an impeller that is pressed and driven by the brake rotor. According to the patent description, the intake opening is located very close to the pad and the rotor, and a distance of less than 5 mm is preferred.

FIGURE 10.7 Brake particle collecting concept of Tallano Technologie SAS, (n.d.).

In addition, the suction device should be integrated as close as possible to the pad. Fig. 10.7 pictures a possible prototype design according to (Tallano Technologie SAS); the device is located on the trailing side of the brake caliper for a better collecting efficiency. The idea behind is to position the capturing device close to the friction zone where all the wear debris is generated, to avoid a spread particle emission and thus to ensure an increased collecting efficiency. The inventors expect an efficiency of 90% of the emitted particles and state that the concept is applicable to different caliper designs and the use as an aftermarket solution is feasible. However, it must be mentioned that the stated efficiency is related to the absolute mass loss of the friction partners; the positive effect on particle numbers was not discussed yet.

In addition, a concept that aimed to collect the wear debris generated from the friction brake is presented in Brake Pad Waste Collection Systems Inc. (n.d.).

The brake pad waste collection system, shown in Fig. 10.8, represents a more or less closed system around the brake system; in other words, it encapsulates the brake. No rotating parts are used; the rotational air flow of the rotor leads the particles to a filter channel.

A detailed description of the system as well as very first results of experimental testing is given in Fieldhouse and Gelb (2016). The authors introduce a system comprising filters that collect particles in the range of 40 μm and above. Additionally, magnets are implemented to collect particles of diameters of 40 μm and below. This combination should allow to cover the entire particle size range and capture also nanoparticles. First, brake dynamometer tests were carried out, and then the gravimetric wear was observed. However, it must be remarked that no particle measurements using solid particle counting systems or particle size analyzers were realized. As a result of the gravimetric studies, the authors state a dust collecting efficiency of 92%, which is even slightly better than (Tallano Technologie SAS) is explaining.

The experimental investigations revealed also an improved wear behavior using the collection system. The brake rotor and pad thickness loss of a baseline test without the collecting system (baseline test) and a test with it are analyzed. Interestingly, the brake rotor and both pads have shown a lower wear when the collecting device was applied than the baseline test. This surprising behavior is explained by a better ventilation and the fact that the wear debris is immediately removed after the friction zone. This means that no debris adheres to the disc surfaces, reenters the frictions zone, and acts as a grinding compound, which leads to less

FIGURE 10.8   Brake pad waste collection system according to Brake Pad Waste Collection Systems Inc. (n.d.).

abrasion in the end. However, this explanation means also that a third body layer cannot be established properly—future investigations will have to clarify this interesting phenomenon.

Briefly summarized, it can be stated that brake particle collecting system is not only considered as an idea or patent only; first experimental studies were already carried out. These results are promising, even impressive in terms of the gravimetric wear debris that is feasible to be collected. However, the efficiency with regard to ultrafine nanoparticles is still to be investigated and some question marks do exist when it comes to the transfer to serial production. Because the space in the wheel and wheel house is very limited, it seems to be very unlikely to implement additional components such as fans, impellers, pipes, and filter boxes into the wheel. Additionally, possible interactions of the collecting devices with the brake are not discussed finally: especially the thermal behavior with high temperatures up to 500°C, 600°C, or 700°C could mean some concerns for the collecting devices. Last but not least, possible consequences for the noise vibration harshness (NVH) behavior should be considered. Rotating parts, especially with direct (frictional) contact to the brake rotor, might cause or at least promote noise and vibration complaints such as brake judder or brake squeal. In the future it is expected to concentrate more on improvements of the friction partners to follow concepts for the prevention of wear debris as it is discussed in the following chapter.

4. PREVENTION OF BRAKE PARTICLE GENERATION

Because BPEs are very complicated to handle because of the "open" character of the brake system, it seems to be very promising to avoid the generation of wear debris. However, the presented approaches represent only first, very interesting, and promising ideas that have not been introduced in a significant way to the market for passenger cars worldwide.

4.1 Role of Regenerative Braking

Thanks to the introduction of hybrid and full-electric vehicles, it might be feasible to reduce or even avoid the use of the traditional friction brake. The electric motor is not only dedicated for traction purposes but can also be used, in the generator mode, for deceleration maneuvers. However, for such a scenario it is obvious that the electric power train has to be dimensioned in a sufficient way or, in other words, powerful motors, power electronics, and battery packs have to be considered. The simple comparison of the required brake power to the traction power in the case of a sports car underlines the challenging demands for the electric power train: although such an exemplary vehicle of 1.500 kg has an engine power output of 330 kW to accelerate up to 310 km/h, it needs ~700 kW to decelerate with 1.1 g down to zero, which means that for at least an emergency braking the required power is almost doubled.

Additionally, it must be mentioned that a so-called highly available brake device has to be implemented. Highly available means a redundancy in the system to ensure its overall functionality also in the case that one subsystem fails. For the brake system, it means that at least two electric motors, which can act as a generator, are installed or a frictional brake as a backup system is considered.

Because the requirements are that demanding, the combined use of friction brake and the electric motor/generator is probably the preferred way for the next few years. A possible brake application scenario is depicted in Fig. 10.9.

The displayed brake step considers a constant deceleration, desired by the driver who actuates the brake pedal. Because the electric machine exhibits a limited operating range in terms of velocity and deceleration, a combination with the friction brake is desired. Especially for higher velocities and close to the so-called instop, when the speed is very low and goes down to 0 km/h, a support from the friction brake is required. In summary it can be stated that during a first phase, the brake initiation, the deceleration is realized by the friction brake, which is later on ramped down. During a second phase, the full recuperation, the electric generator realizes 100% of the deceleration. Finally, in phase 3 (near standstill), the generator is ramped down and the friction brake is used again. It might be remarked that this combined

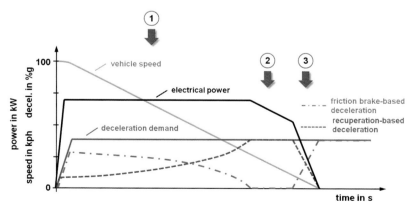

FIGURE 10.9 Brake application with friction brake and electric power train (AUDI AG, 2016a).

braking action is very challenging for the designer of a brake system because a so-called brake blending has to be implemented. This means that the driver only has to press the brake pedal and the brake system controls the friction brake and the electric generator in such a manner that the driver does not recognize any switching effects.

A closer look to the operating parameters with and without regenerative braking reveals the potentials. For the presented example a plug-in hybrid electric sports utility vehicle was instrumented to acquire the operating parameters such as brake disc temperature, brake pressure, deceleration, and vehicle speed. The test vehicle was equipped with a switch to turn on/off the regenerative braking, which allows a direct comparison and shows the potentials of a recuperation system. In terms of the driving profile a customer-oriented cycle that consists of rural, city, and motorway sections was used.

It can be stated that with regard to the brake pressure lower levels were observed: almost 90% of all stops have revealed brake pressures lesser than 30% of the maximum brake pressure p_{max}; without the recuperation only 60% of the brake applications were below 30% of p_{max}. In addition, the temperature level has decreased. In Fig. 10.10 the temperature profiles of the front brake disc, obtained from a thermocouple applied to the friction ring, are depicted. Two signals and their average values are shown: the curves of the higher level represent the study without regenerative braking, while the lower level lines are linked to a test sequence with recuperation.

As it can be seen easily, the gap of about 48% is very impressive and could also lead directly to lower wear losses because higher temperatures lead, in most of the cases, to higher wear rates.

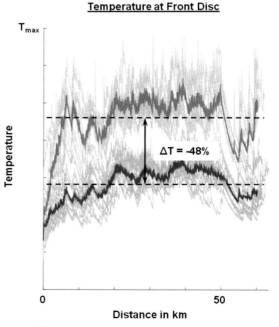

FIGURE 10.10   Temperature profiles of the front disc with and without regenerative braking (AUDI AG, 2016a).

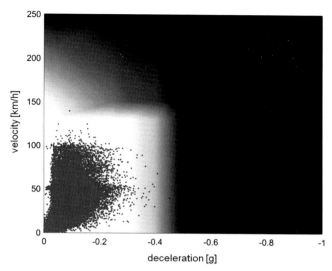

FIGURE 10.11 Velocity—deceleration behavior for a regenerative brake system (AUDI AG, 2016a).

Besides lower levels of brake pressure and temperature, the velocity—deceleration was investigated as well. In Fig. 10.11 each brake application of a test drive with regenerative braking is characterized by its correlating vehicle speed and deceleration. Additionally, the ranges of full recuperation (white), combined friction braking/regenerative braking (gray) and full friction braking (black), are displayed.

It can be seen that almost all of the brake applications of this example (using a moderate-driving customer) could be carried out by the electric generator. Only in some cases, the friction brake is required.

In summary it can be stated that regenerative braking has a huge potential to reduce the use of the friction brake and is a promising measure to reduce brake wear and particle emissions. However, regenerative braking is strictly linked to hybrid and full-electric power train technologies. In 2015, according to (KBA, n.d.), the fleet of hybrid and battery electric vehicle (126,702) had a share of only ≈0.3% of all registered vehicles (44,403,124) in Germany, as an example. These numbers show impressively how necessary and important a significant increase in the market share of electric vehicles is to exploit the potentials of regenerative braking.

4.2 Role of Driver Assistance Systems

Another approach of reducing BPEs and brake wear deals with the consideration of DAS.

It can be stated that mobility is more eco-friendly and resource-conserving when the traffic routing ensures a certain traffic stream without any stop and go, which leads, similar to reduced exhaust emissions, also to a decrease of non-exhaust emissions, according to Gruden (2008). For this objective, some influencing factors exist and are shown in Fig. 10.12.

It is clear that factors such as traffic density and routing cannot be influenced for every use-case; especially in urban regions a high traffic density exists nearly all day and building

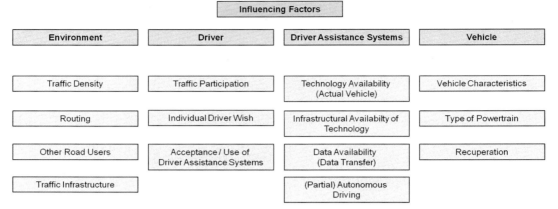

FIGURE 10.12   Influencing factors on traffic flow and brake particle emissions.

additional roads is also not a feasible solution. However, the infrastructure can be improved by considering sensor and transmitter systems to realize a so-called environment-to-car communication. Such an environmental traffic monitoring was investigated successfully by the INTEGREEN project (Joint Research Project: INTEGREEN, n.d.). The project included static monitoring stations for the real assessment of traffic conditions and also traffic-induced emissions. As a consequence of the monitoring, an environmental traffic management was implemented including speed control and adaptive traffic light cycles.

In terms of the driver, a crucial influencing factor is represented by the acceptance and use of driver assistance systems. Because today's vehicles do not drive entirely autonomous, the DAS is only supporting the driver and can offer a help to him/her. Only when the driver is accepting the DAS, all its beneficial advantages can be exploited.

As one example, the possibilities of DAS are claimed in Rychlak (2010), where the driving through a curve is optimized by a DAS. The curve characteristics are determined and a drive scenario is calculated and carried out in such a way that, namely, CO_2 emissions, tire abrasion, and also brake dust are reduced. Several devices are necessary for the realization of the introduced claim. In addition to the already mentioned "environment-to-car" communication, meant by the infrastructural availability of technology, also "car-to-car" communication is required to optimize the driving and completes the data transfer factor. Last but not least, the DAS technologies must be available in the vehicle. These systems include not only a simple cruise control, they have to be much more sophisticated as well. Already existing adaptive cruise control (ACC) systems with integrated stop and go assists, which are combined with lane departure assists, will be accompanied by active curve assists. It is expected that such a combination of sophisticated assistance systems is very close to the requirements of autonomous driving and an enabler for a decrease in driving emissions in general and BPEs in particular.

Finally, the vehicle itself influences its emission behavior. The role of factors such as the vehicle mass is quite clear, but also the type of the power train and possibilities of regenerative braking (recuperation) are relevant as the joint research project UR:BAN shows in Joint Research Project: Urban Space (n.d.). It deals with the development of user-dedicated assistance systems, and one of its objectives is to define a cooperative traffic management for cities to ensure economic and energy-efficient driving. One of the innovative approaches considers the guidance of the traffic flows taking into account the different types of powertrains and possibilities of regenerative braking, which means a management that optimizes the amount of regenerative brake energy.

In summary it can be stated that some approaches of using driver assistance systems in terms of driving emission reduction already exist, and the required vehicle technologies such as ACC, lane departure assists, and adaptive navigation systems are already available or will be ready in the near future, accompanied by additional features such as active curve assists. In combination with a sophisticated data transfer ("environment-to-car" and "car-to-car" communication) the technical requirements are mentioned. Finally, the driver has also to accept that kind of assisted driving and use the advantages of DAS to improve driving safety and reduce driving emissions.

5. SUMMARY

The presented approaches of remediation measures for brake wear and particle emissions can be divided into three columns. The first one takes the friction partner (brake disc and lining) into account, starting with minor changes to commonly used compositions and ending up with complete new design and material approaches. So far, only the FNC-disc treatment can be considered as success story with regard to market penetration, at least for passenger cars in North America. Most of the other proposals could not find the right way to the markets yet, mostly due to critical additional costs, but also in some cases caused by technical difficulties and question marks. The second column is represented by brake particle collection systems. Several different design proposals exist, but no system was introduced to the passenger car market yet. This might be changed in the future, especially when particle emissions in general and emitted particle mass in particular could become part of the legislation. The third and last column deals with the avoidance of wear and particle emissions. Approaches such as regenerative braking and the use of driver assistance systems seem to be very promising. On the other hand, it is obvious that the market penetration of electric vehicles is still very low compared with vehicles with conventional powertrains, and the positive effect of regenerative braking therefore is quite poor. Moreover the use of driver assistance systems is up to now limited, because these systems are, in most of the cases, expensive and innovative communication possibilities such as "car to x" are not available yet.

As a conclusion it can be stated that many approaches for wear and particle emission improvements already exist, but the efforts to introduce them into the market have to be enhanced.

Glossary

ACC Adaptive cruise control
AK Master International dynamometer standard test procedure to determine brake performance and wear
Al-MMC Aluminum metal matrix compound
BPEs Brake particle emissions

DAS Driver assistance system
ECE A common term in brake industry for European brake lining materials with low steel content; derived from ECE regulation program
FNC Ferritic nitro-carburizing
GCI Gray cast iron
NAO Nonasbestos organic
NVH Noise vibration harshness
OEM Original equipment manufacturer
SPCS Solid particle counting system

References

AUDI AG. (1990). *Development foundation brake*. Internal investigations.

AUDI AG. (2016a). *Development foundation brake*. Internal investigations of braking systems for plug-in hybrid electric vehicles.

AUDI AG. (2016b). *Development foundation brake*. Internal investigations of FNC-treated cast iron rotors.

Automotive Parts Co. Ltd Dongying Baofeng. (2014). *Patent No. CN104179860 A*. China.

Bowei, L. E. (2009). *Patent No. CN101592204A*.

Brake Pad Waste Collection Systems Inc. (n.d.). Retrieved from http://www.cleanbrakeperformance.com/index.html.

Dupuis, V. (2013). *Patent No. US2013161132A*. USA.

Fieldhouse, J., & Gelb, J. (2016). New developments of an on-vehicle brake pad waste collection system. In *SAE brake colloquium and exhibition*. Scottsdale: SAE international.

Greil, P., Hammer, J., Rosenlöcher, J., Steinau, M., Travitzky, N., & Zipperle, T. (2011). *Patent No. DE102009053737 A1*. Deutschland.

Gruden, D. (2008). *Umweltschutz in der Automobilindustrie*. Wiesbaden: Vieweg+Teubner.

Hanna, M. E. (2009). *Patent No. US20110079326A1*. USA.

Hino, H., & Miyashita, Y. (1984). *Patent No. JP59173234A*. Japan.

James, M., Murdie, N., & Walker, B. (2002). *Patent No. US2002170787A*. USA.

Jessberger, T. (2012). *Patent No. US8167098 B2*. USA.

Johnson, K. E. (2009). *Patent No. WO 2009037437 A1*.

Joint Research Project: INTEGREEN. (n.d.). Retrieved from http://www.integreen-life.bz.it/home.

Joint Research Project: Urban Space. (n.d.). Retrieved from http://urban-online.org/de/urban.html.

KBA. (n.d.). Retrieved from http://www.kba.de/DE/Statistik/Fahrzeuge/Bestand/Umwelt/b_umwelt_z.html;jsessionid=8179A55B55773F9E69DC21B918DDE746.live11292?nn=663524.

Keith, H. (2009). *Patent No. US2009026025A*. USA.

Khambekar, S., Baumgartner, H., & Pahle, W. (2004). *Patent No. US20040031652 A1*. USA.

Kienzle, A., & Kratschmer, I. (2007). *Patent No. US20080125306 A1*. USA.

Lampke, T., & Özer, I. (2012). *Patent No. DE102012000487 B3*. Deutschland.

Lampke, T., & Özer, I. (2013). *Patent No. DE102011121292 A1*. Deutschland.

Lembach, O., & Mayer, R. (2012). *Patent nr. DE102008053637B4*. Deutschland.

Matsumoto, R., & Forsythe, G. (2002). *Patent No. US6376431 B1*. USA.

Mayer, R., & Lembach, O. (2012). *Patent No. DE102010048075A1*. Deutschland.

Mayer, R., & Lembach, O. (2013). *Patent No. US2013008748A*. USA.

Muthukannan, D., Valarmathi, A., Shariff, S., & Padmanabham, G. (2014). Laser surface nitrided Ti-6Al-4V for light weight automobile disk brake rotor application. *Wear*, 269–274.

Nakanishia, H., Kakiharaa, K., Nakayamaa, A., & Murayamab, T. (2002). Development of aluminum metal matrix composites (Al-MMC) brake rotor and pad. *JSAE Review*, 365–370.

Ning Guofei Eagle Auto Parts Co. Ltd. (2014). *Patent No. Inorganic nano modified phenolic resin brake pad and preparation process thereof*. China.

Ono, M. (1999). Patent No. JP11311272A. Japan.

Özer, I. (2009). Patent No. DE102009059806A1. Deutschland.

Özer, I., & Lampke, T. (2012). Patent No. DE102011120989B3. Deutschland.

Patterson, W., Standke, K., & Kocheisen, K. (1967). *Ursachen für Unterschiede in den mechanischen Eigenschaften und der Gefügeausbildung von Gußeisen mit Lamellengraphit*. Wiesbaden: Springer Fachmedien.

Piwowarsky, E. (1942). *Hochwertiges Gusseisen*. Berlin: Springer Verlag.

Qihong, C. E. (2011). *Patent No. CN201944152U*. China.

Rasim, Y. (2011). *Patent No. DE102009050024 A1*. Deutschland.

Riefe, M., Holly, M., & Learman, C. (2016). An overview of ferritic nitrocarburized (FNC) cast iron brake rotor performance and field experience. In *SAE brake colloquium and exhibition. Scottsdale: SAE international*.

Rocca Serra, C. (2014). *Patent No. WO2014072234 A2*.

Rychlak, S. (2010). *Patent No. DE102009001332 A1*. Deutschland.

Santamaria Ranzo, D., & Persoon, F. (2016). Bio-soluble chemical composition for complementary mineral fibres: An enhanced tribological effect and its influence on disc wear. *SAE International Journal of Materials and Manufacturing*.

Shangdong Hong MA Construction Machinery Co. (2014). *Patent No. CN104195423A*. China.

Sichuan Province Vanadium and Titanium Brake Ltd. (2014). *Patent No. CN104073712A*. China.

Smolen, G. E. (1991). *Patent No. US005224572A*. USA.

Tallano Technologie SAS. (n.d.). Retrieved from http://www.tallano.eu/en.

Turani, S. E. (2011). *Patent No. US20110198170 A1*. USA.

Yuan, W. E. (2009). *Patent No. CN101514251 A*. China.

Zeuner, T., Ruppert, H., & Engels, A. (1998). Developing trends in disc brake technology for rail application. *Materialwissenschaften und Werkstofftechnik, 29*, 726–735.

Zhao, H. (2013). *Patent nr. CN103470666A*. China.

Non-technological Measures on Road Traffic to Abate Urban Air Pollution

Xavier Querol[1], Fulvio Amato[1], Francesc Robusté[2], Claire Holman[3], Roy M. Harrison[4,5]

[1]Institute of Environmental Assessment and Water Research (IDÆA), Spanish National Research Council (CSIC), Barcelona, Spain; [2]Civil and Environmental Engineering, ETESCCP, Universitat Politècnica de Catalunya, Barcelona, Spain; [3]Brook Cottage Consultants Ltd, Bristol, United Kingdom; [4]Department of Environmental Sciences/Centre for Excellence in Environmental Studies, King Abdulaziz University, Jeddah, Saudi Arabia; [5]School of Geography, Earth & Environmental Sciences, University of Birmingham, Birmingham, United Kingdom

O U T L I N E

1. Introduction   230
   1.1 WHO Air Quality Guidelines and EU Legislation and Implications for Air Quality Plans   231
   1.2 Air Quality Plans and Source Attribution for Pollutants   231
     1.2.1 NOx Source Contributions   231
     1.2.2 Particulate Matter Source Contributions   232
     1.2.3 EU Emission Inventories, Source Contributions, and Air Quality Measures   233
     1.2.4 Air Quality Measures for Road Traffic   234

2. Non-technological Road Traffic Measures in Urban Areas   234
   2.1 Public Transport   236
   2.2 Measures That Reduce the Number of In-Use Vehicles   236
   2.3 Measures Favoring the Transformation of the Fleets to More Ecological Options   239
   2.4 Measures on City Logistics and Taxis   241
   2.5 Measures for the Urban Transformation of the City   245
   2.6 Examples of Reremediation Measures   247

Non-Exhaust Emissions
https://doi.org/10.1016/B978-0-12-811770-5.00011-X

229

2.6.1 Road Sweeping and Washing   247

2.6.2 Application of Photocatalytic
Products   252

2.7 Other Nontechnological Measures   253

Acknowledgments   255

References   255

1. INTRODUCTION

Air pollution is one of the major environmental problems, both in developed and developing regions of the world (Lim et al., 2012; World Bank, 2016). In Western Europe, and according to the last evaluation of the WHO's Global Burden of Disease (Lim et al., 2012), air pollution is the 11th greatest avoidable cause of mortality. Thus according to EEA (2016), the premature deaths attributable to atmospheric particulate matter finer than 2.5 μm ($PM_{2.5}$), NO_2, and O_3 exposure in the EU-28 in 2013 reached 438,000, 68,000, and 16,000 deaths/year, respectively, and the years of life lost were 4,668,000, 723,000, and 179,000 years/year, respectively.

These large effects are still evident after a large improvement in air quality for most of the pollutants observed in European urban areas over the last decades (Guerreiro, Foltescu, & de Leew, 2014; EMEP, 2016; Querol et al., 2014). Thus according to EEA (2016), there is a statistically significant decreasing averaged trend of the annual mean of PM_{10} (PM finer than 10 μm) at rates reaching averages of -0.6 and -0.9 μg/m^3 PM_{10}/year from 2000 to 2014 (2σ = 0.03 and 0.05 μg/m^3 PM_{10}/year) in urban background and road traffic sites of the EU-28, and even more pronounced in some regions of Southern Europe (-1.0 and -1.5 μg/m^3 PM_{10}/year in Spain, 2σ = 0.14 and 0.19 μg/m^3 PM_{10}/year). The above EEA evaluation also shows, however, that the EU-28 averaged $PM_{2.5}$ 2006—14 trend is not statistically significant mainly due to different trends observed in various countries, but in a large number of them there is a clearly decreasing trend. The other critical urban pollutant in terms of compliance with EU limit values is NO_2, and concentrations remain at relatively high ambient levels. In this case, decreasing trends are also evident, but less pronounced than for particulate matter (PM). EEA (2016) reported averaged 2000—14 decreasing annual averaged trends of -0.5 and -0.6 μg/m^3NO_2/year (2σ = 0.02 and 0.05 μg/m^3 PM_{10}/year) from 2000 to 2014 for the urban background and urban traffic sites, also with the largest decreases in Southern Europe.

Also according to EEA (2016), 8%—12% and 16%—21% of the EU-28 population are exposed to levels that exceed the annual $PM_{2.5}$ limit value (25 μg/m^3) and the PM_{10} daily limit value (50 μg/m^3 PM_{10} as a 90.4th percentile of the daily concentrations in a year), as well as 7%—9% to levels that exceed the annual NO_2 limit value (40 μg/m^3).

In the case of NO_2, the annual limit value from the EU standard (2008/50/EC) coincides with the health protection guideline from WHO (2006), but the discrepancy is very large for $PM_{2.5}$ and PM_{10}. Thus, the European annual limit values of $PM_{2.5}$ and PM_{10} are higher by a factor of $\times 2$ and $\times 2.5$ than the respective WHO health protection guidelines.

1.1 WHO Air Quality Guidelines and EU Legislation and Implications for Air Quality Plans

As stated above and in Chapter 1 of this book the current PM_{10} and $PM_{2.5}$ EU Limit Values exceed by far the respective WHO air quality guidelines. In the case of PM_{10}, this has remained unchanged since the first daughter air quality directive (EC, 1999). The most recent air quality directive (EC, 2008) did not change the PM_{10} limit value and fixed a $PM_{2.5}$ one 2.5-fold the WHO guidelines. Finally, the 2013–14 Clean Air for Europe Legislation Package announced that the 2008/50/EC directive, containing the PM_{10} limit values (fixed at the 1999 criteria), will continue in force until 2020, when a revision of the directive will take place. The current situation is that if we take as a reference the WHO health protection thresholds the percentages of EU-28 population exposed to levels exceeding the guidelines increase to 85%–91% for the WHO $PM_{2.5}$ and 50%–63% for the daily PM_{10} (EEA, 2016).

Why do we discuss this population exposure and air quality standards issue? Because the current EU legislation states that action plans should be implemented when the limit values are exceeded. In the cases of PM_{10} and $PM_{2.5}$, these are so lax that they are exceeded only in a few regions and reduce the pressure for abatement of ambient air PM. However, it is made clear by all stakeholders and research results that the pollutant having the greatest impact on human health in urban areas is PM (causing 6.4 and 27.5 times more premature deaths in the EU than NO_2 and O_3, according EEA, 2016). According to this, State Members should decide whether to reinforce stricter PM national standards or implement PM abatement plans in areas in compliance with standards.

It is for these reasons that in this chapter we will describe abatement strategies for reducing both PM_{10} and NO_2 ambient levels, even in regions where the EU limit values are not exceeded.

1.2 Air Quality Plans and Source Attribution for Pollutants

Focusing now on air quality plans (AQPs) or measures, it is important to clarify that (1) of course the basis for identifying the sources and abatement needs of a given region to attain ambient air limit values are the emission inventories; but (2) the conclusions of evaluating the source contribution to a given emission inventory does not necessarily match with the true source contribution to the ambient air PM or NO_2 levels; (3) for $PM_{2.5}$ the emission inventories only include the "primary" $PM_{2.5}$, whereas in most urban environments the "secondary" fraction (generated in the atmosphere from gaseous pollutants such as SOx, NOx, NH_3, and voltaile organic compounds (VOCs)) reaches 70%–80% of the ambient air $PM_{2.5}$ mass concentrations (Amato et al., 2016a); and (4) emission inventories have deficiencies for specific sources that might make an important contribution to the exceedance of a limit value (especially for PM). The directive 2008/50/EC specifies in annex XV, Sections 7 and 8, that during the development of an AQP member states should demonstrate quantitatively, not only that the reductions of emissions are relevant, but also that these are sufficient to avoid exceedances in a given target year.

1.2.1 NOx Source Contributions

According to LRTAP (2017) emission inventories for the EU-28 (Fig. 11.1), road traffic accounts for nearly 40% of the EU-28 NOx emissions. However, due to the proximity of urban

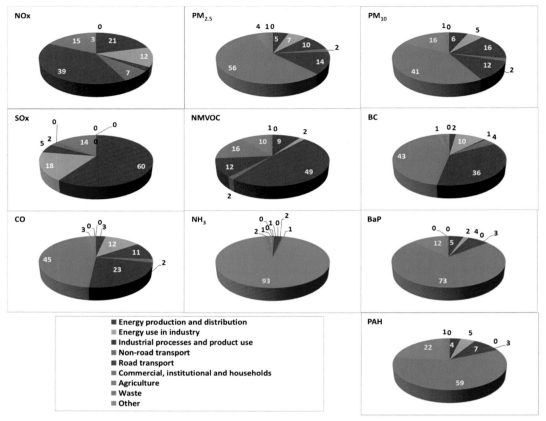

FIGURE 11.1   Contribution (in %) of source sectors to the 2014 EU-28 emission inventories for selected pollutants according to LRTAP (2017) as accessed on January 24, 2017.

road traffic to exposed citizens when compared with other emission sources (industries, power generation, etc.), the road traffic contribution to urban ambient NO_2 is estimated to exceed 60%–80% or to cause a very large proportion of the noncompliance with the air quality standards in many EU cities (AIRPARIF, 2013; Borge, de la Paz, Lumbreras, & Vedrenne, 2014; DEFRA, 2015). Furthermore, the percentage of primary NO_2 is nearly 30% in most diesel passenger car exhausts (Carslaw, Murrells, Andersson, & Keenan, 2016). Thus there is a mismatch here. Although emissions are quantified in terms of NOx (NO + NO_2), ambient air regulations refer to NO_2 and, furthermore, different NOx emission sources have different NO_2/NO ratios. Although most of NO will be converted to secondary NO_2, this conversion can take time and have an effect outside or on the outskirts of a city.

1.2.2 Particulate Matter Source Contributions

For PM_{10} and $PM_{2.5}$, the source attributions and determination of contributions are even more complex because these are made by a mix of pollutants.

As stated previously, secondary PM accounts for around 70% of urban background $PM_{2.5}$. From this, and according to numerous PM speciation studies, approximately 1/5 might be ammonium sulfate (from SO_2 oxidation and reaction with NH_3), another 1/3 be ammonium nitrate (from NO_x oxidation and reaction with NH_3), and the remaining 1/2 may be secondary organic compounds (from the complex oxidation of VOCs) (Amato et al., 2016a; Jimenez et al., 2009). The levels of this secondary $PM_{2.5}$ fraction tend to be relatively homogeneous across urban areas because time is required for the production of PM from gaseous pollutants, and because a relevant fraction of PM has an external (to the city) origin. To decrease levels of this secondary fraction, abatement of its gaseous precursors is required, but these are not necessarily linearly related to the secondary constituents; because, for example, a decrease of SO_2 emissions might clearly decrease levels of ammonium sulfate, but increase those of ammonium nitrate due to more available NH_3. The remaining 30% of urban $PM_{2.5}$ is mainly comprised of soot particles (that include black carbon (BC)) emitted directly from road traffic and biomass burning, with a minor contribution from fine road dust, brake and tire wear, fly ash, soil dust, and sea salt. The emission inventories for $PM_{2.5}$ only refer to this 30% primary fraction. The temporal and spatial variability of levels of primary $PM_{2.5}$ are expected to be high due to the direct influence of the PM emission sources, the latter decreases with distance from the emission sources (the curbside for traffic).

The $PM_{2.5-10}$ fraction, known as the coarse fraction, is mostly primary in origin, and it is mainly made up of soil dust, road dust, brake and tire wear, sea salt, and biological debris. Its primary origin, coarse size, and relatively high density account for very high temporal and spatial variations, also with a marked increase as we approach the emission sources. The emission inventories usually contain data on PM_{10} and $PM_{2.5}$, but not on $PM_{2.5-10}$, but it can be estimated by difference.

1.2.3 EU Emission Inventories, Source Contributions, and Air Quality Measures

According to the LRTAP (2017) emission inventories for the EU-28 countries (Fig. 11.1), road traffic accounts for 14% of the primary $PM_{2.5}$ emissions, whereas domestic, commercial, and institutional (heating) emissions account for 56%. For primary PM_{10}, these contributions reach 12% and 41%, 23% and 45% for CO, 3% and 73% for BaP; 36% and 43% for BC. Very small contributions from road traffic are reported for SO_2 (<0.5%, with dominant emissions from energy production, with a 78% contribution, but taking into account that shipping emissions are not included here) and NH_3 (2% and predominantly emitted by agriculture, with a 93% contribution). Of course, and as previously stated for NO_2, all these contributions do not necessarily reflect the actual source contributions to the ambient air levels of PM_{10} and $PM_{2.5}$ and the other pollutants, because for all of them we should make adjustment to the estimate according the proximity of the sources to the citizens; and for $PM_{2.5}$ and PM_{10}, we should take into account that the dominant proportion of PM in ambient air is not primary (at least for $PM_{2.5}$) but formed from gaseous precursors that may have different sources to the primary PM of the emission inventories.

Once we know the (1) emission inventories and (2) the source contributions to NO_2, PM_{10}, $PM_{2.5}$, and $PM_{2.5-10}$, we can clearly identify the sources responsible from the exceedances of the health protection thresholds, as well as quantify the abatement targets needed to avoid further exceedances of these thresholds. However, before we proceed to focus on the air quality measures, we need to obtain the source contributions for the averaging periods relevant to the health protection limit value being exceeded. This may be for long-term (annual mean) or the short-term (hourly or daily) periods, or during specific pollution events. As one can expect, the type and contribution of dominant sources for long- and short-term exposure of pollutants might differ considerably. As an example, road dust in Scandinavian countries decisively contributes to the exceedances of the PM_{10} daily limit value due to winter sanding of roads and the use of studded tires, but much lower contributions are expected when the annual average is concerned.

1.2.4 Air Quality Measures for Road Traffic

Focusing on the measures in an AQP for road traffic, we can group them into (1) technological and (2) nontechnological measures. The first one includes the application of technology to abate (mostly exhaust) emissions from existing vehicles (retrofitting) or to replace existing high emitting vehicles with newer vehicles equipped with exhaust emission abatement technologies, or use less polluting fuels.

The main objective of this chapter is reviewing the nontechnological measures available for implementation in AQPs for road traffic in urban areas.

2. NON-TECHNOLOGICAL ROAD TRAFFIC MEASURES IN URBAN AREAS

The spectrum of nontechnological measures is much wider than the technological ones. Usually the nontechnological measures are classified according to the tools or policies applied. Thus, it is common to identify groups of measures such as fuel and vehicle taxation policies, congestion taxes or urban tolls, economic incentives for commuters of public and active transport systems, urban planning and road traffic management policies (low emission zones, regulation of parking areas, speed regulations, vehicle restrictions based on number plate restrictions, etc.) and remediation measures, among others. However, we will use in this chapter another grouping of these measures, which is based not on the tools but on the objectives to be met. Thus we classify measures into those devised to the following (Fig. 11.2):

1. attract commuters to public transport instead of private transport;
2. reduce the number of vehicles in use;
3. accelerate the transformation of the vehicle fleet to more eco-friendly options;
4. include environmental criteria in the city logistics (urban freight distribution (UFD) and taxi fleets);
5. transform the physical structure of the city to reduce human exposure to pollutants by increasing the distance between road traffic and citizens (such as green and pedestrian areas, cycle lanes, controlling the traffic around schools, hospitals, medical care centers, playgrounds, etc.), and decreasing the areas used by road traffic;
6. remediation measures, such as road sweeping and washing, application of dust suppressants to abate resuspension, or the application of photocatalytic products; and
7. other nontechnological measures, such as speed reduction to abate resuspension or enhancing traffic flow to reduce congestion.

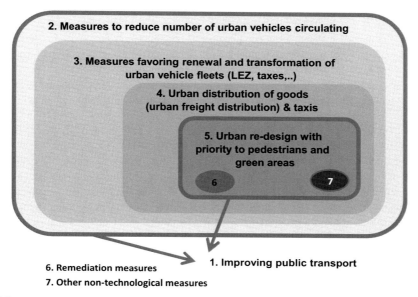

FIGURE 11.2 Classification of nontechnological measures to abate urban air pollution according to their specific aims.

Fig. 11.3 shows a chronogram that is proposed for the implementation of such groups of measures. Following this sequence is of vital importance because in some cases the implementation of one or several of these measures (for example, having a good public transport system, reducing the cars, and the eco-transformation of the fleet) is essential to the success of others (such as urban transformations).

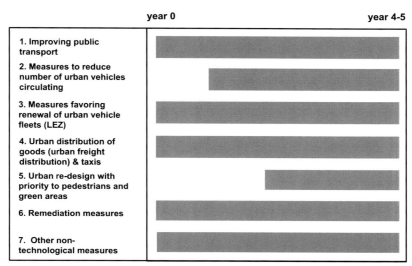

FIGURE 11.3 A typical chronogram for implementing nontechnological measures to abate urban air pollution.

2.1 Public Transport

Indeed this is one of the most relevant groups of measures, because to allow implementing a large proportion of other actions we need the public transport to absorb the commuters from the shift of transport modality. Obviously, strategies favoring this shift are the increase in frequency, spatial coverage, comfort, and reducing costs of the public transport. In many metropolitan areas, the intraurban public transport is relatively good, whereas the interurban transport is the one requiring important improvements.

It is also important to incentivize public transport and active commuting (pedestrian and cycling) through measures such as free public transport during air pollution episodes, implementing public information campaigns, multimodal transport cards that avoid paying during connections among different public transport systems, including public transport costs as tax deductible expenses, or involving private companies and administration in plans to cofinance the use of public transport or cycling for their staff. The Metropolitan Authority of Barcelona recently launched the T-Verda card (T-Green) that gives 3 year allows free use public transport during 3 years if an old car owner decides to bring the car for dismantling and gets the compromise of not purchasing another one.

2.2 Measures That Reduce the Number of In-Use Vehicles

According to the air quality actions implemented in European cities, two major structural measures can be identified in this group, namely the application of congestion charges or city tolls and parking restrictions. Furthermore, the alternating daily restriction of vehicles (typically cars) with odd and even number plates has been applied in a number of cities, but in this case mostly during pollution events.

- According to the European traffic restriction measures website (UAR, 2017), there are in Europe 16 cities (mostly Scandinavian, Italian, or British) applying congestion charges. One of the most successful stories is the congestion charge in Stockholm, which has been applied to the whole city since 2007. Its enforcement is undertaken with automatic number plate recognition (ANPR) cameras on the six bridges that provide access to the city. The charge in peak traffic hours is 3.5 € and is 1 € out of traffic rush hours, to a maximum of 11 € per vehicle/day. According to the reviews by Johansson, Burman, and Forsberg (2009) and Johansson (2016), since its implementation in 2007, the congestion charge has reduced with respect to prior years by around 30% the number of vehicles entering into the city up to 2015 (evaluation reported in 2016), even though the

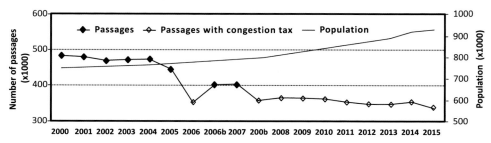

FIGURE 11.4   2000—15 trend of total car passage counts and population in Stockholm city. *Data obtained from Johansson (2016).*

city population increased (Fig. 11.4). It is evident from the various studies on the reduction of vehicles due to congestion charges that it is important that this measure is implemented over a large proportion of the city and not only for specific and relatively small parts of the conurbation. This measure has been excluded in some AQPs because of its potentially socially discriminatory nature; however, it can be implemented with exemption for high-occupancy vehicles, to reduce this collateral effect.

- Parking restriction policies include the prohibition of parking for nonresidents' vehicles in specific areas of the city (the residents from one area cannot park in other areas, as is the case of specific areas in Madrid and Barcelona), or the expansion of payment parking areas (high parking fees for nonresidents and reduced fees for residents for the first car and a high fee for all cars from the second one per family) across most of the city (as occurs in Barcelona). A time—trend analysis of this parking system versus the number of circulating vehicles/working day in central Barcelona has shown a decrease of vehicles from 2005 to 2015 close to 10% (Fig. 11.5).

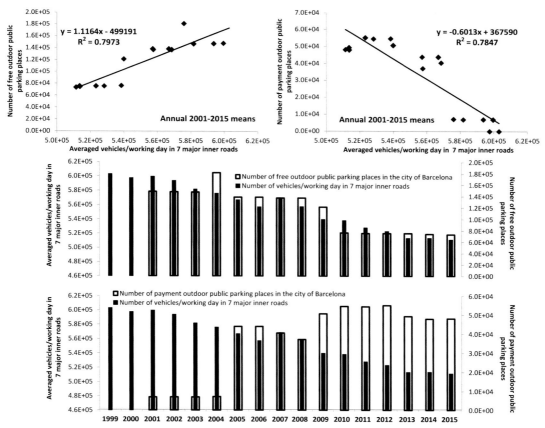

FIGURE 11.5   1999—2015 evolution of the number of outdoor free and restricted parking place types in Barcelona and correlation with the number of daily cars circulating on working days in seven major inner streets of Barcelona. *Source: Statistical data on urban mobility of the city of Barcelona (Ajuntament de Barcelona, 2017).*

- Even/odd number plate restrictions. This has been and is still applied episodically in a number of European cities such as several Italian cities and Paris, and more recently in Madrid, during pollution episodes. The analysis of the daily number of cars during the recent NOx pollution episode (26/12/2016 to 02/01/2017) in Madrid revealed an average decrease in traffic volume for 29/12/2016, when the restriction was in force, of 17% in a large central Madrid area, and an average decrease of the NO_2/SO_2 ratio (the latter mostly arising from domestic heating) of 13%. On Christmas Day, without any traffic restriction, the decreases reached 47% and 35% for the number of journeys and the NO_2/SO_2 relative decrease, respectively (Fig. 11.6), but with lower traffic flows and higher use of domestic heating.
- There are of course other complementary measures to reduce the number of cars, such as incentivizing telecommuting, as a structural policy or during pollution episodes; favoring public transport and private cycling, and high occupancy lanes, among others. However, it is also important to clarify that Annex XV of the Air Quality Directive 2008/50/EC (Section A, paragraphs 7 and 8) clearly states that, for AQPs prepared under Article 23, the impact of measures included in the plan should be quantified, and that the set of measures implemented must be sufficient to reduce ambient air levels of pollutants sufficiently to reach the objectives.

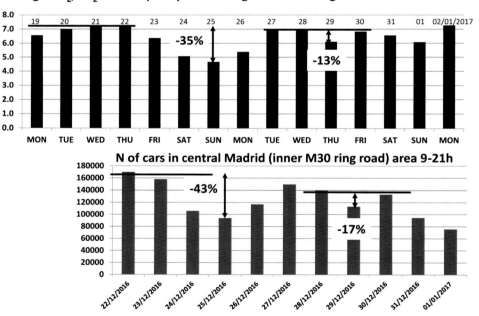

FIGURE 11.6   Evolution of daily rates of ambient air levels of NO_2 and SO_2 and counts of number of cars, both in the periods 09–21 h LT in central Madrid, during an intense NO_2 pollution episode, at the end of December 2016, when parking and even and odd plate restrictions were applied on December 29,2016. Air quality and vehicle counts supplied by the Ajuntamiento de Madrid (Madrid City Council).

2.3 Measures Favoring the Transformation of the Fleets to More Ecological Options

In this group of measures, we include policy actions aiming at facilitating (1) the electrification, gasification, and hybridization of vehicle fleets; (2) the elimination of highly polluting cars from the urban fleet and their substitution by "cleaner" vehicles; and (3) decreasing the number of diesel vehicles using fuel and vehicles taxation policies.

- According to AIRUSE (2015a), although policy actions for conversion to gas and electrification of the vehicle fleets can be easily identified in most EU countries, only Norway, the Netherlands, and Italy have reached a significant proportion of electric (the two first) and gas (the latter) vehicles in their national fleets. Thus, Norway and the Netherlands reached an electrification rate of 5%–6% of their new car fleet in 2013 (0.4% in the EU-28). Italy has reached a 12%–14% proportion of gas-fueled vehicles in the national fleet in recent years. According to ICCT (2016), for 2015, new car sales in the EU-28 were 1% electric (plug-in and battery electric vehicles), the United Kingdom reached 1.2%; the Netherlands 9.7%; Norway 22.4%, and gas-fueled cars in Italy 11.1%. According to the same AIRUSE (2015a) report, major actions leading to this achievement in Norway started in 1990, which included (1) long-term fiscal incentives, including the exemption from vehicle registration tax, road tolls, and VAT (25%), reduced annual tax and reduced rates on the main coastal ferries; (2) bus lane access; and (3) incentives were added sequentially until the market responded, this happened when the price difference between battery electric and gasoline cars reached only about €1000. In the case of Italy, the long-term willingness of car manufacturers and the associated transformation industry to convert gasoline into gas engines, as well as long-term, nationwide support for using and distributing gaseous fuels, seems to have facilitated the rise in the proportion of gas-fueled vehicles, although this increase seems to have slowed in the last few years. Although it seems that the current technology available for hybrid, plug-in hybrid, and electric vehicles will probably facilitate an increase in their proportion of national fleets, this should be considered as a medium- to long-term strategy. Furthermore, it is important to state that efforts and incentives to support this "cleaner fleet" transformation should focus particularly on taxis and UFD fleets (see below) because the daily distance traveled may be around 5- to 12-fold higher than private commuting cars and as a consequence the emission reduction achieved is proportionally much higher. Furthermore, there is a good opportunity for reducing PM and NOx emissions by electrifying motorcycle fleets, especially in some Italian and Spanish cities where these vehicles might reach 15%–30% of the circulating urban fleets.
- The most widely used action to ban urban use of highly polluting cars, as part of a strategy to support the shift to a cleaner urban fleet, is the implementation of low emission zones (LEZs). This measure implements a control system for vehicles, usually based on camera readings of vehicle plates (ANPR) or by eco-labeling of vehicles, allowing the identification of the age and fuel type of the vehicle, and the restriction of the most polluting ones, such as pre-EURO 4 diesel light-duty vehicles, pre-Euro III heavy-duty diesel vehicles, and pre-EURO 3 light-duty gasoline vehicles. Nowadays, around 280 LEZs have been implemented in Europe (UAR, 2017). However, there is still a

technical/scientific debate on the actual efficacy of these in abating air pollution. For detailed information of their efficiency we refer the reader to Morfeld, Groneberg, and Spallek (2014a and b), Holman, Harrison, and Querol (2015), and AIRUSE (2015b). In any case one should consider that the real-world emission of PM from a <2.0 L EURO 5 or 6 diesel passenger car at 30 km/h is only 2% of those from an equivalent EURO 1 cars according to Copert (Copert-EEA, 2017a, 2017b) emission factors. Thus, removing one EURO 1 diesel car would have a similar effect to eliminating 50 EURO 5 or 6 diesel cars, as far as PM is concerned. Also we have to take into account that if, as stated above, 70% of the urban ambient air $PM_{2.5}$ is made up of secondary material (PM formed in the atmosphere from gaseous precursors from regional and local origins), by implementing LEZs we are reducing the primary PM emissions (directly emitted particulate material) from the engines, the magnitude of the LEZ effect on bulk ambient air $PM_{2.5}$ concentrations cannot be not very high (Holman et al., 2015). For PM_{10} the effect would be also reduced due to the lack of effectiveness on non-exhaust (coarse) emissions. The impact of the LEZs can be evaluated specifically for the abatement of diesel soot by monitoring BC or EC (indirectly representing the levels of diesel soot and PM from biomass burning emissions in urban areas, Reche et al., 2011). In spite of the technical difficulties of evaluating the efficacy of LEZs, and the fact that many published studies did not adopt strict statistical analysis (Morfeld et al., 2014a and b), Holman et al. (2015) reported that the largest effect of LEZs was evidenced for EC/BC with decreases of ambient air concentrations up to 9% of ambient concentrations in 19 German LEZs, whereas very little effect or no effect was evidenced for ambient air NO_2, PM_{10}, and $PM_{2.5}$. One should also take into account that these studies were conducted with relatively old air quality data and that EURO 5 (2009−15) and EURO 6 diesel cars emit very low soot levels. Thus any LEZ restricting pre-Euro 5 vehicles is likely to have a greater impact than the earlier LEZs assessed in Holman et al. (2016). It should be noted that diesel soot was declared a Grade 1 carcinogen by WHO in 2014 (IARC, 2012). For NO_2 the LEZ effect is expected to be very low because EURO 5 diesel cars (2009−15) emit around 67% the NOx of an equivalent EURO 1 car (from 1992) (Copert-EEA, 2017a), but probably higher primary NO_2, because the primary NO_2 proportion in NOx has increased by a factor close to 2.5 in the newer diesel cars (Carslaw et al., 2016). There is still an uncertainty regarding the real-world emissions of NOx and NO_2 of EURO 6 diesel cars (Tate, 2016). In any case, Table 11.1 shows the main features of the existing LEZs in Europe according to Holman et al. (2015), updated using UAR (2017) data. Holman et al. (2015) reported that probably the more effective LEZs were those applying restrictions to all vehicles, and not only to buses and heavy- and light-duty commercial vehicles.

- Taxation of fuels and vehicles in a number of EU member states has been applied to support the growth of the diesel share in the national fleets as demonstrated by Fig. 11.7, which shows examples of fuel duties and diesel/gasoline/other car sales in selected countries (AIRUSE, 2015c). It is evident that this policy had a clear influence in countries such as Greece, Portugal, Spain, Luxemburg, and Belgium, but not in others such as the Netherlands, Denmark, and Finland, where, in spite of the high taxes for gasoline, the diesel proportion of new car sales did not increase proportionally. The United Kingdom has the same tax on both gasoline and diesel, and the highest diesel

TABLE 11.1   Main Features of the Existing Low Emission Zone (LEZ) in Europe According to Holman et al. (2015), Updated Using UAR (2017) Data

Country	Number of LEZs	Applicable Vehicles	National Framework
Austria	7	Heavy goods vehicles (HGVs)	Yes
Denmark	4	HGVs + buses	Yes
Finland	1	Buses/refuse trucks	No
France	2	HGVs	No
Germany	73	All four or more wheeled vehicles	Yes
Greece	1	All vehicles	No
Italy	Approximately 100[a]	Various	No
The Netherlands	13	HGVs	Yes
Portugal	1	Cars and HGVs	No
Sweden	8	HGVs + buses	Yes
UK	5	Various	No

[a]Excludes large number of LEZs in communities in Lombardy region.
http://urbanaccessregulations.eu and Holman et al. (2015).

tax in the EU compared with other countries. Italy, Norway, and the Netherlands have devised successful policy actions to favor the growth of the gas (Italy) and electric fleets (Norway and the Netherlands). To reduce the growth of the proportion of diesel cars, avoiding preferential duty reductions for diesel fuel as well as for diesel cars is strongly recommended, at least until the technology allows production of low-emission diesel cars. Also the annual city taxes for vehicles followed in the policies of many countries favor diesel cars based on the relatively lower CO_2 emissions/km compared with gasoline. It is recommended to use not only CO_2 but also locally acting urban pollutants for evaluating the eco-efficiency of cars for annual taxation.

2.4 Measures on City Logistics and Taxis

While the contribution of UFD traffic (in terms of vehicle—km) can be of 16% (London) or 18% (Italy) of the total urban traffic, the contribution of emissions can be magnified by a factor of three; and a large proportion of ambient concentrations of pollutants can arise from this sector (see for example, France in Table 11.2, LET & Aria Technologies, 2000).

Major logistic operators have newer and cleaner fleets and have been shown to be more efficient in unitary emissions (Antún, 1998). Only in the last decade UFD fleets have started going green, especially with the introduction of electric tricycles in European city centers. Their productivity is higher than regular motorized vans or trucks because the tricycles park at the door and they can drive in pedestrian zones regardless of the traffic direction

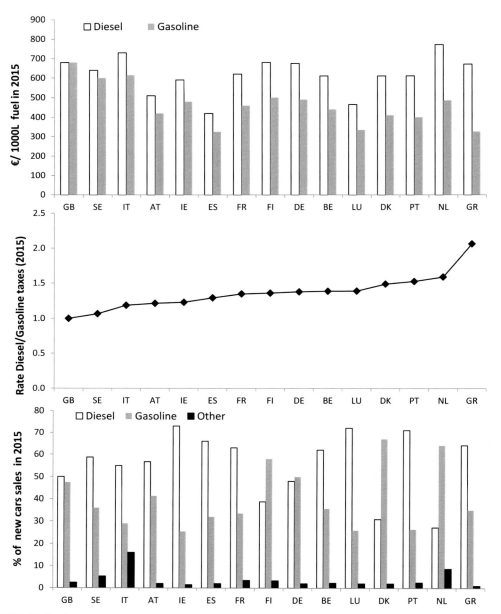

FIGURE 11.7   Comparison of taxes for gasoline and diesel vehicles (2015) and percentages of types of new car sales (2014) in selected countries from EU-28, according to AIRUSE (2015c).

TABLE 11.2 Contribution (%) of Good Transport to the Emission Inventory of France and That of the Urban Freight Distribution to Urban Emissions of Motorized Traffic of Three French Cities (LET & Aria Technologies, 2000)

	CO	NOx	HC	PM	SO2	CO2
PERCENTAGE OF EMISSIONS FROM GOOD TRANSPORT TO NATIONAL EMISSIONS						
France	13	37	16	48	33	27
PERCENTAGE OF EMISSIONS FROM URBAN FREIGHT DISTRIBUTION TO URBAN EMISSIONS OF MOTORIZED TRAFFIC						
Bordeaux	7	15	9	2	18	15
Dijon	9	24	12	42	25	20
Marseille	13	35	16	32	44	26

of the streets. As their capacity is rather small (about 200 kg or 1 m³), they need a mobile depot or logistics platform to feed them.

Although freight logistics policy in cities is pretty old (Julius Caesar passed the *Lex Iulia Municipalis* in 45 BC in Ancient Rome), it has been the "Cinderella" of urban mobility. Because operating costs decrease with the average speed (for the low speeds expected in a city), congestion, street directions, and parking regulations make very expensive to overcome the "last mile." Moving a container from New Jersey to New York (1 mile) has a cost of $150, the same cost as a journey between Connecticut and Ohio (500 miles) (Holguin et al., 2012).

The city logistics sector has many stakeholders, all of them with their particular view, needs, and constraints, which make it difficult to plan and regulate the sector. The main actors include the following:

- City planners. They are usually "offside of game" when considering service, functionality, flows, and economic issues that shape stakeholder's behavior. In general, streets were not conceived for UFD and loading/unloading (L/U) zones are a must.
- Traffic engineers. They are traditionally concerned with traffic flow and street safety. Only recently have they started considering a "democratization of streets" and to learn the causes (not only the effects) of UFD behavior.
- Manufacturers. They usually subcontract transport and use it as a commercial "gift." They like to ship just in time (JIT) without minimum economic order quantity (EOQ) and are against consolidating suppliers (due to strategic or commercial interests).
- Commerce. They undertake an "improvised" *outsourcing* of logistics (time windows, "JIT," zero stock, EOQ = 0). The small and medium enterprises are against UFD during off-peak hours or at night, but with the unattended off-hour delivery implemented in New York (Holguin et al., 2012), their concerns can be overcome.
- Police. There are not enough resources to control short stops. Because logistics operators are usually paid per delivery, they have the maximum interest in not spending more time at each stop than that necessary. That is why in Barcelona, city logistics parking time in L/U zones has been regulated in the last decade with two cardboard circles showing the arrival time and allowing for 30 min stop time. Control of the time parked

is undertaken by the regular on-street parking regulators. The Institute of Transport Studies (University of Leeds) has substituted the primary but effective parking policy regulation by an app that controls the parking time (in the future, that time will be charged).

- Neighbors and street users. They bear the effects of L/U stops and UFD operation: noise, barrier effect, pollution, accidents. They are usually against UFD at night because of the noise, but now we have silent electric vehicles.
- City Hall. Until recently, many cities had vague municipal ordinances, low training and internship, and concentrated on law enforcement. Logistic platforms (fixed or infrastructural in every market and quarter, or mobile) need to be considered as public facilities.
- Vehicle manufacturers. In Europe, they have focused on ergonomics, driving, and fuel consumption (and emissions). But only recently are they considering handling tools and possible intermodality.
- Lawyers. They are used to 0–1 decisions that need to be proved, but for an agile UFD we need global vision and adequate laws for expected behaviors. We also need a functional interpretation of laws and modern laws (e.g., "delivery receipt signed"… w/o checking the delivery content).
- Transportation operators. The sector is considered a victim of the system. They want L/U at the door by productivity in time, comfort in handling (moving goods), and because the existing L/U zones or spots are scarce and may be crowded or full. The sector uses an improvised outsourcing of logistics when transportation is considered as a commercial gift. As most of the operators are small (atomization), there is limited sector coordination and public intervention is recommended.

Geroliminis and Daganzo (2005) review implementations of the "green" city logistics around the world, classifying the policies as follows:

- Restrictions zones (Copenhagen, the United Kingdom, Sweden, Brussels).
- Clean vehicles (Rotterdam, Osaka, Zurich).
- Coordinated transport (Berlin, Stockholm).
- Congestion mitigation (Barcelona, Paris, Rome).
- Charging (London, Germany).
- Information systems (New York, Tokyo, Vancouver).
- Water use (Amsterdam, Venice).

Macário, Galelo, and Martins (2008) propose a different classification scheme for city logistics policies:

- Legislative and organizational measures: Cooperative logistic systems, encouraging night deliveries, public–private partnerships, and intermediate delivery depots.
- Access restriction measures: Access restrictions according to vehicle characteristics (weight or volume), conditional access to pedestrian areas, urban tolls, and periodic restrictions.
- Territorial management measures: Creation of loading and unloading areas, load transfers, and mini logistic platforms.
- Information-assisted measures: GPS, tracking and tracing systems, route planning software, intelligent transport systems, adoption of nonpolluting vehicles, and vehicles adapted to urban characteristics (size and propulsion).

- Infrastructural measures: Construction of urban distribution centers and peripheral storing facilities, use of urban rail for freight (freight trams), and underground freight solutions.

This sector's measures follow the general policy actions (cleaner vehicle fleets, elimination of highly polluting cars of the circulating urban fleet, and favoring the decrease of current circulating diesel vehicles), but are enhanced by "hot topics" such as drop boxes for asynchronous reception and delivery, cargo drones, modal change (when feasible), mobile depots, collaborative deliveries to share the "last mile," off-hour deliveries, etc.

Many of the problems, limitations, and measures listed above for UFD are also true for the taxi sector. Electrification and hybridization, reduction of the mileage of empty taxis, increase in number of the taxi stations, and apps for the easy booking of taxis or for sharing taxis might have also very positive effects in reducing emissions.

2.5 Measures for the Urban Transformation of the City

These are measures not only intended to expand the pedestrian, cycling, and green areas and to reduce the areas used by road traffic, but also to divert citizens from highly trafficked roads to reduce their exposure to urban pollutants; especially those more vulnerable to atmospheric pollution, such as elderly people and children. It is important to note that if these measures are applied before those described above, they are liable to move the pollution from one place to another, or, even worse, might concentrate traffic in specific streets and cause exceedances of the limit values in areas where these were not recorded before.

These measures include all kinds of transformation of the urban structure following the objectives listed above. These include measures such as (1) improving urban cycling networks and extending cycle lanes, including their width and separation from traffic lanes; (2) transforming road areas into pedestrian and green areas; (3) modifying traffic circulation patterns to restrict the number of vehicles on routes that cross a specific area of the city; (4) implementing air quality policies for the location of new hospitals, medical care services, playgrounds, schools, retirement homes, and other facilities used by citizens vulnerable to the effects of air pollution. This could be through regulating a minimum distance (50–150 m, for example) from highly trafficked roads (for example, those >10,000 vehicles/day); and (5) reducing traffic intensities around the above facilities; among others. As an example of the effect of the greening of cities on decreasing air pollution, Fig. 11.8 displays results of the BREATHE-ERC-AG project (Sunyer et al., 2015), which demonstrates a high and positive cross-correlation between 15 days average levels of BC recorded indoors in classrooms of 39 schools of Barcelona with the percentage of area used by road traffic in the neighborhood of each school, as well a clearly negative cross-correlation with the percentage of the same sector occupied by parks (Reche et al., 2015). Fig. 11.8 shows also an example of measurements performed at different distances from the curbside of highly trafficked streets in Barcelona displaying a sharp decrease of BC levels recorded in the first 10 m from the curbside and the relatively slow decrease from 50 m onward, in the case of two major streets. This has relevant implications for urban planning, concerning the location of safe school paths, cycling lanes, and other logistics in a city. The shape of these BC exposure/distance to curb trends depends mainly on traffic density and fleet composition, canyon or open

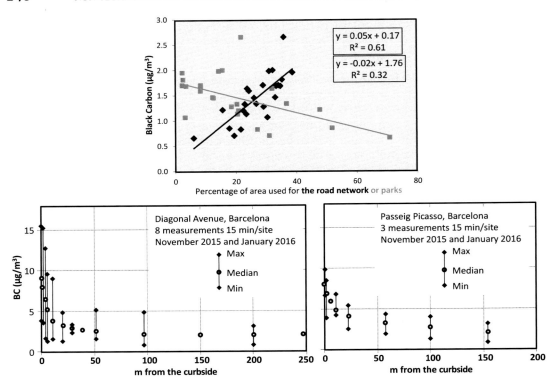

FIGURE 11.8   Top: Cross-correlation between 15 days indoor EC (elemental carbon) concentrations in 39 Barcelona schools and the percentage of space used by traffic and parks in their respective neighborhoods (data from Reche et al., 2015). Bottom: Measurements of BC (black carbon, equivalent to EC) at different distances of the curbside of two major inner streets in Barcelona.

patterns of the street, wind speed and direction, and the width of the street; and might vary considerably across the city.

One of the major transformations of an urban neighborhood is represented by the introduction of a "superblock" system. This strategy has been implemented in cities such as Barcelona and Vitoria in Spain and Vancouver in Canada (Rueda, 2017). In the case of Barcelona, the strategy is based on the definition of a new urban cell, defined by a primary road network that connects the origins and destinations of the whole city, without compromising the functionality of the urban system. The inner streets constitute a local or secondary road network with traffic speed limited to 10 km/h. Traffic cannot drive through the superblock, only around or within it. This means that the majority of vehicle movements inside the superblock are local trips, and overall traffic levels are reduced. Therefore, around 70% of the space that was previously occupied by traffic is now converted into pedestrian and cycle lanes. Fig. 11.9 shows schematically the strategy of the superblock urban cells. Currently measurement studies are being devised to evaluate their effect on air quality.

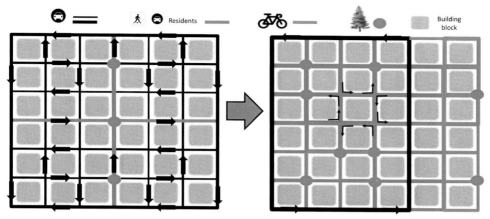

FIGURE 11.9 The concept of the "superblock transformation." *Based on Rueda (2017).*

2.6 Examples of Reremediation Measures

2.6.1 Road Sweeping and Washing

Application of dust suppressants to inhibit road dust suspension: Besides driving speed, traffic volume, and composition, other factors such as road surface materials and conditions heavily affect road dust emissions. In addition, in Scandinavian countries, road sanding and the use of studded tires in winter–spring months provoke exceedances of the PM_{10} daily limit value. In Southern Europe, the small covering by vegetation, urban works, and outbreaks of desert dust can increase road dust loadings and therefore emissions. Therefore, due to the difference in climatic and road conditions, appropriate measures to reduce road dust emissions differ from one region to another.

Measures can be either preventative or mitigating. Preventative strategies aim to avoid the build-up of particles on road surfaces in the first place, such as reducing wear (see Chapter 12), covering truck loads, or road traffic restrictions. Mitigating measures attempt instead to remove or bind those particles already deposited; and these include (1) road sweeping, (2) road water washing; (3) use of dust suppressants; (4) use of porous asphalt; and (5) limitations in the use of studded tires and road sanding.

Road sweeping is generally performed to improve the esthetic appearance of a street by eliminating debris, litter, and dirt so that, generally, neither road sweepers nor their operational procedures are designed to reduce ambient PM_{10} concentrations. However, some studies have evaluated possible beneficial effects on air quality of existing sweeping technology. The sweeper technology has changed relatively little over the past few decades with the exception of the use of bag filters to control fugitive PM_{10} emissions from the dust collected and diesel particulate filters to control exhaust emissions from the vehicle engine, which have recently been introduced to the market (Wiemann, 2013). In the market, there are three main categories of sweepers (mechanical, vacuum, and regenerative air vacuum, AIRUSE, 2014).

There is evidence that the use of some road sweepers can increase local PM_{10} concentrations close to an operating vehicle. Water sprays can be used with vacuum sweepers to reduce the resuspension of dust. A comparison of different road sweepers suggests that they may have different impact on PM_{10} emissions as much as three orders of magnitude (Geddes, 2011). Innovative vehicles include a roots vacuum pump, specially developed nozzles for road surfaces, and a unique filter technology. Dust-free performance is achieved due to brushes being under vacuum throughout the whole sweeping and cleaning operation (DISAB Group, 2013). Other manufacturers have developed a system which creates a cyclone effect within the hopper, for the most efficient filtering via mesh screens, of dust and debris particles prior to discharge to the atmosphere. This range was the first European-manufactured sweeper to achieve full PM_{10} test compliance under the Californian regulations.

There exist a number of PM_{10} certifications and tests for road sweepers (but no ISO standard) aimed at testing the performance of road sweepers in reducing road dust loading, minimizing the dust reentrainment and/or outlet emissions during operation. Details can be found in AIRUSE (2014).

The studies existing in the literature show no evidence of immediate benefits of sweeping on air quality. After sweeping procedures, some of these studies registered an increase or no difference in PM_{10} levels in the United States, Swedish, Norwegian, and German studies (Chow, Watson, Egami, Frazier, & Lu, 1990; Norman & Johansson, 2006). In other studies, although all quite dated, a reduction of PM concentrations was observed (Amato, Querol, Johansson, Nagl, & Alastuey, 2010; Cuscino, Muleski, & Cowherd, 1983; Cowherd, 1982; Fitz, 1998; Fitz & Bumiller, 2000; Hewitt, 1981; Kantamaneni et al., 1996), but none of these studies conclusively demonstrated the effectiveness of sweeping on reducing suspended PM_{10} in the short term. However, this does not mean that street sweeping is not beneficial for air quality in the long term.

Water washing can reduce the mobility of dust load deposited on street surfaces in the same way as rain and is therefore a potentially effective measure for mitigating dust resuspension. When water adheres to the deposited particles, it increases their mass and surface tension forces, decreasing the possibility of suspension and transport, especially as cohesion of wetted particles often persists after the water has evaporated due to the formation of aggregates. It is less likely that water flushing could carry particles below 10 μm into the sewage. Water flushing was found to reduce the road dust mobility by >90% in Spain and 60%—80% in Germany depending on the particle size and methodology applied (Amato et al., 2009; Ang et al., 2008).

Water flushing can be integrated in a street sweeper or manually applied by means of hosepipes. Phreatic and nondrinking water should ideally be used for street washing purposes. The amount of water may vary according to street dirtiness (road dust loading). Although reported in only a few studies, an application rate of at least 1 L/m^2 is recommended. The recovery of road dust emissions to atmosphere after street washing reaches on average 99% after 24 h in Spain and 72 h in the Netherlands (Amato et al., 2012, Fig. 11.10), as the moistening effect of rain is more important than actual particle removal (wash off). Moisture evaporation is the main process controlling the recovery of road dust emissions. Based on these results, it can be concluded that the maximum benefit occurs

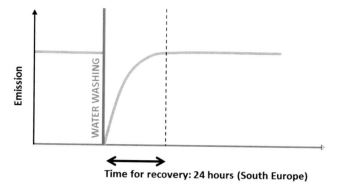

FIGURE 11.10 Schematic effect and duration of street washing on road dust emissions.

during the morning peak emission period (07–09 h) when road washing is performed in the early morning hours (5–6 a.m.) (AIRUSE, 2014).

Ideally, washing should be performed whenever road dust emissions increase significantly. This is closely related to the degree of road moisture. Frequency should therefore be optimized taking into account (1) local resources, (2) rain forecast (washing should be avoided during and 24 h after rain), and (3) episodic intrusion of dust (from the Sahara desert, for example, or local sources).

For paved roads, the higher the road dust loading, the higher the efficiency of street washing. Within an urban environment, road dust emissions may vary widely. In Barcelona, for example, a large variation of road dust loadings was found, from 2 to 23 mg/m^2, indicating that road characteristics (pavement, fleet composition, proximity to unpaved areas or construction areas) substantially affect local emissions. However, traffic volume is also crucial for the total emissions (per day, for example). Therefore, if not all roads can be washed due to a lack of resources, a careful selection of roads should be performed to optimize the cost–benefit ratio.

The efficacy of road washing depends highly on several factors such as (1) climatic conditions, (2) road dust loadings, (3) frequency of washing, (4) road surface material, (5) portion and length of the road that is washed, and (6) relevance of other sources of PM.

The efficiency of water washing has been quantified in absolute or relative terms in a number of studies worldwide. A PM$_{10}$ reduction of about 2 µg/m^3 of the daily mean was found in Düsseldorf, where water flushing was performed twice a week on a busy road (John, Hugo, Kaminski, & Kuhlbusch, 2006). In Stockholm, street washing was performed to the verge next to the carriageway of a highway and only on days with favorable weather conditions (Norman & Johansson, 2006). The mean decrease was 6%, which occurred during the morning hours. Two out of 12 exceedances of 50 µg/m^3 measured at the control station were not measured at the street washing site. Also in the Netherlands (Nijmegen) a beneficial effect was found even though its effectiveness varied depending on local conditions such as road pavement, meteorology, and solubility of particles (Keuken, Denier van der Gon, & van deer Valk, 2010). In Madrid, Karanasiou et al. (2011) found that during a 1-month campaign, the mass contribution from the road dust source was ∼2 µg/m^3 lower during the days that street washing was implemented, corresponding to a reduction of 15% of road dust mass contribution during the days that the road surface was left untreated. The effect of pressurized washing on PM_{10} emissions from the street surface was tested within the KAPU project in Finland (Kupiainen et al. 2011). Emission levels were found to be 15%—60% lower after washing than before. The effect was found to be highest immediately after treatment and to be dependent on the water pressure, volume of water used, and the orientation of the nozzles in the pressure washer. The KAPU project found that the efficacy of cleaning is not only dependent on the efficiency of the cleaning equipment, but also on the frequency of cleaning and the amount of road dust. They concluded that street cleaning would only be effective where there are high street dust levels, and street dust is a dominant source of PM_{10}. In the Barcelona city center, the effect of a cleaning procedure consisting of manual washing after a vacuum assisted broom sweeping was tested. Averaging the daily mean concentration of PM_{10} over the 24 h following each cleaning event, and on the rest of the dry days, a mean decrease of 3.7—4.9 $\mu g/m^3$ (7%—10% of curbside concentrations) was found (Amato et al., 2014).

Within the AIRUSE LIFE + project, street washing was also tested on industrial roads (AIRUSE, 2014). The interest in industrial sites was related to the higher PM emissions due to high road dust loadings. The paved road was located within the ceramic industrial cluster of L'Alcora (Spain). Road dust loadings are within 20—40 mg/m^2, much above the general range in European cities of 1—6 mg/m^2. Given such high road dust loadings, street washing was intensive (27 L/m^2 phreatic water flow was used). Daily mean PM_{10} concentrations decreased by 18.5% on the day of the cleaning (Fig. 11.11). However, this reduction was short lived, being reduced to only 2.2% during the day after cleaning. The main PM_{10} decrease occurred from 7 to 11 a.m., which corresponds to 5 h after the start of cleaning activities. Although short lived, this decrease is sufficient to produce a significant reduction in daily mean PM_{10} concentrations (Amato et al., 2016b).

Chemical dust suppressants have been used on unpaved roads and in the minerals industry to suppress dust for a long time in some countries. Since the 1990s, they have also been used on paved roads in Norway, both in tunnels and on open roads. It has only been more recently that their effectiveness at reducing road dust emissions has been more widely investigated in a number of European cities. There is very little published peer-reviewed literature

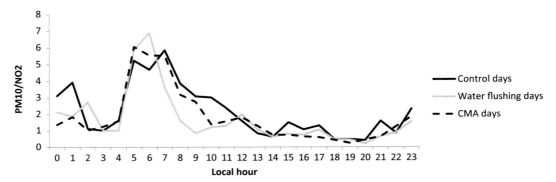

FIGURE 11.11 PM_{10}/NO_2 reduction due to street washing and calcium magnesium acetate (CMA) spraying at industrial paved road.

on the effect on PM_{10} concentrations of applying dust suppressants to paved roads in urban environments.

Consideration needs to be given to the potential environmental effects of the dust suppressants themselves as well as their benefits in reducing ambient PM_{10} concentrations. These include damage to vegetation and human health, and contamination of soil and ground water. Other potential effects include reducing road friction and corrosion to highway infrastructure such as bridges and the road surface.

Dust suppressants are sprayed onto the road surface, which binds the particles that come into contact with it and prevents them from becoming airborne when agitated by the wind, tire action, or vehicle turbulence. Dust suppressants also lower the freezing point of precipitation. In studies performed in the Nordic countries, it has been noted that efficient dust suppression requires repeated application and treatment over large areas. In general, tests of their efficacy have been undertaken on relatively short stretches of road.

Many different types of dust suppressants have been tested around the world, but in Europe the focus has been on salts, which absorb water when relative humidity exceeds 50% (i.e., hygroscopic compounds): $MgCl_2$, $CaCl_2$, calcium magnesium acetate (CMA), and potassium formate (referred to as KF in some publications). Both CMA and potassium formate have been used for deicing at airports because sodium chloride (NaCl) can corrode metal aircraft parts. Sugar has also been trialed as a dust suppressant in Scandinavia. CMA application on paved roads was recently tested in Sweden (Norman & Johansson, 2006), Austria (www.life-cma.at), Germany (Reuter, 2010) and the United Kingdom (Barratt, Carslaw, Fuller, Green, & Tremper, 2012). Other studies (though not in English) are available in Sweden, Finland, and Norway (Gustafsson, Blomqvist, Johansson, Norman, 2013; Gustafsson, Blomqvist, Jonsson, & Ferm, 2010). Positive results were obtained in Sweden and Austria (daily mean PM_{10} decreased by up to 35%) where road dust emissions represent a major contributor to PM_{10} levels, while studies in Germany and the United Kingdom could not detect a significant PM_{10} decrease on typical urban roads. When a PM_{10} decrease was found, this was short lived (a few hours), suggesting that the effectiveness of CMA in binding deposited particles is closely related to the degree of road moisture. In CMA and $MgCl_2$ applications on a typical urban road in Barcelona at a rate of $15-20 \, g/m^2$, no clear air quality benefit could be ascribed to the CMA or $MgCl_2$ application in the city of Barcelona. This inefficacy is likely due to the higher solar radiation (and consequent faster moisture decrease due to evaporation) of the Mediterranean climate in contrast with Sweden, Austria, Norway (for $MgCl_2$), and industrial sites in the United Kingdom. Böhner, Niemann-Delius, and Hennig (2011) stated that a prerequisite for this is that the air relative humidity should be at least 35%. In addition, it is noteworthy that where those studies that found a reduction induced by CMA or $MgCl_2$, the sites were characterized by very high road dust loads (studded tires, road sanding, and industrial dust), which increase the local emissions from the test road.

The discrepancy found with respect to Scandinavian/Alpine regions suggests that effectiveness of dust suppressants can be strongly influenced by local features such as the road dust emission strength (dustiness of the road, type of pavement, etc.), and the relative contribution of "treated" and "untreated" road dust emissions having impact on the receptor point. Solar radiation and road moisture are also believed to play a key role in determining the effectiveness of dust suppressants. For this reason, it is crucial to comprehensively evaluate the effectiveness in more severe scenarios in the Mediterranean region, such as industrial sites, where road dust loadings are comparable to those in countries where studded tires and road sanding are involved.

To discriminate whether road dust loading is a critical parameter in determining the effectiveness of CMA in the Mediterranean region, a paved industrial road with high road dust loadings (40 mg/m^2 for particles below 10 μm) allowed testing the effectiveness of CMA under highly polluted scenarios. Again, CMA efficacy was found to be lower than street washing (water only), which was estimated as 19%.

Although more research is needed, recent studies suggest that besides the wear resistance of mineral materials used for the road surface, other factors can reduce the road wear emission potential, including (1) increasing macrotexture, (2) road pavement construction type (e.g., incorporating rubber into the asphalt, porous asphalt), and (3) good operation and maintenance (See Chapter 10).

2.6.2 Application of Photocatalytic Products

Titanium dioxide (TiO$_2$)-based photocatalytic surfaces have been developed and tested over the two last decades, both at a laboratory scale and in the real world as a mitigation measure for reducing curbside NOx concentrations (Chen & Poon, 2009; Chen, Nanayakkara, & Grassian, 2012; Maggos et al., 2008; Ohama & Van Gemert, 2011; Schneider et al., 2014), but also for VOCs, PM, and O$_3$ (De Richter & Caillol, 2011; Pichat et al., 2000; Strini, Cassese, & Schiavi, 2005).

Two field trials in artificial model street canyons showed high NOx remediation of 25% to 30% (Fraunhofer, 2010) and 40% to 80% (Maggos et al., 2008; PICADA, 2006), respectively. These studies are in general agreement with two experiments in real urban street canyons for which NOx reductions of 19% (Ballari & Brouwers, 2013) and 26% to 66% (Guerrini & Peccati, 2007) were observed. In Italy, photocatalytic cementitious coating materials were applied on the side walls and on the ground surface of an artificial model street canyon test section, but no significant reduction of NOx, O$_3$, and VOCs and no impact on particle mass, size distribution, and chemical composition were observed in the field campaign.

The measured NOx reduction was below the measurement precision errors in recent field projects using photocatalytic noise protection barriers at Putten, the Netherlands (IPL, 2010), and Grenoble, France (Tera, 2009), in agreement with a recent study using active pavement blocks on the sideways of an urban street in Fulda, Germany (Jacobi, 2012). Reasons for these contradictory results are still under discussion, and further field studies are necessary to better assess the impact of this technology on air quality (Gallus et al., 2015). A theoretical study by the UK Air Quality Expert Group (AQEG, 2016) has shown that due to mass transfer limitations, the efficacy of photocatalytic paints in removing NO$_2$ is very low.

Besides the reduction of primary pollutants by photocatalytic surfaces, the potential formation of harmful by-products is another controversial issue. Although it is generally assumed that, for example, NOx is converted into nitrate (Laufs et al., 2010), some recent studies on pure photocatalysts and on self-cleaning window glass also observed the formation of the intermediate nitrous acid (HONO) (Gustafsson, Orlov, Griffiths, Cox, & Lambert, 2006; Monge, D'Anna, & George, 2010; Ndour et al., 2008), which is even more harmful than the primary reactants NO and NO$_2$ (Pitts, 1983). In addition, renoxification and O$_3$ formation originating from photocatalytic decomposition of adsorbed nitrate was recently observed in laboratory experiments (Monge, D'Anna, et al., 2010; Monge, George, et al., 2010). Harmful oxygenated reaction products, for example, aldehydes (Gallus et al., 2015) were also detected during the photocatalytic degradation of VOCs besides the expected end product CO_2.

2.7 Other Nontechnological Measures

Traffic speed management has often been used as a policy measure to improve safety as well as to reduce emissions from vehicles both on highways and in residential areas. The effect on air quality has studied using monitoring and/or modeling data focusing on exhaust missions.

In the Netherlands, speed management has achieved a PM_{10} emissions reduction by 5%–25% (Keuken et al., 2010). Reducing speed limits in residential areas from 50 to 30 km/h, modeled using real-life urban drive cycles, showed a significant decrease in exhaust PM emissions and a smaller reduction in NOx emissions from a small sample of light-duty vehicles, although it might cause light increases in CO and hydrocarbons (Int Panis et al., 2011). In Berlin, lowering the speed limit from 50 to 30 km/h, combined with effective enforcement, resulted in a measured drop in the traffic-related PM increment of 25%–30% (5% of total PM concentration). The traffic NO_2 increment, at the curbside, also fell by 15%–25% and the total concentration by 7%–12% (Lutz, 2013). It should be noted that these results were achieved by synchronizing the traffic lights to ensure smooth traffic flow at 30 km/h (i.e., 30 km/h as instantaneous vehicle speed, rather than average (stop-and-go) speeds) (Casanova & Fonseca, 2012). According to the Copert emission model, which uses average vehicle speeds, reducing traffic speed from 50 to 30 km/h increases both PM and NO_2 emissions (Int Panis et al., 2011).

Traffic flow optimization may, however, in the long term, lead to an increase of traveled kilometers due to a possible encouragement of the use of the private car. This issue needs to be managed in an integrated manner with mobility plans to ensure that speed limits deliver the expected mitigation of the air pollution also in the long term. In Barcelona, a speed limit of 80 km/h was implemented on highways entering the city in 2008, with a modeled emissions reduction up to 5%–8% and 3% for NO_2 and $PM_{10,}$ respectively; although the local effects in the adjacent area to the roadways reached a higher decrease (Baldasano, Gonçalves, Soret, & Jiménez-Guerrero, 2010). However, due to the low acceptation of this measure by the community, in 2011, a variable speed limit, between 40 and 100 km/h, was introduced as a function of pollution levels, traffic density, and meteorology (de Miguel, Minguillón, Viana, & Querol, 2013).

Fewer studies have been undertaken on the effect of speed management on non-exhaust emissions, and these are limited to road dust suspension. Norman et al. (2016) and Hagen, Larssen, and Schaug (2005) studied the effect of vehicle speed reduction on a Norwegian highway in 2004–06. Norman et al. (2016) applied an emission model to quantify the PM_{10} reduction, but the first year after the implementation was much drier than the previous, hampering the quantification of the effectiveness of the measure. Previous studies explored the variation of emission strength of a deposited tracer with the speed of one traveling vehicle, suggesting that reducing the vehicle speed may significantly reduce road dust emissions (Nicholson & Branson, 1990; Sehmel, 1967), but no studies have demonstrated experimentally the effectiveness of such a measure in real conditions. Lee, Kwak, Kim, and Lee (2013) used a mobile laboratory to track emissions to evaluate the concentration of roadway particles (not the emission factors, EF) at different speeds of the vehicle, finding an increase in the mean concentrations only from 80 to 110 km/h, but with high standard deviations. Pirjola et al. (2010) and Pirjola, Kupiainen, Perhoniemi, Tervahattu, Vesala (2009) found a linear increase in dust concentrations, measured behind the rear tire of two testing vehicles, exploring the range 50—80 km/h, but no EFs were calculated. Hussein, Johansson, Karlsson, and Hansson (2008) found an important dependence on vehicle speed, the particle mass concentrations behind a studded tire at 100 km/h being about 10 times higher than that at 20 km/h. However, the speed dependence could also be attributed to the use of studded tires. Etyemezian et al. (2003) and Zhu et al. (2009) found that on the same road (i.e., same dust loading) emissions increase with vehicle speed to the power of approximately 3. More recently, in Milan, Amato et al. (2017) investigated the impact of traffic speed on road dust emissions and calculated the emission factors at three sites along the same road with different instantaneous traffic speeds. EF values were 24.6, 40.9, and 48.4 mg/VKT for traffic speeds of 36, 47, and 57 km/h, respectively, which suggests that road dust resuspension increased with a power of 1.5 of the vehicle speed (Fig. 11.12). As result of this lack of information, most emission models do not enable non-exhaust emissions to be estimated as a function of vehicle speed (EPA, 2006; Kauhaniemi et al., 2011; Omstedt, Johansson, & Bringfelt, 2005) with the exception of the model of Denby et al. (2013), who specified a quadratic dependence on speed for spray emissions but did not specify it for road dust suspension, stating that the model remains uncertain and requires further refinement based on experimental studies.

Another measure for reducing non-exhaust PM emissions is to limit the number of heavy-duty vehicles in vulnerable areas, as there is a direct relationship between vehicle weight and emission potential (EPA, 2006); based on a literature review, trucks have 9—10 times higher emission factors for road dust resuspension than passenger cars (Schaap et al., 2009). This factor may also be applicable for the direct emissions from brake and tire wear, due to the higher weight and friction forces. Note that this is no longer the case for NOx. There is evidence that NOx emissions from Euro 6 HDVs are lower than those from Euro 6 diesel cars. Seems surprising but is due to Euro VI HDVs having to meet real-world driving NOx limits but the equivalent has not yet been introduced for new cars (Transport for London, 2015; UK Department of Transport, 2017a, 2017b).

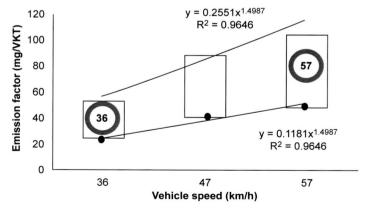

FIGURE 11.12   Relationship between fleet traffic speed and road dust emission factor in Milan. *Adapted from Amato et al., 2017.*

Acknowledgments

The present work has been supported by the LIFE Program (AIRUSE-LIFE + ENV/ES/584project) and by the Generalitat de Catalunya (AGAUR 2015 SGR33 and the DGQA). We would also like to express our gratitude to the city councils of Barcelona and Madrid for supplying data evaluated in specific parts of this chapter.

References

AIRPARIF. (2013). *Air quality in Paris region, 2012.* París: AIRPARIF, 26 pp. https://www.airparif.asso.fr/_pdf/publications/bilan-2012-anglais.pdf

AIRUSE. (2014). *LIFE+ project on testing and development of air quality mitigation measures in Southern Europe.* Mitigation measures in Southern Europe. Report B7-17, 128 pp. http://airuse.eu/wp-content/uploads/2013/11/13_B7-Report-on-mitigation-measures-in-South-Eur.pdf

AIRUSE. (2015a). *LIFE+ project on testing and development of air quality mitigation measures in Southern Europe.* Strategies to encourage use of electric, hybrid and gas vehicles in Central and Northern Europe. Report B8-18, 26 pp. http://airuse.eu/wp-content/uploads/2013/11/18_B8_Strategies-encourage-use-electric-hybrid-and-gas-vehicles-in-Cent-and-North-Europe-FINAL.pdf

AIRUSE. (2015b). *LIFE+ project on testing and development of air quality mitigation measures in Southern Europe.* Low Emissions Zones in Central and Northern Europe. Report B8-16, 35 pp. http://airuse.eu/wp-content/uploads/2013/11/16_B8_Low-Emissions-Zones-in-Central-AND-Northern-Europe.pdf

AIRUSE. (2015c). *LIFE+ project on testing and development of air quality mitigation measures in Southern Europe.* Taxation and pricing strategies to discourage the use of diesel vehicles in Central and Northern Europe. Report B8-17, 20 pp. http://airuse.eu/wp-content/uploads/2013/11/17_B8_Taxation-and-pricing-strategies-discourage-the-use-diesel-vehicles-in-Cent-and-Nort-Eur-FINAL.pdf

Ajuntament de Barcelona. (2017). *Statistical data on urban mobility of the city of Barcelona.* http://mobilitat.ajuntament.barcelona.cat/ca/documentacio?field_tipologia_documentacion_tid=12.

Amato, F., Alastuey, A., Karanasiou, A., Lucarelli, F., Nava, S., Calzolai, G., et al. (2016a). AIRUSE-LIFE+: A harmonized PM speciation and source apportionment in five southern European cities. *Atmospheric Chemistry and Physics, 16,* 3289–3309. https://doi.org/10.5194/acp-16-3289-2016.

Amato, F., Bedogni, M., Padoan, E., Querol, X., Ealo, M., & Rivas, I. (2017). Impact of vehicle fleet speed on road dust emissions. *Aerosol and Air Quality Research* (under review).

Amato, F., Escrig, A., Sanfelix, V., Celades, I., Reche, C., Monfort, E., et al. (2016b). Effects of water and CMA in mitigating industrial road dust resuspension. *Atmospheric Environment, 131,* 334–340.

Amato, F., Karanasiou, A., Cordoba, P., Alastuey, A., Moreno, T., Lucarelli, F., et al. (2014). Effects of road dust suppressants on PM levels in a Mediterranean urban area. *Environmental Science and Technology, 48*(14), 8069–8077.

Amato, F., Querol, X., Alastuey, A., Pandolfi, M., Moreno, T., Gracia, J., et al. (2009). Evaluating urban PM_{10} pollution benefit induced by street cleaning activities. *Atmospheric Environment, 43*(29), 4472–4480.

Amato, F., Querol, X., Johansson, C., Nagl, C., & Alastuey, A. (2010). A review on the effectiveness of street sweeping, washing and dust suppressants as urban PM control methods. *Science of the Total Environment, 408*(16), 3070–3084.

Amato, F., Schaap, M., Denier van der Gon, H. A. C., Pandolfi, M., Alastuey, A., Keuken, M., et al. (2012). Effect of rain events on the mobility of road dust load in two Dutch and Spanish roads. *Atmospheric Environment, 62,* 352–358.

Ang, K. B., Baumbach, G., Vogt, U., Reiser, M., Dreher, W., Pesch, P., et al. (2008). Street cleaning as PM control method. In *Poster presentation, better air quality, Bangkok.*

Antún, J. P. (1998). Escenarios de mitigación de emisiones del transporte de carga en el área metropolitana de la ciudad de México mediante operadores logísticos. In *Memorias del X Congreso Panamericano de Ingeniería de Tránsito y Transporte, Santander, Spain. Proceedings Published by Centro de Publicaciones, Secretaría General Técnica, Ministerio de Fomento, Madrid, Spain, M 34290-1998, 809 pp.*

AQEG. (2016). *Paints and surfaces for the removal of nitrogen oxides.* London: Air Quality Expert Group, Department for Environment, Food and Rural Affairs.

Baldasano, J. M., Gonçalves, M., Soret, A., & Jiménez-Guerrero, P. (2010). Air pollution impacts of speed limitation measures in large cities: The need for improving traffic data in a metropolitan area (2010). *Atmospheric Environment, 44*, 2997–3006.

Ballari, M. M., & Brouwers, H. J. H. (2013). Full scale demonstration of airpurifyingpavement. *Journal of Hazardous Materials, 254–255*, 406–414.

Barratt, B., Carslaw, D., Fuller, G., Green, D., & Tremper, A. (2012). *Evaluation of the impact of dust suppressant application on ambient PM$_{10}$ concentrations along 'hot spot' highway corridors*. London: Kings College London, University of London, 56 p, https://kclpure.kcl.ac.uk/portal/en/publications/evaluation-of-the-impact-of-dust-suppressant-application-on-ambient-pm10-concentrations-along-hot-spot-highway-corridors(1a7771b3-c030-49fc-acdb-c58b01625b4e)/export.html.

Böhner, R., Niemann-Delius, C., & Hennig, A. (2011). The use of salt solutions to reduce PM$_{10}$ on paved roads in open-pit mining [Der einsatz von salzlösungen zur PM$_{10}$-Minderung auf befestigten strafien im tagebau] *World of Mining - Surface and Underground, 63*, 275–279.

Borge, R., de la Paz, D., Lumbreras, J., & Vedrenne, M. (2014). Analysis of contributions to NO$_2$ ambient air quality levels in Madrid city (Spain) through modeling. Implications for the development of policies and air quality monitoring. *Journal of Geoscience and Environment Protection, 2*(1), 6–11.

Carslaw, D. C., Murrells, T. P., Andersson, J., & Keenan, M. (2016). Have vehicle emissions of primary NO$_2$ peaked? *Faraday Discussions, 189*, 439–454.

Casanova, J., & Fonseca, N. (2012). Environmental assessment of low speed policies for motor vehicle mobility in city centres. *Global NEST Journal, 14*(2), 192–201.

Chen, H., Nanayakkara, C. E., & Grassian, V. H. (2012). Titanium dioxide photocatalysis in atmospheric chemistry. *Chemical Reviews, 112*, 5919–5948.

Chen, J., & Poon, C.-S. (2009). Photocatalytic construction and building materials: From fundamentals to applications. *Building and Environment, 44*, 1899–1906.

Chow, J. C., Watson, J. G., Egami, R. T., Frazier, C. A., & Lu, Z. (1990). Evaluation of regenerative air vacuum street sweeping on geological contributions to PM$_{10}$. *Journal of Air and Waste Management, 40*, 1134–1142.

Copert-EEA. (2017a). *Emission factors for transport. The technical development of COPERT is financed by the European Environment Agency (EEA)*. http://naei.defra.gov.uk/data/ef-transport.

Copert-EEA. (2017b). *Software program which is developed as a European tool for the calculation of emissions from the road transport sector*. http://emisia.com/content/download-copert-v5.

Cowherd, C. (1982). Particulate emission reductions from road paving in California oil fields. In *75h annual meeting of the Air & Waste Management Association, Nashville, TN, 1982*.

Cuscino, T., Muleski, G. E., & Cowherd, C. (1983). Determination of the decay in control efficiency of chemical dust suppressants. In *Proceedings-symposium on iron and steel pollution abatement technology for 1982*. Research Triangle Park, NC: US Environmental Protection Agency.

De Richter, R., & Caillol, S. (2011). Fighting global warming: The potential of photocatalysis against CO$_2$, CH$_4$, N$_2$O, CFCs, tropospheric O$_3$, BC and other major contributors to climate change. *Journal of Photochemistry and Photobiology C: Photochemistry, 12*, 1–19.

DEFRA. (2015). *Draft plans to improve air quality in the UK*. Tackling nitrogen dioxide in our towns and cities UK overview document. UK, Department for Environment, Food & Rurakl Affairs September 2015, 51 pp. https://consult.defra.gov.uk/airquality/draft-aq-plans/supporting_documents/Draft%20plans%20to%20improve%20-air%20quality%20in%20the%20UK%20%20Overview%20document%20September%202015%20final%20-version%20folder.pdf

Denby, B. R., Sundvor, I., Johansson, C., Pirjola, L., Ketzel, M., Norman, M., et al. (2013). A coupled road dust and surface moisture model to predict non-exhaust road traffic induced particle emissions (NORTRIP). Part 1: Road dust loading and suspension modeling. *Atmospheric Environment, 77*, 283–300.

DISAB Group. (2013). *DISAB's street fighter promises a cleaning revolution*. http://disab.com/wp-content/uploads/2012/03/DISAB-s-street-fighter-promises-a-cleaning-revolution1.pdf.

EC. (1999). *1999/30/EC of the Council of 22 April 1999 relating to limit values for sulphur dioxide, nitrogen dioxide and oxides of nitrogen, particulate matter and lead in ambient air OJ L 163/41-60, 29.6.1999*.

EC. (2008). *2008/50/EC of the European Parliament and of the Council of 21 May 2008 on ambient air quality and cleaner air for Europe OJ L 152, 11.6.2008*. pp. 1–44.

EEA. (2016). *Air quality in Europe 2016*. European Environmental Agency, EEA Report 28/2016, ISBN 978-92-9213-824-0. https://doi.org/10.2800/413142. http://www.eea.europa.eu/publications/air-quality-in-europe-2016.

EMEP. (2016). *Air pollution trends in the EMEP region between 1990 and 2012*. EMEP/CCC-Report 1/2016, O-7726. ISBN : 978-82-425-2833-9 (printed), ISBN : 978-82-425-2834-6 (electronic) https://www.unece.org/fileadmin/DAM/env/documents/2016/AIR/Publications/Air_pollution_trends_in_the_EMEP_region.pdf.

EPA. (2006). *AP-42. Compilation of air pollutant emission factors* (5th ed.). Research Triangle Park, NC: U.S. Environmental Protection Agency.

Etyemezian, V. H., Kuhns, J., Gillies, Chow, J., Hendrickson, K., McGown, M., et al. (2003). Vehicle-based road dust emission measurement (III): Effect of speed, traffic volume, location, and season on PM_{10} road dust emissions in the Treasure Valley, ID. *Atmospheric Environment, 37*, 4583−4593.

Fitz, D. R. (1998). *Evaluation of street sweeping as a PM_{10} control method*. South Coast Air Quality Management District, Contract No. US EPA-AB2766/96018, 15−19.

Fitz, D. R., & Bumiller, K. (2000). Determination of PM_{10} emission from street sweepers. *Journal of the Air and Waste Management Association, 50*, 181−187.

Fraunhofer. (2010). *Clean air by airclean®*. http://www.ime.fraunhofer.de/content/dam/ime/de/documents/AOe/2009_2010_Saubere%20Luft%20durch%20Pflastersteine_s.pdf.

Gallus, M., Ciuraru, R., Mothes, F., Akylas, V., Barmpas, F., Beeldens, A., et al. (2015). Photocatalytic abatement results from a model street canyon. *Environmental Science and Pollution Research, 22*, 18185−18196.

Geddes, V. (2011). Particulate matter and fine dusts situation in Europe. In *Presentation to Eventopro, 26 and 27th May 2011, Lago di Garda*.

Geroliminis, N., & Daganzo, C. F. (2005). *A review of green logistics schemes used in cities around the world. Working paper. Center for Future Urban Transportation. UCB-ITS-VWP-2005-5*. UC Berkeley Centre for Future Urban Transport, 21 pp. http://www.its.berkeley.edu/sites/default/files/publications/UCB/2005/VWP/UCB-ITS-VWP-2005-5.pdf

Guerreiro, C. B. B., Foltescu, V., & de Leew, F. (2014). Air quality status and trends in Europe. *Atmospheric Environment, 98*, 376−384.

Guerrini, G. L., & Peccati, E. (2007). Photocatalytic cementitious roads for depollution. In *International RILEM symposium on photocatalysis, environment and construction materials, Florence* (pp. 179−186).

Gustafsson, M., Blomqvist, G., Johansson, C., & Norman, M. (2013). *Driftåtgärder mot PM_{10} på Hornsgatan och Sveavägen i Stockholm - utvärdering av vintersäsongen 2011−2012*. VTI−Rapport 767 (in Swedish with English abstract).

Gustafsson, M., Blomqvist, G., Jonsson, P., & Ferm, M. (2010). *Effekter av dammbindning av belagda vägar*. VTI-Rapport 666 (in Swedish with English abstract).

Gustafsson, R. J., Orlov, A., Griffiths, P. T., Cox, R. A., & Lambert, R. M. (2006). Reduction of NO_2 to nitrous acid on illuminated titanium dioxide aerosol surfaces: Implications for photocatalysis and atmospheric chemistry. *Chemical Communications*, 3936−3938.

Hagen, L. O., Larssen, S., & Schaug, J. (2005). *Environmental speed limit in Oslo. Effects on air quality of reduced speed limit on RV4*. NILU, Kjeller (OR 41/2005) (In Norwegian) http://www.nilu.no/Publikasjoner/tabid/62/language/en-GB/Default.aspx.

Hewitt, T. R. (1981). *The effectiveness of street sweeping for redicing particulate matter background concetrations*. Research Triangle Park, NC: Sirine Evironmental Consultants.

Holguin, J., Ozbay, K., Kornhauser, A., Ukkusuri, S., Brom, M. A., Iyer, S., et al. (2012). Overall impacts of off-hour delivery programs in the New York city metropolitan area: Lessons for European cities. In *Association for European Transport and Contributors European Transport Conference, Glasgow, UK*. https://abstracts.aetransport.org/paper/index/id/3879/confid/18.

Holman, C., Harrison, R. M., & Querol, X. (2015). Review of the efficacy of low emission zones to improve urban air quality in European. *Atmospheric Environment, 111*, 161−169.

Hussein, T., Johansson, C., Karlsson, H., & Hansson, H.-C. (2008). Factors affecting non-tailpipe aerosol particle emissions from paved roads: On-road measurements in Stockholm, Sweden. *Atmospheric Environment, 42*, 688−702.

IARC. (2012). IARC: Diesel engine exhaust carcinogenic. *Central European Journal of Public Health, 20*(2), 120, 138.

ICCT. (2016). *European vehicle market statistics, Pocketbook 2016/17*. International Council on Clean Transportation Europe, 2017 http://www.theicct.org/european-vehicle-market-statistics-2016-2017.

Int Panis, L., Beckx, C., Broekx, S., De Vlieger, I., Schrooten, L., Degraeuwe, B., et al. (2011). PM, NOX, and CO_2 emission reductions from speed management policies in Europe. *Transport Policy, 18*, 32−37.

IPL. (2010). *Dutch air quality innovation programme concluded: Improved air quality with coating of titanium dioxide not demonstrated*. http://laqm.defra.gov.uk/documents/Dutch_Air_Quality_Innovation_Programme.pdf.

Jacobi, S. (2012). *NO$_2$−Reduzierung durch photocatalytisch wirksame Oberflächen? Modellversuch Fulda*. Hessisches Landesamt für Umwelt und Geologie. http://www.hlug.de/fileadmin/dokumente/das_hlug/jahresbericht/2012/jb2012_059-066_I2_Jacobi_final.pdf.

Jimenez, J. L., Canagaratna, M. R., Donahue, N. M., Prevot, A. S. H., Zhang, Q., Kroll, J. H., et al. (2009). Evolution of organic aerosols in the atmosphere. *Science, 326*, 1525−1529.

Johansson, C. (2016). The effectiveness of urban air quality actions in Stockholm. In *AIRUSE final Conference. Barcelona, 18−19 April, 2016*. http://airuse.eu/wp-content/uploads/2016/01/13_C-JOHANSSON_Stockholm.pdf.

Johansson, C., Burman, L., & Forsberg, B. (2009). The effects of congestions tax on air quality and health. *Atmospheric Environment, 43*, 4843−4854.

John, A., Hugo, A., Kaminski, H., & Kuhlbusch, T. (2006). *Ntersuchung zur Abschätzung der Wirksamkeit von Nassreinigungsverfahren zur Minderung der PM$_{10}$-Immissionen am Beispiel der Corneliusstraße, Düsseldorf*. Duisburg (in German): Institut für Energie und Umwelttechnik e. V.

Kantamaneni, R., Adams, G., Bamesberger, L., Allwine, E., Westberg, H., Lamb, B., et al. (1996). The measurement of roadway PM$_{10}$ emission rates using atmospheric tracer ratio techniques. *Atmospheric Environment, 30*(24), 4209−4223.

Karanasiou, A., Moreno, T., Amato, F., Lumbreras, J., Narros, A., Borge, R., et al. (2011). Road dust contribution to PM levels - evaluation of the effectiveness of street washing activities by means of Positive Matrix Factorization. *Atmospheric Environment, 45*, 2193−2220.

Kauhaniemi, M., Kukkonen, J., Härkönen, J., Nikmo, J., Kangas, L., Omstedt, G., et al. (2011). Evaluation of a road dust suspension model for predicting the concentrations of PM$_{10}$ in a street canyon. *Atmospheric Environment, 45*, 3646−3654.

Keuken, M., Denier van der Gon, H., & van deer Valk, K. (2010). *Science of the Total Environment, 408*, 4591−4599.

Kupiainen, K., Pirjola, L., Ritola, R., Väkevä, O., Viinanen, J., Stojiljkovic, A., et al. (2011). *Street dust emissions in Finnish cities − summary of results 2006−2010*. Helsinki: City of Helsinki, Environment Centre, Publication 5/2011.

Laufs, S., Burgeth, G., Duttlinger, W., Kurtenbach, R., Maban, M., Thomas, C., et al. (2010). Conversion of nitrogen oxides on commercial photocatalytic dispersion paints. *Atmospheric Environment, 44*, 2341−2349.

Lee, S., Kwak, J., Kim, H., & Lee, J. (2013). Properties of roadway particles from interaction between the tire and road pavement. *International Journal of Automotive Technology, 14*, 163−173.

LET & Aria Technologies. (2000). Balance of Urban Goods Movement in Bordeaux, Marseilles. 70 p. + annexes. 120 pp. (Cited in BESTUFSII (2006). Best Urban Freight Solutions II, Deliverable 5.1. Quantification of effects, EC 6th FP. 72 pp. http://www.bestufs.net/download/BESTUFS_II/key_issuesII/BESTUF_Quantification_of_effects.pdf).

Lim, S. S., Vos, T., Flaxman, A. A. D., Danaei, G., Shibuya, K., Adair-Rohani, H., et al. (2012). A comparative risk assessment of burden of disease and injury attributable to 67 risk factors and risk factor clusters in 21 regions, 1990−2010: A systematic analysis for the global burden of disease study 2010. *The Lancet, 380*, 2224−2260.

LRTAP. (2017). *The air pollutant emissions data viewer (LRTAP Convention) providing access to the data contained in the EU emission inventory report 2000-2014*. Under the UNECE Convention on Long-range Transboundary Air Pollution (LRTAP) as donwloaded on 24/01/2017 from http://www.eea.europa.eu/data-and-maps/data/data-viewers/air-emissions-viewer-lrtap.

Lutz. (2013). Mitigation strategies: Berlin, Germany. Chapter 5. In *Particulate matter. Environmental monitoring and mitigation*. Future sciences Ed.

Macário, R., Galelo, A., & Martins, P. M. (2008). Business models in urban logistics. *Ingeniería & Desarrollo, 24*, 77−96.

Maggos, T., Plassais, A., Bartzis, J. G., Vasilakos, C., Moussiopoulos, N., & Bonafous, L. (2008). Photocatalytic degradation of NOx in a pilot street canyon configuration using TiO$_2$-mortar panels. *Environmental Monitoring and Assessment, 136*, 35−44.

de Miguel, E., Minguillón, M. C., Viana, M., & Querol, X. (2013). Mitigation strategies: Barcelona, Spain. In *Particulate Matter. Environmental monitoring and mitigation* (Future Sciences ed.) (Chapter 7).

Monge, M. E., D'Anna, B., & George, C. (2010a). Nitrogen dioxide removal and nitrous acid formation on titanium oxide surfaces − an air quality remediation process? *Physical Chemistry Chemical Physics: PCCP, 12*, 8991−8998.

Monge, M. E., George, C., D'Anna, B., Doussin, J.-F., Jammoul, A., Wang, J., et al. (2010b). Ozone formation from illuminated titanium dioxide surfaces. *Journal of the American Chemical Society, 132*, 8234−8235.

Morfeld, P., Groneberg, D. A., & Spallek, M. (2014a). Effectiveness of low emission zones: Analysis of the changes in fine dust concentrations (PM_{10}) in 19 German cities. *Pneumologie, 68*, 173—186.

Morfeld, P., Groneberg, D. A., & Spallek, M. F. (2014b). Effectiveness of low emission zones: Large scale analysis of changes in environmental NO_2, NO and NOx concentrations in 17 German cities. *PLoS One, 9*, e102999.

Ndour, M., D'Anna, B., George, C., Ka, O., Balkanski, Y., Kleffmann, J., et al. (2008). Photoenhanced uptake of NO_2 on mineral dust: Laboratory experiments and model simulations. *Geophysical Research Letters, 35*, L05812.

Nicholson, K. W., & Branson, J. R. (1990). Factors affecting resuspension by road traffic. *The Science of the Total Environment, 93*(C), 349—358.

Norman, M., & Johansson, C. (2006). Studies of some measures to reduce road dust emissions from paved roads in Scandinavia. *Atmospheric Environment, 40*, 6154—6164.

Norman, M., Sundvor, I., Denby, B. R., Johansson, C., Gustafsson, M., Blomqvist, G., et al. (2016). Modelling road dust emission abatement measures using the NORTRIP model: Vehicle speed and studded tyre reduction. *Atmospheric Environment, 134*, 96—108.

Ohama, Y., & Van Gemert, D. (2011). *Application of titanium dioxide photocatalysis to construction materials.* State-of-the-Art Report of the RILEM Technical Committee 194-TDP, 48 pp. Springer, XII.

Omstedt, G., Johansson, C., & Bringfelt, B. (2005). A model for vehicle induced non-tailpipe emissions of particles along Swedish roads. *Atmospheric Environment, 39*, 6088—6097.

PICADA. (2006). *European PICADA Project, GROWTH Project GRD1-2001-40449.* Retrieved from http://www.picada-project.com/domino/SitePicada/Picada.nsf?OpenDataBase.

Pichat, P., Disdier, J., Hoang-Van, C., Mas, D., Goutailler, G., & Gaysse, C. (2000). Purification/deodorization of indoor air and gaseous effluents by TiO_2 photocatalysis. *Catalysis Today, 63*, 363—369.

Pirjola, L., Johansson, C., Kupiainen, K., Stojiljkovic, A., Karlsson, H., & Hussein, T. (2010). Road dust emissions from paved roads measured using different mobile systems. *Journal of the Air and Waste Management Association, 60*, 1422—1433.

Pirjola, L., Kupiainen, K. J., Perhoniemi, P., Tervahattu, H., & Vesala, H. (2009). Non-exhaust emission measurement system of the mobile laboratory SNIFFER. *Atmospheric Environment, 43*, 4703—4713.

Pitts, J. N., Jr. (1983). Formation and fate of gaseous and particulate mutagens and carcinogens in real and simulated atmospheres. *Environmental Health Perspectives, 47*, 115—140.

Querol, X., Alastuey, A., Pandolfi, M., Reche, C., Pérez, N., Minguillón, M. C., et al. (2014). 2001—2012 trends on air quality in Spain. *The Science of the Total Environment, 490*, 957—969. https://doi.org/10.1016/j.scitotenv.2014.05.074.

Reche, C., Querol, X., Alastuey, A., Viana, M., Pey, J., Moreno, T., Rodríguez, S., González, Y., Fernández-Camacho, R., Sánchez De La Campa, A. M., De La Rosa, J., Dall'osto, M., Prévôt, A. S. H., Hueglin, C., Harrison, R. M., & Quincey, P. (2011). New considerations for PM, Black Carbon and particle number concentration for air quality monitoring across different European cities. *Atmospheric Chemistry and Physics, 11*, 6207—6227.

Reche, C., Rivas, I., Pandolfi, M., Viana, M., Bouso, L., Àlvarez-Pedrerol, M., et al. (2015). Real-time indoor and outdoor measurements of black carbon at primary schools. *Atmospheric Environment, 120*, 417—426.

Reuter U. Mit CMA gegen Feinstaub? Beispiel Stuttgart. In: Vortrag beim internationalen Kongress "Innovativer Winterdienst-Feinstaubreduktion" am 30.09.2010 in Lienz.

Rueda, S. (2017). The superblock: A new urban cell for the construction of a new functional and urban model of Barcelona. In P. de La Cal, & J. Monclús (Eds.), *"Jaca landscapecity" Seie Estrategias urbanas y paisajisticas.* Prensas University of Zaragoza (In press).

Schaap, M., Manders, A. M. M., Hendriks, E. C. J., Cnossen, J. M., Segers, A. J. S., Denier van der Gon, H. A. C., et al. (2009). *Regional modelling of particulate matter for The Netherlands.* www.pbl.nl.

Schneider, J., Matsuoka, M., Takeuchi, M., Zhang, J., Horiuchi, Y., Anpo, M., et al. (2014). Understanding TiO_2 photocatalysis: Mechanisms and materials. *Chemical Reviews, 114*, 9919—9986.

Sehmel, G. A. (1967). Particle resuspension from an asphalt road caused by car and truck traffic. *Atmospheric Environment, 7*, 291—309.

Strini, A., Cassese, S., & Schiavi, L. (2005). Measurement of benzene, toluene, ethylbenzene and o-xylene gas phase photodegradation by titanium dioxide dispersed in cementitious materials using a mixed flow reactor. *Applied Catalysis B: Environmental, 61*, 90—97.

Sunyer, J., Esnaola, M., Alvarez-Pedrerol, M., Forns, J., Rivas, I., López-Vicente, M., et al. (2015). Association between traffic-related air pollution in schools and cognitive development in primary school children: A prospective Cohort study. *PLoS Medecine, 12*(3), e1001792. https://doi.org/10.1371/journal.pmed.1001792.

Tate, J. (2016). Real driving emissions of diesel cars. In *AIRUSE Workshop on urban air Quality: Problems and possible solutions. 4–5 February 2016, Valencia, Spain.* http://airuse.eu/wp-content/uploads/2016/01/7d_JTate_Remote-SensingVehicleEmissions.pdf.

Tera. (2009). *In situ study of the air pollution mitigating properties of photocatalytic coating, Tera Environement, (Contract number 0941C0978).* Report for ADEME and Rhone-Alpe region, France http://www.air-rhonealpes.fr/site/media/telecharger/651413.

Transport for London. (2015). In-service emissions performance of Euro 6/VI vehicles. A summary of testing using London drive cycles. *The Mayor of London,* 18. http://content.tfl.gov.uk/in-service-emissions-performance-of-euro-6vi-vehicles.pdf.

UAR. (2017). *Urban acess regulations. CLARS (charging, low emission zones, other access regulation schemes).* Funded by the European Commission. http://www.urbanaccessregulations.eu/.

UK Department of Transport. (2017a). *Vehicle emissions testing programme, April 2016.* London: Cm 9259, Williams Lea Group, ISBN 9781474131292.

UK-Department of Transport. (2017b). Emissions Testing of Gas-Powered Commercial Vehicles: The results of tests to measure the greenhouse gas and air pollutant emission performance of various gas-powered HGVs. *Low Carbon Vehicle Partnership,* 38. https://www.gov.uk/government/uploads/system/uploads/attachment_data/file/581859/emissions-testing-of-gas-powered-commercial-vehicles.pdf.

WHO. (2006). *Air Quality Guidelines: Global Update 2005. Particulate matter, ozone, nitrogen dioxide and sulfur dioxide.* Copenhagen: World Health Organisation, ISBN 92 890 2192 6. http://www.euro.who.int/__data/assets/pdf_file/0005/78638/E90038.pdf.

Wiemann, J. (July 19, 2013). *Dulevo international spa, personnel communication.*

World Bank. (2016). *The cost of air pollution. Strengthening the economic case for action.* The World Bank and Institute for Health Metrics and Evaluation, University of Washington, Seattle. Published by World Bank Group, 102 pp. https://openknowledge.worldbank.org/bitstream/handle/10986/25013/108141.pdf?sequence=4&isAllowed=y

Zhu, D., Kuhns, H. D., Brown, S., Gillies, J. A., Etyemezian, V., & Gertler, A. W. (2009). Fugitive dust emissions from paved road travel in the lake tahoe basin. *Journal of the Air and Waste Management Association, 59,* 1219–1229.

Non-Exhaust PM Emissions From Battery Electric Vehicles

Victor R.J.H. Timmers, Peter A.J. Achten

INNAS BV, Breda, The Netherlands

O U T L I N E

1. **Introduction**   262
   1.1 *Electrification of Passenger Cars*   262
   1.2 *Electric Vehicle Market Growth*   262

2. **Weight and Non-Exhaust Emissions**   264
   2.1 *Theory*   264
   2.2 *Literature*   264
      2.2.1 Qualitative Evidence   264
      2.2.2 Experimental Data and Models   264
      2.2.3 Emission Inventories   266
   2.3 *Weight Comparison of Electric and Conventional Passenger Cars*   266

3. **Non-Exhaust Emissions of Electric Vehicles Compared With Internal Combustion Engine Vehicles: A Review of Previous Studies**   269
   3.1 *Van Zeebroek and De Ceuster (2013)*   269

3.2 *Soret, Guevara, and Baldasano (2014)*   271
3.3 *Timmers and Achten (2016)*   272
3.4 *Platform for Electro-Mobility (2016) Response*   273
3.5 *Hooftman et al. (2016)*   274

4. **Discussion**   278
   4.1 *Challenges of Comparing Electric Vehicles and Internal Combustion Engine Vehicles*   278
   4.2 *Contribution of Secondary Particulate Matter*   278
   4.3 *Contribution of Resuspension*   279
   4.4 *Future Research*   280

5. **Conclusions**   281
   5.1 *Remaining Uncertainties*   281
   5.2 *Consensus*   282

References   282

1. INTRODUCTION

1.1 Electrification of Passenger Cars

Electric vehicles (EVs), most notably electric passenger cars, have been proposed as a solution to several of the society's problems. Some of the advertised benefits include reduced greenhouse gas emissions, reduction in oil consumption, less noise pollution, as well as improved air quality.

In terms of air quality, the advantage of EVs in terms of nitrous oxides (NOx) reduction is well established, but their effect on particulate matter (PM) emissions is still being investigated. PM emissions are of great importance because of their detrimental effect on human health. Research has shown that traffic is a major contributor to PM levels, especially in urban areas (Charron, Harrison, & Quincey, 2007; Kousoulidou, Ntziachristos, Mellios, & Samaras, 2008; Pant & Harrison, 2013). Therefore, the increase in EVs has been hailed by many as the solution to urban air pollution, offering zero emissions and promising cleaner air for everyone (Dutch Government, 2011; EU, 2005; Murrells & Pang, 2013). However, EV proponents often neglect to consider PM emissions from non-exhaust sources, such as tire wear, brake wear, road wear, and resuspension of road dust.

In fact, as tailpipe emissions standards have become increasingly stringent over the last decade, non-exhaust emissions are already considered to be the dominant source of PM from traffic (see Chapter 6 and Amato et al., 2014; Ketzel et al., 2007). By the end of the decade, non-exhaust sources are predicted to contribute as much as 90% to total PM emissions from traffic (see Chapter 6 and Denier van der Gon et al., 2013; Jörß & Handke, 2007; Rexeis & Hausberger, 2009; Squizzato et al., 2016).

EVs have the benefit of having no tailpipe emissions, but still have tires and brakes. Therefore, just like internal combustion engine vehicles (ICEVs), EVs will emit PM through non-exhaust sources. In addition, EVs tend to weigh more than ICEVs because of their heavy batteries and are therefore significant contributors to PM. This means that the increasing popularity of EVs will not necessarily have a straightforward effect on PM levels and air quality.

This chapter reviews the research done in this area. First, studies will be presented, which investigate the relationship between weight and non-exhaust emissions. Second, the weight of EVs and ICEVs will be compared. Finally, the literature comparing PM emissions from EVs and ICEVs will be analyzed and important points of discussion highlighted.

1.2 Electric Vehicle Market Growth

The impact of EVs on air quality may seem unimportant because of the relatively small number of EVs on the road. However, this is set to change as electric passenger cars are being heavily incentivized throughout the world (EAMA, 2016; Mock & Yang, 2014). For example, in the EU, 21 out of 28 countries have some form of incentives for EVs (EAMA, 2016). In most of these countries, EVs are exempt from circulation and/or registration tax, potentially saving thousands of euros. Some countries, such as Sweden, the United Kingdom, and France offer subsidies or rebates worth thousands as well.

At the same time, numerous countries have begun investing in their charging infrastructure. In 2013, the number of public slow- and fast-charging stations in Europe exceeded 20,000 and 1000, respectively. For reference, the number of petrol stations in Europe is in the order of 131,000 (McKinsey & Company, 2014). The distribution of charging stations is still uneven, with most located in Western Europe. Leading countries in charging infrastructure are Denmark, the Netherlands, Germany, the United Kingdom, and France. The EU's Clean Fuel Directive sets a target to increase the number of public charging stations to 800,000 throughout Europe by 2020—with individual targets being set for each member state (McKinsey & Company, 2014).

As a result, the market share of EVs has seen a dramatic increase each year. Globally, new registration of electric cars increased by 70% between 2014 and 2015 with over 550,000 EVs sold worldwide (IEA, 2016). The total number of EVs on the road during this year increased to one million. Norway currently has the largest EV market share in the world, with 22.4% of car sales coming from EVs (Yang, 2016). See Fig. 12.1. This high figure can be explained by the fact that the Norwegian government has introduced numerous incentives for EVs since the 1990s, including exemption from registration tax, parking fees, and road tolls; reduced company car tax and annual car tax; and access to bus lanes (AIRUSE, 2016).

The Electric Vehicle Initiative (EVI), a multigovernment policy forum comprising 16 nations, has set a goal of 20 million electric cars by 2020. This represents an annual increase of 60% and a global market share of 1.7% (IEA, 2016).

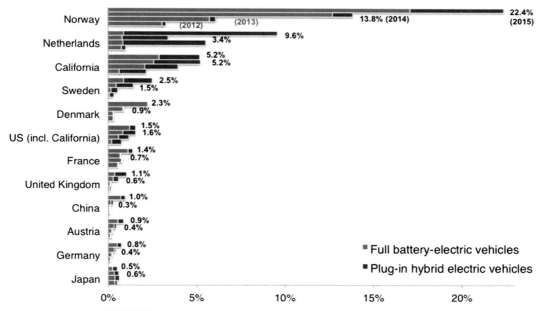

FIGURE 12.1   Market share of electric passenger cars (Yang, 2016).

2. WEIGHT AND NON-EXHAUST EMISSIONS

2.1 Theory

The idea that vehicle weight has an influence on non-exhaust PM emissions is not new or far fetched. Salminen (2014) suggests that the logical cause of tire wear is the load acting on the tire—road contact surface. Greater loads result in larger tire deformations and higher strain on the tire's side wall and tread. This in turn results in increased fatigue and wear.

Road surface wear, including surface cracking, rutting (permanent deformation), and raveling (loss of stones) are caused by shear loadings of pavement by vehicle tires (NVF, 2008). Higher loading therefore likely causes higher wear. This effect of vehicle weight on road wear was tested by the AASHO between 1958 and 1960. From these experiments, it was concluded that road wear is related to the fourth power of vehicle weight. This so-called fourth power law is commonly used as a rule of thumb for heavy-duty vehicles. Authors warn, however, that the law cannot simply be applied to light-duty vehicles (LDVs), such as passenger cars, because only heavy-duty vehicles were included in the study (NVF, 2008).

Brake wear is caused by the friction between the brake pads and the wheels. Garg et al. (2000) note that a variety of factors contribute the differences in wear rates including brake composition and the inertia weight being stopped. Because heavier vehicles will have more inertia, they will also have higher rates of brake wear.

Resuspension is caused by vehicle-induced turbulence and tire shear (Kupiainen, 2007). Vehicle-induced turbulence is determined by the size, weight, and aerodynamics of the vehicle, whereas tire shear is mainly dependent on vehicle weight. Therefore, heavier vehicles result in increases in both these factors, which will lead to increased rates of resuspension.

2.2 Literature

2.2.1 Qualitative Evidence

Some authors have speculated about the possible influence of weight but not directly measured it. Barlow (2014) mentions that vehicle weight is likely to be one of the factors affecting tire wear. He also says that, in general, larger vehicles produce larger non-exhaust emissions. Similarly, Garg et al. (2000) mention that the inertia weight being stopped is one of the factors contributing to brake wear rate, but they do not perform any tests with varying weights to confirm this.

Carslaw (2006) mentions that experimental work has shown that non-exhaust emissions increase with vehicle weight but not by what amount. He notes that robust relationships between vehicle weight and emission factors of non-exhaust PM have not yet been established. Even so, Carslaw suggests that vehicle weight appears to be an important factor in non-exhaust emissions and reducing vehicle weight may also contribute to lower non-exhaust PM.

2.2.2 Experimental Data and Models

In more recent years, experimental data and models related to weight and non-exhaust emissions have started to appear. In his paper, Simons (2013) presents new life cycle data for emissions of passenger cars in the eco-inventv3 model. He distinguishes between vehicle exhaust and non-exhaust emissions and is one of the first to define non-exhaust emissions as a factor of vehicle weight, with the intention of being applied to studies on hybrid and EVs. Simons suggests that PM_{10} emission factors could be scaled directly to vehicle weight and provides emission factors for tire, brake, and road wear per kilogram of vehicle weight. For example, tire, brake, and road wear increase by around 50% when comparing a medium (1600 kg) and small (1200 kg) car. Compared with a small car, large cars (2000 kg) emitted more than double the amount of PM_{10} (see Fig. 12.2).

There are numerous other studies that link non-exhaust emissions to vehicle weight. Wang, Huang, Chen, and Liu (2016) investigated the relationship between contact features of tires and tire wear. They found that vertical load had an approximately linear effect on tire wear. Salminen (2014) obtained similar results, but suggested that tire wear increases exponentially with vertical load.

Several individual studies have found higher non-exhaust emission factors for heavier vehicle categories, indicating a correlation with weight. Despite varying definitions for the weight of vehicle categories, the consensus is that LDVs emit more PM than passenger cars (Lükewille et al., 2001). For example, Garben, Wiegand, Liwicki, and Eulitz (1997) found tire wear of LDVs to be 75% higher than that of passenger cars. Similarly, Gebbe, Hartung, Berthold, and (1997) found tire wear for LDVs to be more than twice that of passenger cars. BUWAL (2001) found that the PM_{10} emissions of passenger cars' brakes were twice as much as those from motorcycles. LDVs on the other hand, emitted over two and a half

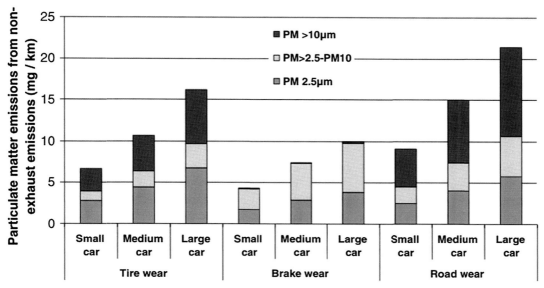

FIGURE 12.2 Non-exhaust particulate matter (PM) emissions by source and car size. *From Simons (2013) based on Ntziachristos and Boulter (2009).*

times more PM_{10} than passenger cars. Research by Garg et al. (2000) distinguishes between brake emissions from small cars, large cars, and large pickup trucks. They found that the brakes of large cars emit 55% more total suspended particles (TSP), PM_{10} and $PM_{2.5}$, than small cars. Large pickup trucks were found to emit more than double the quantity of particulates compared with small cars.

Gillies, Etyemezian, Kuhns, Nikolic, and Gillette (2005) investigated the influence of vehicle characteristics on resuspension of road dust on unpaved roads. They found that emissions had a strong linear relationship with not only vehicle speed but also vehicle weight. This is in line with the results from a study by Amato et al. (2012), which found that PM_{10} resuspension rates were 10 times higher for passenger cars than for motorcycles and three to four times higher for LDVs than for passenger cars.

2.2.3 *Emission Inventories*

Emission inventories are regularly updated reviews of existing literature that provide estimates of emission factors. The European Environmental Agency (EEA) publishes an Emission Inventory Guidebook (Ntziachristos & Boulter, 2013) that outlines emission factors for different vehicle types. In this emission inventory, passenger cars are defined as vehicles carrying up to nine passengers, whereas LDVs are defined as vehicles with a gross weight of up to 3500 kg. LDV emission factors of TSP, PM_{10} and $PM_{2.5}$, were 57% higher than those of passenger cars for both tire and brake wear, but road surface wear was the same for both.

The Pollutant Release and Transfer Register (PRTR) in the Netherlands provide their own emission inventory with emission factor estimates for tire wear based on extensive research (ten Broeke, Hulskotte, & Denier van der Gon, 2008). They consider the average empty weight of a passenger car to be 850—1050 kg and the gross weight of a van to be around 2000 kg. They suggest that the total tire wear, PM_{10} and $PM_{2.5}$ emissions were 40% higher for vans compared with regular passenger cars. The PRTR also has a report on calculating emissions per tire for different vehicle categories (Klein et al., 2014). In this report, wear rate per tire is 10% higher for passenger cars than for motorcycles, 20% higher for delivery vans than for passenger cars, and 130% higher for lorries than for passenger cars.

Finally, the US Environmental Protection Agency (EPA) has an emission inventory called the AP42 Method (EPA, 2011) that contains guidance for estimating resuspension. This includes a factor based on vehicle weight to the power 1.02, suggesting resuspension increases almost linearly with weight. This is in accordance with the results from Gillies et al. (2005).

2.3 Weight Comparison of Electric and Conventional Passenger Cars

To determine the additional non-exhaust emissions that EVs produce, a comparison must be made between the weight of EVs and ICEVs. One way of doing this involves categorizing highway-capable EVs by market segment and comparing their weight to the average weight of that segment. Van Zeebroek and De Ceuster (2013) used this method for two vehicles and

TABLE 12.1 Comparison of Weight Between Electric Vehicles (EVs) and the Segment Average

EV Car Name	Market Segment	Mass in Running Order (kg)	Segment Average (kg)	Weight Difference (%)
BMW i3	Small (B)	1372	1175	+16.8
BMW Brilliance Zinoro 1E	Upper medium (E)	2045	1835	+11.4
Bolloré Bluecar	Mini (A)	1145	978	+17.1
BYD e6	Upper medium (E)	2095	1835	+14.2
Chery QQ3 EV	Mini (A)	1125	978	+15.0
Chevrolet Spark EV	Mini (A)	1431	978	+46.3
Citroën C-Zero	Mini (A)	1155	978	+18.1
Fiat 500e	Mini (A)	1427	978	+45.9
Ford Focus Electric	Lower medium (C)	1719	1402	+22.6
Honda Fit EV	Small (B)	1550	1175	+31.9
JAC J3 EV	Small (B)	1335	1175	+13.6
Kia Soul EV	Lower medium (C)	1617	1402	+15.3
Lightning GT	Sports car (S)	1925	1572	+22.5
Mercedes B-Class Electric Drive	Lower medium (C)	1725	1402	+23.0
Nissan Leaf	Small (B)	1568	1175	+33.4
Renault Fluence ZE	Lower medium (C)	1618	1402	+15.4
Renault Zoe	Small (B)	1543	1175	+31.3
Smart Electric Drive	Mini (A)	1055	978	+7.9
Tesla Model S (Various)	Luxury (F)	2036 to 2314	2074	−1.8% to +11.6
Volkswagen e-Golf	Lower medium (C)	1617	1402	+15.3
Volkswagen e-Up!	Mini	1289	978	+31.8

Based on Manufacturer Information.

found their weight to be 21% heavier. A more complete comparison has been made by the authors for 20 electric passenger cars (Table 12.1) using manufacturer data. This table includes EVs, which are, or have been, available on the market in the last 5 years. Using this method for 20 EVs shows that EVs are 22% heavier than the average of their market segments.

Comparing EVs to the average passenger car in their category has several problems associated with it, however. First, there are various classification systems available. This

TABLE 12.2 Comparison of Weight Between Electric Vehicles (EVs) and Their Internal Combustion Engine Vehicle (ICEV) Counterparts

EV	ICEV	MRO EV (kg)	MRO ICEV (kg)	Difference (kg)	Difference (%)
Ford Focus Electric	Ford Focus	1719	1500	+219	+14.6
Honda Fit EV	Honda Fit	1550	1215	+335	+27.6
Fiat 500e	Fiat 500	1427	1149	+278	+24.2
Smart Electric Drive Coupe	Smart Coupe	1055	820	+235	+28.7
Kia Soul EV	Kia Soul	1617	1306	+311	+23.8
Volkswagen e-Up!	Volkswagen Up	1289	1004	+284	+28.3
Volkswagen e-Golf	Volkswagen Golf	1617	1390	+227	+16.3
Chevrolet Spark EV	Chevrolet Spark	1431	1104	+327	+28.6
Renault Fluence EV	Renault Fluence	1618	1300	+318	+24.4
Mercedes B-Class Electric Drive	Mercedes B-Class Hatchback	1725	1420	+305	+21.5

Adapted from Timmers and Achten (2016).

study used the EURO market segment classification, but vehicles can also be classified using the US EPA Size Class or EURO NCAP Class, which would produce different results. Second, EVs do not always fit well into a specific class. To further complicate the comparison, the exact definition of the segments has been left open by the European Commission (1999). Finally, and most importantly, the vehicles in any market segment will vary significantly. They will have a range of dimensions, engine sizes, and features that makes comparison problematic.

Another method of comparing the weight of EVs to ICEVs is by comparing EVs and their equivalent nonelectric version. This avoids the problem of having to classify vehicles and ensures that the vehicles are as similar as possible. Timmers and Achten (2016) used this method to compare the mass in running order of nine EVs and their ICEV counterparts. On average, they found the difference in weight to be 280 kg. This corresponds to an average increase in weight of 24% when using EVs instead of conventional counterparts (see Table 12.2 for the full list).

However, this method of comparison also has a range of problems associated with it. A fair comparison would have identical vehicle specifications except for the drivetrain. However, EVs and ICEVs often have different specifications in terms of materials used, range, top speed, trailer load, types of tires, brakes, among others. This is discussed in detail in Section 4.1.

The weight of both EVs and ICEVs is likely to change in the future as battery density improves and lightweight materials become the norm in car manufacture. Unfortunately, very few studies investigate the weight difference between EVs and ICEVs and even fewer make future predictions. An analysis by the Burnham (2012) of the Argonne National Laboratory provides updated vehicle specifications for the GREET Vehicle-Cycle model. The gross

TABLE 12.3 Comparison of Electric Vehicle (EV) And Internal Combustion Engine Vehicle (ICEV) Gross Vehicle Weight (GVW) for Regular and Lightweight (LW) Vehicles

Vehicle Type	GVW (kg)		Difference (%)	GVW (kg)		Difference (%)
	ICEV	EV		LW ICEV	LW EV	
Passenger car	1352	1937	+43	826	1216	+47
SUV	1642	2504	+52	1007	1656	+64
Pickup truck	1891	2953	+56	1247	2177	+75

Adapted from Burnham (2012)

vehicle weight of EVs and ICEVs was compared for passenger cars, SUVs, and pickup trucks. This was also done for the lightweight (LW) version of the vehicles. In terms of EV specifications, the author refers to a paper by Moawad, Sharer, and Rousseau (2011) in which EVs are said to have a range of 150 miles (241 km). According to the study, EVs are between 43% and 56% heavier than ICEVs. In the future, the lightweight versions of EVs and ICEVs are predicted to be between 47% and 75% heavier than ICEVs (see Table 12.3).

Another study comparing the weight is that of Bauer, Hofer, Althaus, Del Duce, and Simons (2015). They used a simulation of a midsize vehicle to compare the weight of ICEVs and EVs in 2012 and projected in 2030. They found that in 2012, ICEVs were 1567 kg on average, whereas EVs were 1944 kg or 24% heavier. The projected values for 2030 were 1383 and 1613 kg for ICEVs and EVs, respectively. This corresponds to a difference of 17%.

3. NON-EXHAUST EMISSIONS OF ELECTRIC VEHICLES COMPARED WITH INTERNAL COMBUSTION ENGINE VEHICLES: A REVIEW OF PREVIOUS STUDIES

Non-exhaust emissions have only started to receive the attention of the scientific community relatively recently. Therefore, not many comparisons have been made between the non-exhaust emissions of EVs and ICEVs. There is also an ongoing debate about many important aspects of the process, including which emissions to include, which numbers to use, and how the data are represented. Extrapolating the results to the future introduces even more uncertainty. However, all studies do agree on one thing: something must be done about non-exhaust emissions.

3.1 Van Zeebroek and De Ceuster (2013)

In 2013, two researchers from Transport and Mobility Leuven, Van Zeebroek and De Ceuster (2013), wrote the first paper that investigated the non-exhaust emissions of EVs. The paper, translated as "Electric cars barely reduce particulate matter," presents three reasons why EVs could not deliver the benefits to society, which were expected. The most prominent of these is that they emit almost equal amounts of PM as conventional cars.

Van Zeebroek and De Ceuster consider nitrogen oxides (NO_x) and PM to be the most harmful emissions in urban environments. Based on the methods used by Delhaye,

TABLE 12.4   Particulate Matter (PM) Emissions From Petrol and Diesel Cars in Van Zeebroek and De Ceuster (2013)

Car Type	Exhaust (mg/vkm)			Non-Exhaust (mg/vkm)			Total (mg/vkm)		
	PM_{10}	$PM_{2.5}$	Soot	PM_{10}	$PM_{2.5}$	Soot	PM_{10}	$PM_{2.5}$	Soot
Petrol (EURO 5)	1	1	0.2	30	17	2.7	31	18	2.9
Diesel (DPF, EURO 5)	4	4	0.8	30	17	2.7	34	21	3.5
Diesel (no DPF, EURO 2)	62	62	50	30	17	3	92	79	52

DPF, diesel cars with particulate filter.

De Ceuster, and Maerivoet (2010), they calculate that the costs to society of PM are far greater than those of NO_x. The paper is therefore mainly concerned with PM emissions.

Van Zeebroek and De Ceuster only consider non-exhaust PM emissions from brake, tire, and road wear, neglecting resuspension. They find that each of these contributes about one-third to PM emissions, of which a significant portion is $PM_{2.5}$. Based on the GAINS-Copert IV model (Borken-Kleefeld & Ntziachristos, 2012), they find that total non-exhaust emissions are 20 mg/vkm of PM_{10} and 11 mg $PM_{2.5}$. However, in urban areas, the authors argue, these emission factors increase by about 50% due to the large amount of cornering, stopping, and starting. This is in line with findings of the European Environment Agency (EEA, 2009), which estimates $PM_{2.5}$ emissions due to non-exhaust reach 16 mg/vkm in urban areas.

For tailpipe emissions of conventional cars, Van Zeebroek and De Ceuster used estimates from Graz University (2012) to determine $PM_{2.5}$ emissions of diesel cars to be 4 mg/vkm. Note that this is only for diesel cars with particulate filter (DPF). For petrol cars, this figure was determined to be 1 mg/vkm based on the GAINS-Copert model (Borken-Kleefeld & Ntziachristos, 2012) (see Table 12.4).

Van Zeebroek and De Ceuster do not consider the effect of EVs on secondary PM, which is formed from gaseous precursors normally emitted by the exhaust. EVs are likely to reduce the emission of these precursors, but because secondary PM is currently not considered in emission inventories, this contribution was not included in the study.

As a result, non-exhaust emissions account for 80%—100% of PM emissions from vehicles, say Van Zeebroek and De Ceuster. They do note that there are some exceptions for which these numbers do not apply. For example, petrol vehicles with direct injection emit more PM than regular diesel cars and were excluded from the analysis. In addition, the authors also excluded any effects of reduced performance of DPFs due to aging.

For electric cars, Van Zeebroek and De Ceuster make two adjustments to the non-exhaust emission figures. They cite the Belgian research project Clever, which notes there are two factors that influence the production of non-exhaust PM. These are regenerative braking and the weight of the vehicle. The purpose of regenerative braking is to increase range by recuperating some of the energy normally lost when braking. An additional benefit is that brake wear emissions are reduced significantly. The authors assume regenerative braking should reduce the PM emissions associated with brake wear by 50%. The weight of the vehicle should have a proportional effect on the total non-exhaust emissions.

To calculate the total emissions of EVs, the authors compared the weight of a Volkswagen Golf and Kia Soul to their electric counterparts. Both conventional vehicles have a weight of 1210 kg, whereas the electric versions are 1500 and 1460 kg, respectively. This corresponds to approximately a 21% increase in weight.

Van Zeebroek and De Ceuster assume this has a linear effect on emissions. As a result, they consider EVs to have tire and road emission factors that are 21% higher than ICEVs. The overall brake wear of EVs is calculated to be 39.5% lower when adjusting for regenerative braking. Taking these factors into consideration, the overall PM emissions from EVs and ICEVs are calculated to be equal.

The authors note that the vehicles they chose had conservative weights. The VW Golf and Kia Soul have relatively low weights compared with other vehicles in their weight class, such as the Opel Astra at 1370 kg and Renault Megane at 1280 kg. Finally, the authors compare total PM emission rates when using regenerative braking 75% or 90% of the time. In those cases, the reduction of PM from EVs is in the order of 10%–16%.

3.2 Soret, Guevara, and Baldasano (2014)

Soret et al. (2014) investigated the impact of three different fleet electrification scenarios (13%, 26%, and 40% transport share of EVs) on air quality in Barcelona and Madrid, Spain. Their research used HERMESv2.0 to model the emissions. This high-resolution emission model was developed specifically for Spain and uses a comprehensive database with updated methodologies to determine biogenic and anthropogenic emissions (Guevara, Martínez, Arevalo, Gassó, & Baldasano, 2013).

To determine non-exhaust emission factors, Soret et al. used the COPERT 4 software, which uses the emission factors detailed in the European Emission Inventory (EEA, 2009) for tire and brake wear. They employed the EEA Tier 3 method that calculates emission factors as a function of circulation speed for 256 different vehicle categories, rather than generic vehicles and speeds. Unlike van Zeebroek and De Ceuster, Soret et al. also considered emissions factors for resuspension of road dust. The HERMESv2.0 model has emission factors for resuspension of 88 mg/vkm (Amato et al. 2010), which is adjusted based on rainfall (Pay, Jimenez-Guerrero, & Baldasano, 2011). No emission factors for road wear were included, however.

For exhaust emissions, Soret et al. used the European Emission Inventory, as it provides estimates for all exhaust emissions, except for PM. To determine PM emission factors, the authors conducted their own literature review with, as most notable source, a study by Graham (2005). This resulted in an exhaust PM emission factor of 4 mg/vkm.

For EVs, Soret et al. assumed that they had no exhaust-related emissions and equal non-exhaust emissions to ICEVs. Using these numbers, they find that in all fleet electrification scenarios, PM_{10} emissions from traffic are only reduced by up to 5% in Barcelona and 3% in Madrid. They attribute these low percentages to the importance of non-exhaust emissions and the fact that fleet electrification does not reduce these types of emissions.

They conclude that, in general, fleet electrification presents potential air quality benefits, but the reductions achieved are insufficient to ensure proper air quality levels. They therefore warn that fleet electrification cannot be considered the sole solution, especially regarding PM. Additionally, to acquire a significant level of improvement, a large proportion of the fleet will need to be electrified (26%–40%).

3.3 Timmers and Achten (2016)

The research by Timmers and Achten (2016) compared the PM emissions of EVs with those of gasoline and diesel passenger cars complying with EURO 6 standards. The EURO 6 standards were introduced in 2014 and stipulate that new passenger vehicles must not emit more than 5 mg/vkm of PM. Timmers and Achten assumed that in the next decade, most vehicles would comply with these standards and based their comparison on this near-future scenario.

To determine the actual PM emissions of passenger cars, not just the regulations, the study looked at emission inventories from the EU, United States, Netherlands, and United Kingdom. They found exhaust emission factors were generally below the EURO 6 limits, with an average of approximately 3 mg/vkm. The study used the same approach to determine non-exhaust emission factors. By averaging the abovementioned emission inventories, they found PM_{10} emission factors for tire wear to be 6.1 mg/vkm, brake wear to be 9.3 mg/vkm, road wear to be 7.5 mg/vkm, and resuspension to be 40 mg/vkm.

They calculated EVs to be 24% heavier than ICEVs, based on comparisons between passenger cars with both an electric and equivalent conventional model. This translated to EVs being 280 kg heavier, on average. Using the research by Simons (2013), they found an increase in weight of 280 kg would result in a PM_{10} increase of 1.1 mg per vehicle kilometer (mg/vkm) for tire wear, 1.1 mg/vkm for brake wear, and 1.4 mg/vkm for road wear. However, because of EVs' regenerative braking, they assumed a conservative estimate of zero brake wear emissions. For resuspension, they assumed a linear increase based on the research by Gillies et al. (2005).

From these numbers, they found that EVs emitted approximately the same amount of PM_{10} and only 1%–3% less $PM_{2.5}$ than their conventional equivalents (see Table 12.5).

The results of the study caused significant controversy in the media and have been criticized in some of its aspects, including the importance it places on resuspension and the exclusion of secondary PM. In the study, resuspension indeed plays a large role in the non-exhaust emission factors, accounting for more than 60% of the total. The resuspension figure used was solely based on the UK national emission inventory because no other inventory provides numbers for this emission source. Individual studies on resuspension have exceedingly varied results, with some reporting emission factors of more than 100 mg/vkm (Abu-Allaban, Gillies, Gertler, Clayton, & Proffitt, 2003; Fitz & Bufalino, 2002, pp. 1–18; Omstedt, Bringfelt, & Johansson, 2005) and others reporting emission factors of less than 20 mg/vkm (see Chapter 11 and Bukowiecki et al., 2009; Luhana et al., 2004; Thorpe, Harrison, Boulter, & McCrae, 2007) (see Section 4.3 and Table 12.9 for details). However,

TABLE 12.5 Comparison Between PM_{10} Emissions of Electric Vehicles (EVs), Gasoline and Diesel Internal Combustion Engine Vehicles (ICEVs) (Timmers & Achten, 2016)

Vehicle Technology	Exhaust (mg/vkm)	Tire Wear (mg/vkm)	Brake Wear (mg/vkm)	Road Wear (mg/vkm)	Resuspension (mg/vkm)	Total (mg/vkm)
EV	0	7.2	0	8.9	49.6	65.7
Gasoline ICEV	3.1	6.1	9.3	7.5	40	66.0
Diesel ICEV	2.4	6.1	9.3	7.5	40	65.3

the relative contribution of resuspension in the Timmers and Achten study is similar to that reported in earlier literature by Weinbruch et al. (2014). They found that 56% and 58% of PM_{10} emissions came from resuspension at an urban and curbside measuring station in Germany.

In addition, the notion that resuspension increases linearly with vehicle weight has received criticism as well. Timmers and Achten based this on the study by Gillies et al. (2005) and the US Emission Inventory (EPA, 2011). However, researcher Auke Hoekstra reported this in the Dutch newspaper *De Volkskrant* (Van de Weijer, 2016). He argued that the undercarriage of EVs is flatter than those of ICEs due to the lack of a tailpipe, which should improve the aerodynamics of the car and reduce resuspension.

Other critics argue that including resuspension is not relevant because without ICEVs there would be no road dust. This is unjustified, however, as road dust is largely made up of soil-derived minerals and non-exhaust PM, with only a small fraction coming from exhaust PM (Gunawardana, Goonetilleke, Egodawatta, Dawes, & Kokot, 2012). Therefore, road dust will exist regardless of drivetrain technology.

Timmers and Achten argue that not including resuspension would be a serious omission. Even though there is no consensus on the actual emission factors, multiple studies have shown that it is an important contributor to overall PM emissions. The numbers they used were the best estimate based on the available information at the time.

The other main criticism is that the papers received were due to the authors' decision to exclude secondary PM from their analysis. Some sources (Hooftman, Oliveira, Messagie, Coosemans, & Van Mierlo, 2016) suggest that secondary PM may significantly contribute to total PM levels.

Timmers and Achten defend their decision to exclude secondary PM based on issues pointed out by the previous research. Several articles mention that it is exceedingly difficult to model secondary PM emissions (Air Quality Expert Group, 2012; Hoogerbrugge, Denier van der Gon, van Zanten, & Matthijsen, 2015). Not only do many studies have difficulty determining the fractional contribution vehicles make to secondary PM, but it is also problematic to differentiate between primary and secondary PM (Amato, Schaap, Reche, & Querol, 2013; Bahreini et al., 2012; Viana, 2011). Therefore, there is always the risk of double-counting PM (Humbert, Fantke, & Jolliet, 2015). Timmers and Achten also warn that secondary PM may already be partly included in the resuspension figure, which is why they decided to exclude it.

3.4 Platform for Electro-Mobility (2016) Response

Due to the coverage that the Timmers and Achten study received in the media, the Platform for Electro-Mobility (2016) issued a briefing paper in response. The Platform for Electro-Mobility is a European organization that consists of more than 20 producers, infrastructure managers, operators, transport users, cities, and civil society organizations.

The briefing paper agrees with the overall conclusion of Timmers and Achten that non-exhaust emissions need to be regulated and weight reduction encouraged for all vehicles, but disagrees with the conclusions about PM emissions from EVs. They cite a case study performed by Aria Technologies (Groupe Renault, 2015, p. 170), which concluded that a 20% shift to fleet electrification would reduce PM_{10} levels by 30%. It is not known, however, if this study included non-exhaust emissions at all.

Tkhe briefing paper also mentions that the scope of the Timmers and Achten study is too limited as it only focuses on the emissions while driving. The paper argues that full life cycle assessment (LCA) should be used to objectively compare between fuel technologies. The Mobility, Logistics and Automotive Technology Research Centre (MOBI) of Brussels University performed such an LCA. The results from this study (Hooftman et al., 2016) show that in the full well-to-wheels cycle, EVs emit four times less PM than ICEVs. The differences between this article and Timmers and Achten study are investigated in detail in Section 3.5.

In addition, the briefing paper mentions that EVs come equipped with low rolling resistance tires, which they claim produce fewer emissions from wear. However, no research has been done to prove this. Low rolling resistance tires have not been considered in the Timmers and Achten study. However, even if low rolling resistance tires reduce wear, they do not need to be exclusive to EVs. If non-exhaust emissions standards be set, and if these indeed result in reduced wear, it is likely that ICEVs will start using similar tires to reduce tire wear emissions as well.

Another factor, which the briefing paper mentions, is future improvements in battery energy density. Current battery technology allows for an energy density of 120−170 Wh/kg. According to EUROBAT (2015), this is likely to increase up to 290 Wh/kg in 2030. Because the battery pack is the main source of the increased weight of EVs, higher densities could lead to fewer battery cells needed and therefore lower weight. However, it is also possible that higher energy density will not reduce the amount of battery cells in EVs but rather increase the range of EVs. In this case, energy density will not have an impact on EV weight.

The briefing paper wrongfully comments that the Timmers and Achten study considers braking the most important source of non-exhaust emissions. They mention that EVs come equipped with regenerative braking, which reduces brake wear by 25%−50%. However, in their analysis, Timmers and Achten already use a conservative estimate of zero brake wear emissions. Therefore, this argument would not change the study's conclusions.

The final argument the briefing paper presents is that as a group, EV drivers have better ecodriving behavior than drivers of nonelectric cars. They base this statement on a longitudinal study by Neumann, Franke, Cocron, Bühler, and Krems (2015). In the study, 40 drivers were questioned on what ecodriving strategies they knew before and after driving an EV for 3 months. This argument is not very relevant to the EV versus ICEV debate, however. Although it is true that driving behavior has a direct impact on non-exhaust emission rates, this is dependent on drivers rather than the drivetrain technology of the vehicle. If the aim is to improve driving behavior, then far more effective methods exist compared to replacing ICEVs with EVs.

3.5 Hooftman et al. (2016)

Around the same time as the Timmers and Achten paper, a team of scientists from the Brussels University in Belgium published a paper on the environmental impacts of diesel, gasoline, and electric cars. In their paper, Hooftman et al. (2016) performed a comparative LCA (well-to-wheel) of each vehicle type for emissions including CO_2, NO_x, PM, and many others.

Interestingly, they find similar results for tailpipe PM emission factors as Timmers and Achten. They state that PM emissions have been reducing steadily over the last decade with more stringent standards. In addition, actual PM emissions from passenger cars were

TABLE 12.6 Overview of the Type Approval (TA) Values and Applied Real Driving Emission (RDE) Values of Particulate Matter (PM) for Petrol and Diesel Cars

TA and RDE Values	Petrol		Diesel	
	PM		PM	
(mg/km)	TA	RDE	TA	RDE
Euro 4	—	1.5	25	50
Euro 5	5	2.2	5	2.8
Euro 6	5	2.4	5	1.1

Adapted From Hooftman et al. (2016).

found to be well below the limits set by EURO 6, based on test results from the Netherlands Organisation for Applied Scientific Research (TNO). For EURO 6 petrol cars, real driving emissions were found to be 2.4 mg/vkm and for diesel cars this was determined to be 1.1 mg/vkm (see Table 12.6).

For non-exhaust emissions, Hooftman et al. also reached the same conclusion as Timmers and Achten about the reliability of individual studies. They note that after analyzing several literature studies on emission factors, it was found that the data collection process was too widespread and did not provide overall reliable outputs. In the end, they decided to use the German Informative Inventory Report's (GIIR) data for 2015. This report sets PM_{10} emissions for tire wear at 6.4 mg/vkm, brake wear at 7.35 mg/vkm, and road abrasion at 7.5 mg/vkm. The report does not include any estimates of resuspension emission factors (see Table 12.7).

These numbers are close to those used by Timmers and Achten, who averaged the figures of four emission inventories, rather than using one. However, the major difference is Hooftman et al.'s decision to omit resuspension of road dust from the analysis entirely. They state that resuspension is difficult to quantify and that it is submissive to many external factors such as season, precipitation, and road moisture content. These are also the reasons why resuspension is not dealt with in the emission inventory reporting in the framework of the Convention on Long-Range Transboundary Air Pollution (CLRTAP). Despite this, the authors acknowledge that it dominates PM emissions in some countries.

More differences start to appear between the papers by Hooftman et al. and Timmers and Achten when discussing the non-exhaust emissions of EVs. Hooftman et al. do not investigate the weight difference between EVs and their conventional counterparts, but instead make a general statement about EVs having increased average vehicle weight due to the mass of the

TABLE 12.7 Non-Exhaust Emission Inventory Used in Hooftman et al. (2016)

Species Category	Material	Tire Wear	Brake Wear	Road Abrasion
Particulate matter (mg/km)	$PM_{2.5}$	4.49	2.93	4.05
	PM_{10}	6.4	7.35	7.5
	Total suspended particles	10.7	7.5	15

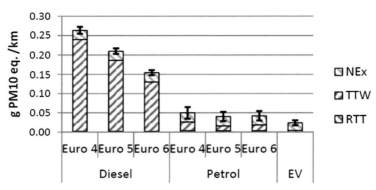

FIGURE 12.3 Particulate matter impact contribution due to non-exhaust (NEx), tank-to-wheel (TTW), and generation to tank (RTT) emissions. *From Hooftman et al. (2016).*

battery pack and its auxiliary systems (e.g., management systems, bus bars, and power electronics).

The effects of this weight on non-exhaust emissions are estimated rather than scientifically quantified. Hooftman et al. assume EVs have higher tire wear due to their increased weight and EV-specific tires, resulting in an increase of 10%. In terms of brake wear, they state that EVs require about two-thirds less braking than ICEVs due to regenerative braking. Their analysis is based on the service times of brake pads on Teslas, BMW i3s, and Nissan Leafs, which demonstrates that on average, the brake pads last roughly two-thirds longer than on diesel/petrol vehicles. They note that this outweighs the additional wear due to the vehicle's mass. In terms of road wear, Hooftman et al. do not consider EVs to emit more PM, stating that pavement quality is the more determining factor than vehicle weight and tire characteristics. As mentioned before, Hooftman et al. completely omit resuspension factors from their analysis and therefore do not differentiate between EVs and ICEVs in this respect.

The results of the study show that EVs reduce emissions compared with both petrol and diesel vehicles throughout all categories, including non-exhaust and total PM emissions. In terms of total PM emissions, Hooftman et al. find that petrol cars and diesel cars emit twice and six times as much PM as EVs, respectively. Their total PM emissions include those from wear, exhaust, and secondary PM due to NO_x, methane, and other hydrocarbons (see Fig. 12.3).

These results are surprising. From Fig. 12.3, it can be seen that tank-to-wheel (TTW) emissions are around 130 mg/km for EURO 6 diesel cars, even though the authors determined earlier that PM_{10} emissions from exhaust were only be 1.1 mg/km. This means that the remaining 129 mg/km (or 99%) of TTW emissions must come from secondary PM. This is far higher than existing estimates (see Section 4.2 and Table 12.8). Unfortunately, Hooftman et al. do not elaborate which numbers they used nor how they obtained their results. The

TABLE 12.8 NO_x and Secondary Particulate Matter (PM) Emissions From Petrol and Diesel Cars

Source	Petrol		Diesel	
	NOx (mg/vkm)	PM$_{10}$ Equivalent[a] (mg/vkm)	NOx (mg/vkm)	PM$_{10}$ Equivalent[a] (mg/vkm)
EMEP/EEA	59	13	210	46
UK DEFRA	49	11	274	60
NL PRTR	40	9	230	51

[a]Based on Goedkoop et al. (2014) calculation.

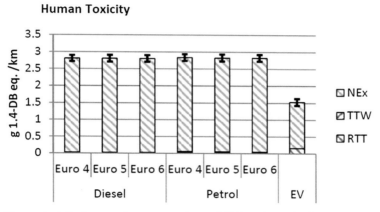

FIGURE 12.4 Human toxicity impacts due to non-exhaust (NEx), tank-to-wheel (TTW), and generation to tank (RTT) emissions. *From Hooftman et al. (2016).*

difference in results between the papers of Hooftman et al. and Timmers and Achten can mainly be explained by these extremely high values for NO_x-related secondary PM.

Also important are Hooftman et al.'s findings that non-exhaust emissions are lower for EVs than ICEVs. This is the opposite of the results by Timmers and Achten, who found an increase in non-exhaust emissions from EVs. As mentioned, this is mostly due to Hooftman et al. neglecting the contributions of additional weight and resuspension. Because non-exhaust is by far the most important contributor to human toxicity, this difference has large implications for human health (see Fig. 12.4).

Not all conclusions are different from Timmers and Achten, however. Hooftman et al. agree that non-exhaust emissions require active regulation. They mention this could be achieved by using alternative materials during production of tires, brakes, and pavements, or by introducing alternative technologies such as regenerative braking in ICEVs. Furthermore, they note that tires should be subjected to technological pushes to mitigate wear and tire composition.

4. DISCUSSION

4.1 Challenges of Comparing Electric Vehicles and Internal Combustion Engine Vehicles

The process of comparing the weight of EVs and ICEVs can be challenging. EVs often do not fit in conventional vehicle classes and if they do they may not share many of the characteristics of the vehicle class. Therefore, the best method to compare EVs and conventional cars is by only comparing EVs, which have an ICEV equivalent. Even then, important differences remain. For example, the weight of the body of EVs is often reduced significantly by using aluminum instead of steel to improve the range of the vehicle (Nealer & Hendrickson, 2015). If this would be done with conventional cars, the weight difference would be even greater than it already is.

Furthermore, EVs have many limitations that ICEVs do not have. For example, the Volkswagen e-Golf has a top speed of 140 km/h, a range of 133 km, and cannot carry any trailer load. The Volkswagen Golf, on the other hand, has a top speed depending on engine size between 179 and 203 km/h, a range of over 1000 km, and can carry a trailer load up to 1100 kg. A fair comparison would need to have identical specifications for all these factors. EVs cannot reach ranges anywhere near those of ICEVs and if they could, they would have to be exceedingly heavy. This all makes direct comparison problematic, especially because only limited data on vehicle specifics are publicly available.

EVs also come equipped with several additional features that are not currently used in ICEVs. One such example is the use of tires with low rolling resistance. Because range is a limiting factor of EVs, a large amount of research is being done into reducing the resistance of tires to save energy. Some authors, such as the Platform of Electro-Mobility (2016), claim that these EV-specific tires also have a lower wear rate than regular tires. However, other sources instead note that low rolling resistance tires have higher tire wear rates and need to be replaced more frequently (Innovative Systems Consultants, 2010; Reed, 2013). As more research is being done, wear rates will likely improve, but, at present, no sources suggest that low rolling resistance tires will have significantly lower wear rates.

Another additional feature of EVs (and hybrid EVs) is their regenerative brakes, which reduce the wear on the brakes and extend their lifetime. Very little experimental research has been done to quantify the reduction, but several authors have provided a range of estimates. Barlow (2014) suggested that regenerative braking produces virtually no brake wear. A report by the TNO (Ligterink, Stelwagen, & Kuenen, 2014) assumes regenerative braking reduces wear by up to 95%. Del Duce, Gauch, and Althaus (2014) reports that brake wear emissions are reduced by 80% for EVs, based on a report by Althaus and Gauch (2010). The study by Van Zeebroek and De Ceuster (2013) suggests a 50% reduction in brake wear emissions, while the study by Hooftman et al. (2016) assumes a 66% reduction. In any of these cases, regenerative braking is an effective abatement measure for brake wear and its application should therefore be encouraged for all vehicles, including ICEVs.

4.2 Contribution of Secondary Particulate Matter

The main point of discussion between the articles is the contribution of secondary PM and resuspension (see Section 4.3) to overall emissions. As can be seen from the comparison between Timmers and Achten (2016) and Hooftman et al. (2016) in Section 3, the inclusion or exclusion of these factors can make an enormous difference to the overall results.

Secondary PM is formed in the atmosphere when gases such as NO_x, sulfur dioxide (SO_2), and ammonia (NH_3) undergo a series of physical and chemical reactions under the influence of sunlight. In terms of traffic, NO_x is the largest precursor of secondary PM. The measurement of secondary PM is very challenging, however. The atmospheric chemical reactions that form secondary PM are highly nonlinear and time dependent, making source apportionment difficult (Mysliwiec & Kleeman, 2002). In addition, the formation process of secondary PM can take hours or even days, which means that the air containing the pollution can travel long distances (AQEG, 2005, p. 3).

At the moment, standards are set for NO_x emissions by the European Commission. The current EURO 6 emission standards specify maximum NO_x emission factors of 60 and 80 mg/km for petrol and diesel cars, respectively (Williams & Minjares, 2016). However, real emissions from diesel cars are currently unlikely to comply with these regulations. Since the Volkswagen emission scandal, it has been shown that manufacturers routinely cheat in NO_x emission tests (Transport & Environment, 2015).

It is possible, however, to make a rough estimate of the contribution of secondary PM by using data from emission inventories and LCA conversion factors. The EMEP/EEA Inventory Guidebook (Ntziachristos & Samaras, 2013) specifies NO_x emission factors in the order of 59 and 210 mg/km for EURO 6 gasoline and diesel cars, respectively. The UK National Emission inventory (Pang & Murrells, 2014) has similar estimates of 49 and 274 mg/km for EURO 6 gasoline and diesel cars, respectively. The Dutch PRTR (Klein et al., 2016) suggests 40 and 230 mg/vkm for EURO 6 gasoline and diesel cars.

Very little research exists, which has determined how much of this NO_x turns into PM. The ReCiPe methodology (Goedkoop et al., 2014) for LCA used in Hooftman et al. (2016) proposes a conversion factor of 22%. This would mean that on average EURO 6 gasoline and diesel cars would produce in the order of 11 and 52 mg/vkm secondary PM, respectively (see Table 12.8). These values are likely to decrease as more representative NO_x test cycles are developed, which diesel vehicles will need to comply with.

4.3 Contribution of Resuspension

Just like secondary PM, the contribution of resuspension is often omitted from PM calculations. As Hooftman et al. (2016) note, resuspension is difficult to quantify and is dependent on several external factors. Thorpe and Harrison (2008) explain that because road dust is made up of a huge array of PM sources, a single tracer element cannot be used to determine resuspension contributions. Further complicating identification is the fact that the composition of road dust is heavily dependent on the local geology. In addition, it is difficult to determine if the resuspended particles have not already been attributed to other sources. This issue of double-counting particles is one of the reasons why so few emission inventories include resuspension (AQEG, 2005, p. 3).

The sampling process for estimating road dust emission factors can be done in various ways. The most popular methods involve taking measurements at the roadside. These can then be used together with measurements of ambient air concentrations of PM_{10} and NOx at two locations to determine emission factors (Abu-Allaban et al., 2003; Bukowiecki et al., 2010; Nicholson, 2000; Omstedt et al., 2005; Thorpe et al., 2007). Alternatively, emission

TABLE 12.9   Results of Resuspension Studies Using Various Measurement Methods

Reference Study	PM$_{10}$ (mg/vkm)	Method	Location
Nicholson (2000)	40	Roadside	UK
Fitz and Bufalino (2002)	64–124	Mobile	Riverside, CA
Abu-Allaban et al. (2003)	40–720	Roadside	Reno, NV; Durham, NC
Luhana et al. (2004)	0.8	Tunnel	Hatfield, UK
Omstedt et al. (2005)	205	Roadside	Stockholm, Sweden
Thorpe et al. (2007)	15–23	Roadside	London, UK
Bukowiecki et al. (2009)	5–76	Mobile	Zurich, Switzerland
Bukowiecki et al. (2010)	9–28	Roadside	Zurich and Reiden, Switzerland
Amato et al. (2011)	35	Roadside	Girona, Spain
Amato et al. (2012)	9.4–36.9	Roadside	Barcelona, Spain
Amato et al. (2016)	5.4–17.0	Roadside	Paris, France

factors can be determined based on the assumption of equilibrium between resuspension and deposition fluxes on roadways (Amato et al., 2016, 2012). Another method is taking samples at the inlet and outlet of a tunnel and comparing these to background PM levels (Luhana et al., 2004). A final method involves attaching a mobile measurement station to a moving vehicle to directly sample emitted particulates (Fitz & Bufalino, 2002, pp. 1–18; Bukowiecki et al., 2009) (see Table 12.9).

4.4 Future Research

Additional research is needed in multiple areas to accurately determine PM emissions from EVs. At the moment, many aspects of the quantification of emissions are based on simple assumptions or individual studies, rather than complete bodies of knowledge.

For example, the effect of additional weight on non-exhaust emissions is simply assumed in the studies of Van Zeebroek and De Ceuster (2013) and Hooftman et al. (2016). In Timmers and Achten (2016), EV tire, brake, and road wear emissions are based on one paper by Simons (2013). More studies, which investigate the influence of vehicle characteristics on emissions, are needed, such as those by Wang et al. (2016) and Salminen (2014). Especially the influence of weight on resuspension is poorly understood with a single dated study by Gillies et al. (2005) and one emission inventory offering recommendations.

The same problem applies to the effects of low rolling resistance tires and regenerative braking on brake wear. Several suggestions have been made by authors, but these tend to differ and none are backed up by concrete experimental data. Predictions range from a 66%–95% reduction in brake wear, which can make a large difference in total emissions. More experimental data are needed on EV driving behavior, as well as tire and brake pad replacement rates to more accurately assess EV non-exhaust emission factors.

The main area of uncertainty and need for clarifying research is in the contribution of resuspension and secondary PM. More accurate measurement methods are needed to reduce the uncertainty in emission factors for both. In the last couple of years, research into resuspension emission factors has increased (Amato et al., 2016), but it remains difficult to draw general conclusions due to the dependence on external factors. For secondary PM, more research is needed to assess what proportion of NO_x and other precursor gases effectively become PM_{10} and their impact on local urban air quality.

A final area of research and development that is needed is in abatement measures for non-exhaust emissions. One way to do this is through applying EV technology to ICEVs. So far, there has not been much interest from the auto industry to equip ICEVs with regenerative braking. However, once limits are introduced for non-exhaust emissions, this will be a viable way to reduce brake wear. The same applies to low rolling resistance tires. These are currently only used for EVs, but could provide further reductions in tire wear for all vehicles.

Another abatement measure where research is ongoing is in weight reduction methods. Keeping weight low is paramount for EVs, but the methods used to this can be applied to all vehicles. Research into the use of aluminum, high-strength steel, and carbon fiber could reduce weight, but several hurdles such as high cost and production difficulty remain (McKinsey & Company, 2012). Additionally, the development of smaller, lower weight vehicles in combination with government incentives would be one of the most effective ways of reducing non-exhaust PM.

5. CONCLUSIONS

This chapter has reviewed the current literature regarding PM emissions from battery EVs. Because the topic is relatively new, there is still much uncertainty and several issues are being debated. Even so, it is possible to draw some important conclusions where the literature has come to a consensus.

5.1 Remaining Uncertainties

The main source of uncertainty in determining the PM emissions from electric cars is the contribution of secondary PM. Still very little is known about how aerosols become secondary PM and their impact on local air quality. Most studies therefore do not take secondary PM into account. Hooftman et al. (2016) did account for this and their results suggest that as much as 99% of total PM from diesel cars is from secondary PM. However, this seems to be an overestimate compared with values proposed by emission inventories.

Further uncertainty exists in the relationship between vehicle weight and resuspension. Very few studies investigate this relationship and more up-to-date measurement data are needed. This is important because resuspension plays a major role in studies where it is included, such as Soret et al. (2014) and Timmers and Achten (2016).

In addition, it is unknown to what extent regenerative braking and especially low resistance tires reduce wear. More experimental data are needed to quantify this influence. Currently, most studies rely on assumptions without justification.

5.2 Consensus

Despite the remaining uncertainties, there are several issues on which the literature agrees. The first of these is that tailpipe PM emissions have seen a strong reduction over the past decades and will likely continue to reduce in the coming years. This has led to non-exhaust emissions becoming the dominant source of PM emissions. In fact, multiple studies suggest that non-exhaust emissions will account for 90% of total PM emissions from traffic by the end of the decade.

All studies agree that EVs reduce NOx, but still produce non-exhaust PM emissions. In addition, research has shown that an increase in weight of the vehicle fleet correlates to an increase in non-exhaust emissions. Studies indicate that a linear relationship exists between weight and non-exhaust emissions such as tire wear and resuspension. Emission inventories also consistently estimate higher non-exhaust emissions for heavier vehicles.

The relationship between weight and non-exhaust emissions is important because the literature shows that EVs are heavier than ICEVs and likely to remain that way in the future. Estimates for additional weight of present-day EVs compared with ICEVs range from 21% to 56%. For future EVs, the additional weight estimates range from 17% to 75%.

Therefore, the claims that EVs are emission free are unjustified. The increase in weight of EVs is likely linked to higher non-exhaust emissions. Results of individual studies vary depending on the methodology used, but it appears that EVs do not result in significant reductions of PM. As a result, EVs cannot be considered the single solution to urban air pollution.

References

Abu-Allaban, M., Gillies, J., Gertler, A., Clayton, R., & Proffitt, D. (2003). Tailpipe, resuspended road dust, and brake-wear emission factors from on-road vehicles. *Atmospheric Environment, 37*(37), 5283–5293. http://dx.doi.org/10.1016/j.atmosenv.2003.05.005.

Air Quality Expert Group. (2012). *Fine particulate matter ($PM_{2.5}$) in the UK.* London: UK Government. Available from https://www.gov.uk/government/uploads/system/uploads/attachment_data/file/69635/pb13837-aqeg-fine-particle-matter-20121220.pdf.

AIRUSE. (2016). *Deliverable B7.5: Technical guide for mitigation measures from the experience of Northern and Central European countries.* Available from http://airuse.eu/wp-content/uploads/2014/02/2016_AIRUSE-B7_5-Tech Guide-mitigation-meas-from-NtoC-Euro.pdf.

Althaus, H. J., & Gauch, M. (2010). *Vergleichende Ökobilanz individueller Mobilität: Elektromobilität versus konventionelle Mobilität mit Bio- und fossilen Treibstoffen.* Empa, Swiss Federal Laboratories for Materials Science and Technology. Report available from http://www.empa.ch/plugin/template/empa/*/104369/l=1.

Amato, F., Cassee, F. R., van der Gon, H. A. D., Gehrig, R., Gustafsson, M., Hafner, W., et al. (2014). Urban air quality: The challenge of traffic non-exhaust emissions. *J. Haz. Mat., 275,* 31–36. http://dx.doi.org/10.1016/j.jhazmat.2014.04.053.

Amato, F., Favez, O., Pandolfi, M., Alastuey, A., Querol, X., Moukhtar, S., et al. (2016). Traffic induced particle resuspension in Paris: Emission factors and source contributions. *Atmospheric Environment, 129,* 114–124. http://dx.doi.org/10.1016/j.atmosenv.2016.01.022.

Amato, F., Karanasiou, A., Moreno, T., Alastuey, A., Orza, J., Lumbreras, J., et al. (2012). Emission factors from road dust resuspension in a Mediterranean freeway. *Atmospheric Environment, 61,* 580–587. http://dx.doi.org/10.1016/j.atmosenv.2012.07.065.

Amato, F., Nava, S., Lucarelli, F., Querol, X., Alastuey, A., Baldasano, J. M., et al. (2010). A comprehensive assessment of PM emissions from paved roads: Real-world emission factors and intense street cleaning trials. *The Science of the Total Environment, 408*(20), 4309–4318. http://dx.doi.org/10.1016/j.scitotenv.2010.06.008.

Amato, F., Pandolfi, M., Escrig, A., Querol, X., Alastuey, A., Pey, J., et al. (2009). Quantifying road dust resuspension in urban environment by multilinear engine: A comparison with PMF2. *Atmospheric Environment, 43*(17), 2770—2780. http://dx.doi.org/10.1016/j.atmosenv.2009.02.039.

Amato, F., Pandolfi, M., Moreno, T., Furger, M., Pey, J., Alastuey, A., et al. (2011). Sources and variability of inhalable road dust particles in three European cities. *Atmospheric Environment, 45*(37), 6777—6787. http://dx.doi.org/10.1016/j.atmosenv.2011.06.003.

Amato, F., Schaap, M., Reche, C., & Querol, X. (2013). Road traffic: A major source of particulate matter in Europe. *Urban Air Quality in Europe*, 165—193. http://dx.doi.org/10.1007/698_2012_211.

AQEG. (2005). *Particulate matter in the United Kingdom*. Prepared for: Department for Environment, Food and Rural Affairs. Available from https://uk-air.defra.gov.uk/assets/documents/reports/aqeg/pm-summary.pdf.

Bahreini, R., Middlebrook, A., Gouw, J., Warneke, C., Trainer, M., Brock, C., et al. (2012). Gasoline emissions dominate over diesel in formation of secondary organic aerosol mass. *Geophysical Research Letters, 39*(6). http://dx.doi.org/10.1029/2011GL050718.

Barlow, T. (2014). *Briefing paper on non-exhaust particulate emissions from road transport*. Wokingham, UK: Transport Research Laboratory. [Cited 18 September 2016]. Available from http://www.lowemissionstrategies.org/downloads/Jan15/Non_Exhaust_Particles11.pdf.

Bauer, C., Hofer, J., Althaus, H., Del Duce, A., & Simons, A. (2015). The environmental performance of current and future passenger vehicles: Life cycle assessment based on a novel scenario analysis framework. *Applied Energy, 157*, 871—883. http://dx.doi.org/10.1016/j.apenergy.2015.01.019.

Borken-Kleefeld, J., & Ntziachristos, L. (2012). *The potential for further controls of emissions from mobile sources in Europe*. IIASA, Monitoring and Assessment of Sectorial Implementation Actions. [Cited 12 October 2016] Available from http://pure.iiasa.ac.at/10161/1/XO-12-014.pdf.

ten Broeke, H., Hulskotte, J., & Denier van der Gon, H. (2008). *Emissies door bandenslijtage afkomstig van het wegverkeer*. Dutch Emission Inventory. [Cited 11 October 2016]. Available from https://goo.gl/37kpXs.

Bukowiecki, N., Gehrig, R., Lienemann, P., Hill, M., Figi, R., & Buchmann, B. (2009). PM_{10} emission factors of abrasion particles from road traffic. *Swiss Federal Department of the Environment, Transport, Energy and Communications*, 1—194. Available from http://www.bafu.admin.ch/luft/00575/00578/index.html?lang=de&download=NHzLpZeg7t,lnp6I0NTU042l2Z6ln1acy4Zn4Z2qZpnO2Yuq2Z6gpJCGeXx8fGym162epYbg2c_JjKbNoKSn6A–.

Bukowiecki, N., Lienemann, P., Hill, M., Furger, M., Richard, A., Amato, F., et al. (2010). PM_{10} emission factors for non-exhaust particles generated by road traffic in an urban street canyon and along a freeway in Switzerland. *Atmospheric Environment, 44*(19), 2330—2340. http://dx.doi.org/10.1016/j.atmosenv.2010.03.039.

Burnham, A. (2012). *Updated vehicle specifications in the GREET vehicle-cycle model*. Argonne National Laboratory.

BUWAL, & German Federal Office for the Environment, Forests and Landscape. (2001). *Measures to reduce PM_{10} emissions*. Final Report. BUWAL Air Pollution Control Division. Cited in Lükewille et al. (2001).

Carslaw, D. (2006). New directions: A heavy burden for heavy vehicles: Increasing vehicle weight and air pollution. *Atmospheric Environment, 40*(8), 1561—1562. http://dx.doi.org/10.1016/j.atmosenv.2005.11.039.

Charron, A., Harrison, R., & Quincey, P. (2007). What are the sources and conditions responsible for exceedences of the 24h PM_{10} limit value ($50\mu gm^{-3}$) at a heavily trafficked London site? *Atmospheric Environment, 41*(9), 1960—1975. http://dx.doi.org/10.1016/j.atmosenv.2006.10.041.

Del Duce, A., Gauch, M., & Althaus, H. J. (2014). Electric passenger car transport and passenger car life cycle inventories in ecoinvent version 3. *International Journal of Life Cycle Assessment*, 1—13.

Delhaye, E., De Ceuster, C., & Maerivoet, S. (2010). *Internalisering van externe kosten van transport in Vlaanderen Eindrapport*. MIRA, Milieurapport Vlaanderen. [Cited 10 October 2016]. Available from http://www.milieurapport.be/Upload/Main/MiraData/MIRA-T/01_SECTOREN/01_06/TRAN_O&O_05.PDF.

Denier van der Gon, H., Gerlofs-Nijland, M., Gehrig, R., Gustafsson, M., Janssen, N., Harrison, R., et al. (2013). The policy relevance of wear emissions from road transport, now and in the future -an international workshop report and consensus statement. *Journal of the Air & Waste Management Association, 63*(2), 136—149. http://dx.doi.org/10.1080/10962247.2012.741055.

Dutch Government. (2011). *Elektrisch Rijden in de versnelling, Plan van Aanpak 2011—2015*. Amsterdam: Rijksoverheid. [Cited 14 September 2016]. Available from https://www.rvo.nl/sites/default/files/bijlagen/Plan%20van%20aanpak%20-elektrisch%20rijden%20in%20de%20versnelling-.pdf.

EAMA. (2016). *European automobile manufacturers association. Overview of purchase and tax incentives for electric vehicles in the EU in 2016*. Brussels: EAMA. [Cited 11 October 2016] Available from http://www.acea.be/uploads/publications/Electric_vehicles_overview_2016.pdf.

EEA. (2009). *EMEP-corinair emission inventory guidebook — 2009*. European Environmental Agency Technical Report N. 9. Available from http://www.eea.europa.eu/publications/emep-eea-emission-inventory-guidebook-2009.

EEA. (2014). *Air quality in Europe — 2014 report*. Copenhagen: EEA. Available from http://www.eea.europa.eu/publications/air-quality-in-europe-2014.

EPA. (2011). *AP42 section 13.2.1 paved roads*. Washington DC: EPA. [Cited 5 November 2016]. Available from https://www3.epa.gov/ttn/chief/ap42/ch13/final/c13s0201.pdf.

EU. (2005). *Thematic strategy on air pollution*. Brussels: European Commission. [Cited 14 September 2016]. Available from http://eur-lex.europa.eu/legal-content/EN/TXT/?uri=CELEX:52005DC0446.

EUROBAT. (2015). Battery technology roadmap 2030. [Cited 13 October, 2016]. Available from http://eurobat.org/sites/default/files/eurobat_emobility_roadmap_lores_2.pdf.

European Commission. (1999). *Regulation (Eec) No 4064/89 merger Procedure*. Brussels, Belgium. [Cited 13 October 2016] Available from http://ec.europa.eu/competition/mergers/cases/decisions/m1406_en.pdf.

Fitz, D., & Bufalino, C. (2002). *Measurement of PM_{10} emission factors from paved roads using on-board particle sensors*. California: EPA. Available from https://www3.epa.gov/ttnchie1/conference/ei11/dust/fitz.pdf.

Garben, M., Wiegand, G., Liwicki, M., & Eulitz, S. (1997). *Automobile traffic emission inventories in Berlin 1993. IVU GmbH Berlin, report commissioned by the Senate Department for Urban Development*. Unpublished. Berlin: Environmental Protection and Technology. Cited in Lükewille et al. (2001).

Garg, B., Cadle, S., Mulawa, P., Groblicki, P., Laroo, C., & Parr, G. (2000). Brake wear particulate matter emissions. *Environmental Science & Technology, 34*(21), 4463–4469. http://dx.doi.org/10.1021/es001108h.

Gebbe, Hartung, & Berthold. (1997). *Quantification of tire wear of motor vehicles in Berlin*. Berlin: TU Berlin, Environmental Protection and Technology. Cited in Lükewille et al. (2001).

Gillies, J., Etyemezian, V., Kuhns, H., Nikolic, D., & Gillette, D. (2005). Effect of vehicle characteristics on unpaved road dust emissions. *Atmospheric Environment, 39*(13), 2341–2347. http://dx.doi.org/10.1016/j.atmosenv.2004.05.064.

Goedkoop, M., Heijungs, R., Huijbregts, M., de Schryver, A., Struijs, J., & van Zelm, R. (2014). *ReCiPe 2008, ReCiPe 2008: A life cycle impact assessment method which comprises harmonised category indicators at the midpoint and endpoint level*. Characterisation and normalisation factors http://www.lcia-recipe.net/file-cabinet/ReCiPe111.xlsx?attredirects=0.

Graham, L. (2005). Chemical characterization of emissions from advanced technology light-duty vehicles. *Atmospheric Environment, 39*, 2385–2398. http://dx.doi.org/10.1016/j.atmosenv.2004.10.049.

Groupe Renault. (2015). *Registration document 2014, including the annual financial report*. [Cited 13 October 2016] Available from https://group.renault.com/wp-content/uploads/2015/04/renault_regitration_doc.pdf.

Guevara, M., Martínez, F., Arevalo, G., Gassó, S., & Baldasano, J. M. (2013). An improved system for modelling Spanish emissions: HERMESv2.0. *Atmospheric Environment, 81*, 209–221. http://dx.doi.org/10.1016/j.atmosenv.2013.08.053.

Gunawardana, C., Goonetilleke, A., Egodawatta, P., Dawes, L., & Kokot, S. (2012). Source characterisation of road dust based on chemical and mineralogical composition. *Chemosphere, 87*(2), 163–170. http://dx.doi.org/10.1016/j.chemosphere.2011.12.012.

Hooftman, N., Oliveira, L., Messagie, M., Coosemans, T., & Van Mierlo, J. (2016). Environmental analysis of petrol, diesel and electric passenger cars in a Belgian urban setting. *Energies, 9*(2), 84. http://dx.doi.org/10.3390/en9020084.

Hoogerbrugge, R., Denier van der Gon, H., van Zanten, M., & Matthijsen, J. (2015). *Trends in particulate matter*. Netherlands Research Program on Particulate Matter, Bilthoven. Available from https://www.researchgate.net/publication/264408355_Trends_in_Particulate_Matter.

Humbert, S., Fantke, P., & Jolliet, O. (2015). Particulate matter formation. *Life Cycle Impact Assessment*, 97–113. http://dx.doi.org/10.1007/978-94-017-9744-3_6.

IEA. (2016). *Global EV outlook 2016: Beyond one million electric cars*. International Energy Agency. [Cited 21 September 2016]. Available from https://www.iea.org/publications/freepublications/publication/Global_EV_Outlook_2016.pdf.

Innovative Systems Consultants. (2010). *Save fuel with lower rolling resistance tyres*. [Cited 1 October 2016] Available from http://www.isc-consultants.co.uk/ISC/Distribution_Resources_1_files/FBP%201032%20WEB%20-%20-TAGGED.pdf.

Jörß, W., & Handke, V. (2007). *Emissionen und Maßnahmenanalyse Feinstaub 2000-2020*. Berlin: Umweltbundesambt. [Cited 14 September 2016]. Available from https://www.umweltbundesamt.de/sites/default/files/medien/publikation/long/3309.pdf.

Ketzel, M., Omstedt, G., Johansson, C., Düring, I., Pohjola, M., Oettl, D., et al. (2007). Estimation and validation of $PM_{2.5}/PM_{10}$ exhaust and non-exhaust emission factors for practical street pollution modelling. *Atmospheric Environment, 41*(40), 9370—9385. http://dx.doi.org/10.1016/j.atmosenv.2007.09.005.

Klein, J., Geilenkirchen, G., Hulskotte, J., Ligterink, N., Fortuin, P., & Molnár-in 't Veld, H. (2014). *Methods for calculating the emissions of transport in The Netherlands — 2014*. Dutch Emission Inventory [Cited 8 October 2016]. Available from https://goo.gl/Yn7Gbn.

Klein, J., Molnár-in 't Veld, H., Geilenkirchen, G., Hulskotte, J., Ligterink, N., Kadijk, G., et al. (2016). *Methods for calculating the emissions of transport in The Netherlands - 2016*. Dutch Emission Inventory [Cited 3 December 2016]. Available from https://goo.gl/zv13bX.

Kousoulidou, M., Ntziachristos, L., Mellios, G., & Samaras, Z. (2008). Road-transport emission projections to 2020 in European urban environments. *Atmospheric Environment, 42*(32), 7465—7475. http://dx.doi.org/10.1016/j.atmosenv.2008.06.002.

Kupiainen, K. (2007). *Road dust from pavement wear and traction sanding* (1st ed.). Helsinki: Finnish Environment Institute.

Ligterink, N., Stelwagen, U., & Kuenen, J. (2014). *Emission factors for alternative drivelines and alternative fuels*. TNO Report for the Dutch Pollutant Release and Transfer Register. Available from goo.gl/vsmnME.

Luhana, L., Sokhi, R., Warner, L., Mao, H., Boulter, P., McCrae, I., et al. (2004). Measurement of non-exhaust particulate matter. *Deliverable 8 of the European Commission*, 1—103.

Lükewille, A., Bertok, I., Amann, M., Cofala, J., Gyarfas, F., Heyes, C., et al. (2001). *A framework to estimate the potential and costs for the control of fine particulate emissions in Europe*. IIASA Interim Report IR-01-023. Laxenburg, Austria: International Institute for Applied Systems Analysis.

McKinsey & Company. (2012). *Lightweight, heavy impact: How carbon fiber and other lightweight materials will develop across industries and specifically in automotive*. [Cited 14 Octover 2016] Available from http://www.mckinsey.com/~/media/mckinsey/dotcom/client_service/Automotive%20and%20Assembly/PDFs/Lightweight_heavy_impact.ashx.

McKinsey & Company. (2014). *Electric vehicles in Europe: Gearing up for a new phase?* Amsterdam Roundtables Foundation. [Cited 11 October 2016]. Available from http://www.mckinsey.com/~/media/McKinsey%20Offices/Netherlands/Latest%20thinking/PDFs/Electric-Vehicle-Report-EN_AS%20FINAL.ashx.

Moawad, A., Sharer, P., & Rousseau, A. (2011). *Light-duty vehicle fuel consumption displacement potential up to 2045*. ANL/ESD/11—14. Argonne, Ill: Center for Transportation Research, Argonne National Laboratory.

Mock, P., & Yang, Z. (2014). *Driving electrification: A global comparison of fiscal incentive policy for electric vehicles*. Washington DC: International Council on Clean Transportation. [Cited 11 October 2016]. Available from http://www.theicct.org/sites/default/files/publications/ICCT_EV-fiscal-incentives_20140506.pdf.

Murrells, T., & Pang, Y. (2013). *Emission factors for alternative vehicle technologies*. London: National Atmospheric Emissions Inventory. [Cited 15 September 2016]. Available from http://naei.defra.gov.uk/resources/NAEI_Emisison_factors_for_alternative_vehicle_technologies_Final_Feb_13.pdf.

Mysliwiec, M. J., & Kleeman, M. J. (2002). Source apportionment of secondary airborne particulate matter in a polluted atmosphere. *Environmental Science & Technology, 36*(24), 5376—5384. http://dx.doi.org/10.1021/es020832s.

Nealer, R., & Hendrickson, T. (2015). Review of recent life cycle assessments of energy and greenhouse gas emissions for electric vehicles. *Current Sustainable/Renewable Energy Reports, 2*(3), 66—73. http://dx.doi.org/10.1007/s40518-015-0033-x.

Neumann, I., Franke, T., Cocron, P., Bühler, F., & Krems, J. F. (2015). Eco-driving strategies in battery electric vehicle use—how do drivers adapt over time? *IET Intelligent Transport Systems, 9*(7), 746—753. http://dx.doi.org/10.1049/iet-its.2014.0221.

Nicholson, K. (2000). *Resuspension from roads: Initial estimate of emission factors*. Internal AEA Technology Report. Cited in Thorpe et al. (2007).

Ntziachristos, L., & Boulter, P. (2009). *EMEP/EEA air pollutant emissions inventory guidebook 2009: Road vehicle tyre and brake wear; road surface wear*. Copenhagen: European Environment Agency. Cited in Simons (2013).

Ntziachristos, L., & Boulter, P. (2013). *EMEP/EEA air pollutant emissions inventory guidebook 2013: Road vehicle tyre and brake wear; road surface wear*. Copenhagen: European Environmental Agency. [Cited 20 September 2016]. Available from http://www.eea.europa.eu/publications/emep-eea-guidebook-2013/part-b-sectoral-guidance-chapters/1-energy/1-a-combustion/1-a-3-b-road-tyre.

Ntziachristos, L., & Samaras, Z. (2013). *EMEP/EEA emission inventory Guidebook 2013: Exhaust emissions from road transport*. Copenhagen: EEA. [Cited 11 December 2016]. Available from http://www.eea.europa.eu/publications/emep-eea-guidebook-2013/part-b-sectoral-guidance-chapters/1-energy/1-a-combustion/1-a-3-b-road-transport.

NVF. (2008). *Road wear from heavy vehicles - an overview*. Report nr. 08/2008, NVF committee Vehicles and Transports. [Cited 24 November, 2016] Available from http://www.nvfnorden.org/lisalib/getfile.aspx?itemid=1586.

Omstedt, G., Bringfelt, B., & Johansson, C. (2005). A model for vehicle-induced non-tailpipe emissions of particles along Swedish roads. *Atmospheric Environment, 39*(33), 6088–6097. http://dx.doi.org/10.1016/j.atmosenv.2005.06.037.

Pang, Y., & Murrells, T. (2014). *NOx speed-related emission functions (COPERT 4v10)*. London: National Emissions Inventory. [Cited 23 November 2016]. Available from http://naei.defra.gov.uk/data/ef-transport.

Pant, P., & Harrison, R. (2013). Estimation of the contribution of road traffic emissions to particulate matter concentrations from field measurements: A review. *Atmospheric Environment, 77*, 78–97. http://dx.doi.org/10.1016/j.atmosenv.2013.04.028.

Pay, M. T., Jimenez-Guerrero, P., & Baldasano, J. M. (2011). Implementation of resus- pension from paved roads for the improvement of CALIOPE air quality system in Spain. *Atmospheric Environment, 45*, 802–807. http://dx.doi.org/10.1016/j.atmosenv. 2010. 10.032.

Platform for Electro-mobility. (2016). *Briefing: Non-Exhaust emissions of electric cars*. Brussels, Belgium. [Cited 28 September 2016] Available from http://www.platformelectromobility.eu/2016/06/28/non-exhaust-emissions-of-electric-cars/.

Reed, P. (2013). *What you need to know about low-rolling-resistance tires*. Edmunds. [Cited 1 October 2016] Available from http://www.edmunds.com/fuel-economy/resistance-movement.html.

Rexeis, M., & Hausberger, S. (2009). Trend of vehicle emission levels until 2020-prognosis based on current vehicle measurements and future emission legislation. *Atmospheric Environment, 43*(31), 4689–4698. http://dx.doi.org/10.1016/j.atmosenv.2008.09.034.

Salminen, H. (2014). *Parametrizing tyre wear using a brush tyre model*. Master's thesis. Sweden: Royal Institute of Technology Stockholm.

Simons, A. (2013). Road transport: New life cycle inventories for fossil-fuelled passenger cars and non-exhaust emissions in ecoinvent v3. *The International Journal of Life Cycle Assessment*, 1–15. http://dx.doi.org/10.1007/s11367-013-0642-9.

Soret, A., Guevara, M., & Baldasano, J. (2014). The potential impacts of electric vehicles on air quality in the urban areas of Barcelona and Madrid (Spain). *Atmospheric Environment, 99*, 51–63. http://dx.doi.org/10.1016/j.atmosenv.2014.09.048.

Squizzato, S., Masiol, M., Agostini, C., Visin, F., Formenton, G., Harrison, R., et al. (2016). Factors, origin and sources affecting PM1 concentrations and composition at an urban background site. *Atmospheric Research, 180*, 262–273. http://dx.doi.org/10.1016/j.atmosres.2016.06.002.

Thorpe, A., & Harrison, R. (2008). Sources and properties of non-exhaust particulate matter from road traffic: A review. *The Science of the Total Environment, 400*(1–3), 270–282. http://dx.doi.org/10.1016/j.scitotenv.2008.06.007.

Thorpe, A., Harrison, R., Boulter, P., & McCrae, I. (2007). Estimation of particle resuspension source strength on a major London Road. *Atmospheric Environment, 41*(37), 8007–8020. http://dx.doi.org/10.1016/j.atmosenv.2007.07.006.

Timmers, V., & Achten, P. (2016). Non-exhaust PM emissions from electric vehicles. *Atmospheric Environment, 134*, 10–17. http://dx.doi.org/10.1016/j.atmosenv.2016.03.017.

Transport & Environment. (2015). *Don't breathe here: Beware the invisible killer. Tackling air pollution from vehicles*. [Cited 25 December 2016] Available from http://www.transportenvironment.org/sites/te/files/publications/Dont_Breathe_Here_report_FINAL.pdf.

Van de Weijer, B. (2016). *Positieve effected elektrische auto op luchtkwaliteit overschat*. De Volkskrant. [Cited 12 October 2016] Available from http://www.volkskrant.nl/wetenschap/positieve-effecten-elektrische-auto-op-luchtkwaliteit-overschat~a4296738/.

Van Zeebroek, B., & De Ceuster, G. (2013). *Elektrische wagens verminderen fijnstof nauwelijks*. Transport & Mobility Leuven. [Cited 4 October 2016] Available from http://www.tmleuven.be/project/fijnstof/belang_niet-uitlaat_fijn_stof_emissies_lang.pdf.

Viana, M. (2011). Urban air quality in Europe. *The Handbook of Environmental Chemistry, 26*. http://dx.doi.org/10.1007/978-3-642-38451-6.

Wang, C., Huang, H., Chen, X., & Liu, J. (2016). The influence of the contact features on the tyre wear in steady-state conditions. *Proceedings of the Institution of Mechanical Engineers, Part D: Journal of Automobile Engineering.* http://dx.doi.org/10.1177/0954407016671462.

Weinbruch, S., Worringen, A., Ebert, M., Scheuvens, D., Kandler, K., Pfeffer, U., et al. (2014). A quantitative estimation of the exhaust, abrasion and resuspension components of particulate traffic emissions using electron microscopy. *Atmospheric Environment, 99,* 175–182. http://dx.doi.org/10.1016/j.atmosenv.2014.09.075 (EPA, 2006).

Williams, M., & Minjares, R. (2016). *A technical summary of Euro 6/VI vehicle emission standards.* International Council on Clean Transportation. [Cited 28 October 2016] Available from http://www.theicct.org/sites/default/files/publications/ICCT_Euro6-VI_briefing_jun2016.pdf.

Yang, Z. (2016). *2015 Global electric vehicle trends: Which markets are up (the most).* International Council on Clean Transportation. [Cited 20 September 2016] Available from http://www.theicct.org/blogs/staff/2015-global-electric-vehicle-trends.

CHAPTER

13

Air Quality in Subway Systems

*Teresa Moreno[1], Vânia Martins[2], Cristina Reche[1],
Maria Cruz Minguillón[1], Eladio de Miguel[3], Xavier Querol[1]*

[1]Institute of Environmental Assessment and Water Research (IDÆA), Spanish National
Research Council (CSIC), Barcelona, Spain; [2]Universidade de Lisboa, Portugal; [3]Transports
Metropolitans de Barcelona, TMB Santa Eulalia, L'Hospitalet de Llobregat, Spain

OUTLINE

1. Particulate Matter Concentrations
   and Size Ranges 301
   1.1 On Platforms 301
   1.2 Inside Trains 307
2. Airborne Particle Characteristics,
   Sources, and Chemical
   Composition 308
3. Bioaerosols in the Subway System 312
4. Health Effects, Mitigation, and
   Future Trends 313

Acknowledgments 316

References 316

The operation of deep-level underground electric railways began (in London) in the late 19th century, and nowadays, over 125 years later, such systems can carry more passengers than any other city transport mode, so that many millions of commuters worldwide regularly spend a proportion of their daily time in subway trains. These systems can be viewed as a transport lifeline, helping to improve the quality of urban infrastructure and relieve road traffic congestion (Vasconcellos, 2001). They are of especial value in aiding the reduction of air pollution above ground: Silva, Saldiva, Amato-Lourenço, Rodrigues-Silva, and Miraglia (2012) concluded that a subway strike in São Paulo led to a 45%–60% increase in daily mean PM_{10} (particulate matter below 10 microns in size) concentrations due to extra road traffic. However, one environmental disadvantage of any underground transport system is that it operates in a confined space that may permit the accumulation of unhealthy concentrations of airborne contaminants (e.g., see the early review by Nieuwenhuijsen,

Non-Exhaust Emissions
https://doi.org/10.1016/B978-0-12-811770-5.00013-3

Gómez-Perales, & Colvile, 2007). Furthermore, much of the inhalable airborne particulate matter (PM) in subway air is sourced underground and so is substantially different in nature from typical PM found outdoors (e.g., Adams, Nieuwenhuijsen, Colvile, McMullen, & Khandelwal, 2001; Furuya et al., 2001; Querol et al., 2012; Salma, Weidinger, & Maenhaut, 2007). In particular, much subway PM is sourced from moving train parts, such as wheels and brake pads, as well as from the steel rails and power supply materials, giving the particles a peculiarly metalliferous character. Dominantly ferruginous particles derived from these materials are mixed with PM from a range of other sources, including rock ballast from the track, bioaerosols, and infiltrating outdoor air, and driven through the tunnel system on turbulent air currents generated by train piston effects and ventilation systems (e.g., Johansson & Johansson, 2003; Kim, Kim, Roh, Lee, & Kim, 2008; Kwon, Jeong, Park, Kim, & Cho, 2015; Loxham et al., 2013; Moreno et al., 2014; Park & Ha, 2008; Perrino, Marcovecchio, Tofful, & Canepari, 2015; Ripanucci, Grana, Vicentini, Magrini, & Bergamaschi, 2006; Salma et al., 2007).

With the exception of a few pioneering studies such as that by Trattner, Perna, Kimmel, and Birch (1975) (on respirable dust on an underground station of the Newark system) or Chan, Spengler, Özkaynak, and Lefkopoulou (1991) (on commuter exposure to volatile organic compounds in Boston), most studies on subway air quality have been published over the last 15 years. A tabulated summary of many of the recent studies focused on PM in subway systems worldwide is provided in Table 13.1, offering comparative information on the kinds of air quality data that have been collected, such as time duration of the study, microenvironment measured (on platforms and/or inside train), identification of the measurement type (real-time or off-line), the characteristics of the measurements (e.g., time resolution, size fraction, number of samples), and chemical species analyzed. Most of these studies have focused on particle mass concentration on platforms (much more rarely inside trains), commonly on a limited number of samples (usually much less than 50). The information offered by such studies is often not directly comparable because some measurement campaigns have used off-line monitors to measure the PM concentrations, whereas others used personal exposure equipment. Furthermore, papers may report on different particle size fractions, or use data uncorrected against reference methods, or not provide samples directly equivalent to personal exposure. Not many studies have collected enough PM sample for detailed chemical analysis (usually excluding organic components) and, although likely sources are routinely mentioned, very few studies show quantitative data on the contribution of each source to the total PM mass using positive matrix factorization analyses (Martins et al., 2016; D. Park et al., 2014; Park, Oh, Yoon, Park, & Lee, 2012).

Despite these limitations, however, there are by now (January 2017) nearly 200 publications dealing with subway air quality, mostly dealing with underground train systems in European, East Asian, or North and South American cities, and offering data on a range of inorganic and organic particles and gases. A comprehensive reference list is provided at the end of this chapter and demonstrates that, in Europe, publications have so far most commonly focused on data from Stockholm (Bigert et al., 2008; Gustafsson, Blomqvist, Swietlicki, Dahl, & Gudmundsson, 2012; Johansson & Johansson, 2003; Karlsson, Nilsson, & Möller, 2005; Midander et al., 2012; Sundh, Olofsson, Olander, & Jansson, 2009) and Barcelona (Martins, Minguillón et al., 2015; Martins, Moreno, Mendes et al., 2016; Martins, Moreno et al., 2015; Martins, Moreno, Minguillón et al., 2016; Moreno, Kelly et al., 2017;

TABLE 13.1 Summarized Literature Review of Subway Studies

| | | Measurements on Platform | | | | | | | Measurements Inside Train | | | | |
| | | Real-Time | | | Off-Line | | | | Real-Time | | | Off-Line | Study |
City	Time Duration	Stations Number	Time Resolution	Size Fraction	Stations Number	Size Fraction	Samples Number	Species Analyzed	Lines Number	Time Resolution	Size Fraction	Samples Number	Reference
Amsterdam (Netherlands)	3 days	—	—	—	1	<0.18, <2.5, 2.5–10 μm	3	NO_3^-, SO_4^{2-}, Cl^-, Fe, Cu, As, Mn, Zr, Mo, Sn, V, Cr, Ni, Nb, Hf, Ca, Mg, Zn, Ba, Sb, Rb, Cd, Pb, Al, Ti, Sr, Sc, La, Hg, Li, B, Na	—	—	—	—	Loxham et al. (2013)
Athens (Greece)	21 days	5	5 s	<1, <2.5, <10 μm	1	<2.5 μm	18	TC, SO_4^{2-}, Cl^-, NO_3^-, NH_4^+, Fe, Ca, Al, Ba, Mg, Cu, Mn, Zn, Na, K, Ti, Cr, Sr, Mo, Sb, Sn, Ni, As, Pb, V, Co, W, Li, Rb, Cd, Bi	2	5 s	<1, <2.5, <10 μm	—	Martins, Minguillón et al. (2015)
Barcelona (Spain)	21 days	2	5 min	<1, <2.5, <10 μm	2	<2.5 (11 days); <10 μm (10 days)	20 (PM_{10}) + 19 ($PM_{2.5}$)	Cl^-, SO_4^{2-}, NO_3^-, NH_4^+, OC, EC, Fe, Ca, Al, Ba, Mg, Cu, Mn, Zn, Na, K, Ti, Cr, Sr, Zr, Mo, Sb, Sn, Ni, As, Pb, V, Co, P, W, Li, Hf, Rb, Nb, Ge, Ga, U, Y, Th, Ta, Cd, Bi	3	30 s	<1, <2.5, <10 μm	—	Querol et al. (2012)
Barcelona (Spain)	13 days	10	15 min	<1, <3, <10 μm	—	—	—	—	—	—	—	—	Moreno et al. (2014)

(Continued)

TABLE 13.1 Summarized Literature Review of Subway Studies—cont'd

City	Time Duration	Measurements on Platform							Measurements Inside Train				Study Reference
		Real-Time			Off-Line				Real-Time			Off-Line	
		Stations Number	Time Resolution	Size Fraction	Stations Number	Size Fraction	Samples Number	Species Analyzed	Lines Number	Time Resolution	Size Fraction	Samples Number	
Barcelona (Spain)	32 days	—	—	—	4	<10 μm	4 (1/station)	C, O, Na, Mg, Al, Si, S, Cl, K, Ca, Ti, V, Cr, Mn, Fe, Co, Zn, As, Sn, Sb, Ba	—	—	—	—	Moreno, Martins et al. (2015) and Moreno, Reche et al. (2015)
Barcelona (Spain)	8 months	24	5 s	<1, <2.5, <10 μm	4	<2.5 μm	236 (55/station)	Cl⁻, SO$_4^{2-}$, NO$_3^-$, NH$_4^+$, TC, Fe, Ca, Al, Ba, Mg, Cu, Mn, Zn, Na, K, Ti, Cr, Sr, Zr, Mo, Sb, Sn, Ni, As, Pb, V, Co, P, W, Li, Hf, Rb, Nb, Ge, Ga, U, Y, Th, Ta, Cd, Bi. PAHs.	6	5 s	<1, <2.5, <10 μm	—	Martins, Moreno et al. (2015) and Martins et al. (2016)
Boston (USA)	—	—	—	0.02–1, <2.5 μm	—	1 μm	—	PAHs	—	—	—	—	Levy et al. (2002)
Budapest (Hungary)	1 day	1	30 s	<10 μm	1	<2, 2–10 μm	4	Na, Mg, Al, Si, P, S, Cl, K, Ca, Ti, V, Cr, Mn, Fe, Ni, Cu, Zn, Ga, Ge, As, Se, Br, Rb, Sr, Y, Zr, Nb, Mo, Ba, Pb	—	—	—	—	Salma et al. (2007, 2009)
Buenos Aires (Argentina)	30 days (5 days/station)	—	—	—	6	TSP	30 (5/station)	Fe, Zn, Cu	—	—	—	—	Murruni et al. (2009)
Cairo (Egypt)	7 days	—	—	—	—	TSP	—	—	—	—	—	—	Awad (2002)

Location												Reference
Helsinki (Finland)	16 days	1	1 min	2	<2.5 µm	12 (6/station)	Al, Ca, Cl, Cr, Cu, Fe, K, Mn, Ni, P, Pb, S, Si, Ti, V, Zn, OC, EC	—	—	<2.5 µm	5 (PM₂.₅)	Aarnio et al. (2005)
Hong Kong (China)	—	—	—	—	—	—	—	—	—	<2.5, <10 µm	—	Chan et al. (2002)
Istanbul (Turkey)	—	—	—	—	—	—	—	—	30 s	<2.5 µm	—	Onat and Stakeeva (2013)
Istanbul (Turkey)	35 days	6	15 min	6	<0.4, 0.4–0.7, 0.7–1.1, 1.1–2.1, 2.1–3.3, 3.3–4.7, 4.7–5.8, 5.8–9, >9 µm	—	Cu, Fe	—	—	—	—	Şahin, Onat, Stakeeva, Ceran, and Karim (2012)
London (UK)	15 days	—	—	—	—	—	—	—	—	—	56 (PM₂.₅)	Adams et al. (2001)
London (UK)	3 days	3	<2.5 µm	3	2.5,<3.5, <10 µm	—	Fe, Cr, Cu, Zn, Mn	3	—	<2.5 µm	—	Seaton et al. (2005)
Los Angeles (USA)	75 days	27	30 s	—	<2.5, 2.5 –10 µm	—	Cl⁻, NO₃⁻, SO₄²⁻, PO₄³⁻, Na⁺, NH₄⁺, K, Mg, Al, K, Ca, Ti, Cr, Mn, Fe, Co, Ni, Cu, Zn, Mo, Cd, Ba, Eu, OC, EC, PAHs, hopanes, steranes	2 (1 UG; 1 AG)	30 s	<2.5, <10 µm	—	Kam, Cheung et al. (2011), Kam, Ning et al. (2011), and Kam et al. (2013)
Mexico city (Mexico)	4 weeks	—	—	—	<2.5 µm	18	Si, S, K, Ca, Ti, V, Cr, Mn, Fe, Ni, Cu, Zn, Se, Br, Pb, TC, OC, EC	—	—	—	—	Gómez-Perales et al. (2004)

(Continued)

TABLE 13.1 Summarized Literature Review of Subway Studies—cont'd

| | | Measurements on Platform | | | | | | | Measurements Inside Train | | | | |
| | | Real-Time | | | Off-Line | | | | Real-Time | | | Off-Line | |
City	Time Duration	Stations Number	Time Resolution	Size Fraction	Stations Number	Size Fraction	Samples Number	Species Analyzed	Lines Number	Time Resolution	Size Fraction	Samples Number	Study Reference
Mexico city (Mexico)	21 days	—	—	—	1	<2.5, <10 µm	42 (1/day – each size)	Al_2O_3, SiO_2, Ca, Fe, Ba, Mg, Cu, Co, Cr, Mn, Ni, Pb V, Zn, Na, K, S	—	—	—	—	Múgica-Alvarez et al. (2012)
Milan (Italy)	54 days (9/station)	—	—	—	6	<10 µm	162 (3/day)	Al, Si, S, Cl, K, Ca, Ti, Mn, Ni, Cu, Zn, Br, Rb, Pb, Fe, Ba, Sb, Sr, Sn	—	—	—	—	Colombi et al. (2013)
Naples (Italy)	5 days	7 (1 AG; 6 UG)	120 s	<2.5, <10 µm	—	—	—	—	1	120 s	<2.5, <10 µm	—	Carteni et al. (2015)
New York (USA)	8 h	—	—	—	—	<2.5 µm	2 (5h each)	Fe, Mn, Cr	—	—	—	2 ($PM_{2.5}$, 3h each)	Chillrud et al. (2004)
New York (USA)	—	—	1 min	<2.5 µm	—	—	—	—	—	—	—	—	Morabia et al. (2009)
New York (USA)	4 months	—	—	—	—	<2.5 µm	39	Fe, Cr, Mn	—	—	—	—	Grass et al. (2010)
Oporto (Portugal)	19 days	5	5 s	<1, <2.5, <10 µm	1	<2.5 µm	15	TC, SO_4^{2-}, Cl^-, NO_3^-, NH_4^+, Fe, Ca, Al, Ba, Mg, Cu, Mn, Zn, Na, K, Ti, Cr, Sr, Mo, Sb, Sn, Ni, As, Pb, V, Co, W, Li, Rb, Cd, Bi	2	5 s	<1, <2.5, <10 µm	—	Martins et al. (2016)
Paris (France)	15 days	—	—	—	2	<10 µm	—	Fe, Mn, Ca, Cu, S, Si	—	—	—	—	Bachoual et al. (2007)

Paris (France)	11 days	1 min	1	<2.5, <10 μm	1	<2.5, 2.5–10 μm	18	Na⁺, NH₄⁺, Cl⁻, NO₃⁻, SO₄²⁻, Mg, Ca, K, BC, OC	–	–	Raut et al. (2009)
Prague (Czech Republic)	1 year	1 min	–	<10 μm	–	–	–	–	–	–	Braniš et al. (2006)
Prague (Czech Republic)	24 h	3 min	1	14–637 nm 0.5–20 μm	1	<1, <2.5, <10 μm	–	TC, SO₄²⁻, Cl⁻, NO₃⁻, NH₄⁺, Fe, Ca, Al, Ba, Mg, Cu, Mn, Zn, Na, K, Ti, Cr, Sr, Mo, Sb, Sn, Ni, As, Pb, V, Co, W, Li, Rb, Cd, Bi	–	–	Cusack et al. (2015)
Rome (Italy)	–	–	–	–	–	<10 μm	–	Fe, Cu, Zn, Mn, Pb, Sb, Ni, Cr, rare earth, Sn, Th, Zr, Co, As, Y, Bi, Cd	–	–	Ripanucci et al. (2006)
Rome (Italy)	25 days	–	1	0.3–0.5, 0.5–0.7, 0.7–1, 1–2, 2–3, 3–5, 5–10, >10 μm	1	<10 μm	5	Al, Ca, Fe, K, Mg, Na, Si, Cl, NO₃⁻, SO₄²⁻, Na, NH₄⁺, Mg, Ca, As, Ba, Be, Cd, Ce, Co, Cu, Fe, La, Mn, Ni, Pb, Sb, Sn, Sr, Ti, V, Zn, OC, EC (*)	–	10 (PM₁₀, with & without AC)	Perrino et al. (2015)
Rotterdam (Netherlands)	1 day	–	–	–	1	<0.18, 0.18–2.5, 2.5–10 μm	1	Al, Cu, Fe, Ni, V, Zn, NH₄⁺, Cl⁻, NO₃⁻, SO₄²⁻, EC, OC, PAHs	–	–	Steenhof et al. (2011)
Seoul (Korea)	–	–	–	–	5	–	90	–	–	–	Cho et al. (2006)

(Continued)

TABLE 13.1 Summarized Literature Review of Subway Studies—cont'd

| | Measurements on Platform | | | | | | | | Measurements Inside Train | | | | |
| | Real-Time | | | | Off-Line | | | | Real-Time | | | Off-Line | |
City	Time Duration	Stations Number	Time Resolution	Size Fraction	Stations Number	Size Fraction	Samples Number	Species Analyzed	Lines Number	Time Resolution	Size Fraction	Samples Number	Study Reference
Seoul (Korea)	—	22 (8 AG; 14 UG)	—	<2.5, <10 μm	—	—	—	—	4	—	<2.5, <10 μm	—	Kim et al. (2008)
Seoul (Korea)	8 days	—	—	—	1	2–1, 2–4, 4–8, 8–16, >16 μm	8	Fe, C, N, Mg, Al, Si, S, Ca, Ti, Cr, Mn, Ni, Cu, Zn, Ba	—	—	—	—	Kang et al. (2008)
Seoul (Korea)	3 days	—	—	—	2	1–2.5, 2.5–10 μm	6 (3/ station)	Fe, C, Ca, Mg, Si, Ba, Al, NO$_3^-$, SO$_4^{2-}$, Na$^+$, NH$_4^+$, Cu	—	—	—	—	B. W. Kim et al. (2010)
Seoul (Korea)	6 days	—	—	—	6	TSP	3–5/ station	Fe, Cu, K, Ca, Zn, Ni, Na, Mn, Mg, Cr, Cd, Ba, Pb	—	—	—	—	C. H. Kim et al. (2010)
Seoul (Korea)	—	—	—	—	—	TAB	67	—	—	—	—	—	Hwang et al. (2010)
Seoul (Korea)	—	—	—	—	—	<10 μm	—	PAHs	—	—	—	—	H. J. Jung et al. (2012) and M. H. Jung et al. (2012)
Seoul (Korea)	—	—	—	—	1	<2.5, <10 μm	—	—	—	—	—	—	Kim et al. (2012)
Seoul (Korea)	80 h (40h/ tunnel)	—	—	—	2 (tunnel samples)	<10 μm	23	—	—	—	—	—	Eom et al. (2013)
Seoul (Korea)	4 days	—	—	—	—	—	—	—	2	30 s	<2.5, <10 μm	—	Kim et al. (2014)

City	Duration	No.	Time	Particle size	Sample description	No. samples	Elements	No.	Time	Size	Standard	Reference
Seoul (Korea)	32 h	1 (tunnel)	—	<1, <2.5, <10 μm	—	—	—	—	—	—	—	Son, Dinh et al. (2014) and Son, Jeon et al. (2014)
Seoul (Korea)	3 weeks (3 seasons)	6	1 min	<1, <2.5, <10 μm	—	—	—	—	—	—	—	Kwon et al. (2015)
Seoul (Korea)	—	16	—	—	Platform floor samples	66	—	—	—	—	—	Hwang and Park (2014)
Seoul (Korea)	12 days	4	—	—	1–2.5, 2.5–10, <10 μm	12 (3/station)	Fe, C, N, Mg, Al, Si, S, Ca, Ti, Cr, Mn, Ni, Cu, Zn, Ba	—	—	—	—	Jung et al. (2010)
Seoul (Korea)	30 days	—	—	—	—	—	Mg, Al, Si, Ti, V, Cr, Mn, Fe, Ni, Cu, Zn, Ba, Pb, NO_3^-, SO_4^{2-}, Cl^- (*)	1	6 s	<10 μm	30 (PM_{10})	Park, Oh, Yoon, Park, and Lee (2012)
Seoul (Korea)	12 days	4 points in a tunnel	—	—	<10 μm	44	Mg^{2+}, Al, Si, Ti, Cr, Mn, Fe, Ni, Cu, Zn, Ba, Pb, Cl^-, NO_3^-, SO_4^{2-}, Na^+, K^+, Ca^{2+}	—	—	—	—	D. Park et al., 2014
Shanghai (China)	—	16	5 s	<1,<2.5,<10 μm	—	—	—	—	—	—	—	Ye et al. (2010)
Shanghai (China)	—	56 (29 UG; 27 AG)	—	—	Total dust samples	56 (1/station)	Al, Fe, Ca, Na, Ti, Mn, Cr, Cu, Ni	—	—	—	—	Zhang et al. (2011)
Shanghai (China)	—	5	—	—	—	45	VOCs (monocyclic aromatic hydrocarbons, PAHs)	—	—	—	—	Zhang et al. (2012)

(Continued)

TABLE 13.1 Summarized Literature Review of Subway Studies—cont'd

| City | Measurements on Platform | | | | | | | | Measurements Inside Train | | | | Study Reference |
| | Real-Time | | | | Off-Line | | | | Real-Time | | | Off-Line | |
	Time Duration	Stations Number	Time Resolution	Size Fraction	Stations Number	Size Fraction	Samples Number	Species Analyzed	Lines Number	Time Resolution	Size Fraction	Samples Number	
Shanghai (China)	—	—	—	—	—	<1, <2.5 µm	4	Mg, Al, Na, K, Ca, Ti, V, Cu, Cr, Mn, Fe, Ni, Zn, Sr, Mo, Ba, Pb	—	—	—	—	Guo et al. (2014)
Shanghai (China)	46 days	—	—	—	3	<2.5 µm	15 (5/station)	Li, Be, Na, Mg, Al, K, Ca, V, Cr, Mn, Fe, Co, Ni, Cu, Zn, Ga, As, Se, Rb, Sr, Ag, Cd, Cs, Ba, Tl, Pb, Bi, Th, U	—	—	—	—	Lu et al. (2015)
St. Petersburg (Russia)	4 months	—	—	—	—	—	200	—	—	—	—	—	Bogomolova and Kirtsideli (2009)
Stockholm (Sweden)	2 weeks	1	—	<2.5, <10 µm	—	—	—	—	—	—	—	—	Johansson and Johansson (2003)
Stockholm (Sweden)	—	—	5 min	14–330 nm	—	<2.5, <10 µm	—	EC, OC, Fe, C, Ca, Al, Mg, Cl, Tl, Cu, Ag, Cr, Ni, Zr, Ce, Si, K, S, Na, P, VOCs	—	—	—	—	Midander et al. (2012)
Stockholm (Sweden)	—	—	—	—	—	<10 µm	—	—	—	—	—	—	Karlsson et al. (2008)
Stockholm (Sweden)	—	—	—	—	—	<2.5, <10 µm	—	Fe, Ba, Mn, Cu	—	—	—	—	Klepczynska-Nyström et al. (2012)

Location														Reference
Taipei (Taiwan)	–	5	1 min	<2.5, <10 μm	–	–	–	–	–	4 routes	1 min	<2.5, <10 μm	–	Cheng et al. (2008)
Taipei (Taiwan)	84 days	3 (1 UG, 2 AG)	1 min	<2.5, <10 μm	–	–	–	–	–	–	–	–	–	Cheng and Yan (2011)
Taipei (Taiwan)	–	1	–	0.23–20 μm	–	–	–	–	–	–	–	–	–	Cheng and Lin (2010)
Taipei (Taiwan)	–	–	–	–	–	–	–	–	–	2 routes (1 AG and 1 UG)	1 min	<2.5, <10 μm	–	Cheng, Liu, and Yan (2012)
Tokyo (Japan)	4 days	–	–	–	3	TSP	–	Fe, Si, Ba, Cl, Ca, S, K, Cu, Zn, Cr, Ni, Mg, P, PAHs	–	–	–	–	–	Furuya et al. (2001)
Washington, D.C. (USA)	6 days	–	–	–	–	<1 μm	–	–	–	–	–	–	–	Birenzvige et al. (2003)

AC, air-conditioning system; AG, above ground; PAHs, polycyclic aromatic hydrocarbons; TAB, total airborne bacteria; TSP, total suspended particles; UG, underground; (*) species analyzed in the PM_{10} samples collected on the platform and inside the trains.

Moreno, Martins et al., 2015; Moreno et al., 2014; Moreno, Reche et al., 2017; Moreno, Reche et al., 2015; Querol et al., 2012; Reche et al., 2017; Triadó-Margarit et al., 2016). Other European subways cited include those in Athens (Martins et al., 2016), Berlin (Fromme, Oddoy, Piloty, Krause, & Lahrz, 1998), Budapest (Salma, 2009; Salma et al., 2009, 2007), Helsinki (Aarnio et al., 2005), Istanbul (Onat & Stakeeva, 2013), London (Adams et al., 2001; Seaton et al., 2005), Milan (Colombi, Angius, Gianelle, & Lazzarini, 2013), Naples (Cartenì, Cascetta, & Campana, 2015), Oporto (Martins et al., 2016), Paris (Bachoual et al., 2007; Raut, Chazette, & Fortain, 2009; Tokarek & Bernis, 2006), Prague (Braniš, 2006; Cusack et al., 2015; Sysalova & Szakova, 2006), Rome (Perrino et al., 2015; Ripanucci et al., 2006), Rotterdam (Steenhof et al., 2011), St. Petersburg (Bogomolova & Kirtsideli, 2009), and Zurich (Bukowiecki et al., 2007). In East Asia many publications have emanated from research on the Seoul subway system (Cho, Hee Min, & Paik, 2006; Eom et al., 2013; Han, Kwon, & Chun, 2016; Heo & Lee, 2016; Hwang & Park, 2014; Hwang, Yoon, Ryu, Paik, & Cho, 2010; H. J. Jung et al., 2012, 2010; Jung, Kim, Park et al., 2012; Kang, Hwang, Park, Kim, & Ro, 2008; B. W. Kim et al., 2010; C. H. Kim et al., 2010; Kim et al., 2014; Kim, Ho, Jeon, & Kim, 2012; Kim et al., 2008, 2015; Y. Kim et al., 2010; Kwon, Namgung, Jeong, Park, & Eom, 2016; Lee et al., 2016; Park & Ha, 2008; D. Park et al., 2014; Park et al., 2012; Park, Woo, & Park, 2014; Song, Lee, Kim, & Kim, 2008), with other contributions including those from Beijing (Cui, Zhou, & Dearing, 2016), Fukuoka (Ma, Matuyama, Sera, & Kim, 2012); Hong Kong (Chan, Lau, Lee, & Chan, 2002; Leung, Wilkins, Li, Kong, & Lee, 2014; Zheng, Deng, Cheng, & Guo, 2017), Shanghai (Gong et al., 2017; Guo et al., 2014; Lu et al., 2015; Wang, Zhao et al., 2016; Xu et al., 2016; Ye, Lian, Jiang, Zhou, & Chen, 2010; Zhang et al., 2011, 2012), Taipei (Cheng, Lin, & Liu, 2008; Cheng, Liu, & Lin, 2009, Cheng, Liu, & Yan, 2012; Cheng & Yan, 2011; Tsai, Wu, & Chan, 2008), Taiwan (Chen, Sung, Chen, Mao, & Lu, 2016), Tianjin (Wang, Liu, Ren, & Chen, 2016), and Tokyo (Furuya et al., 2001). In the American continent there have been studies on the subways of Boston (Chan et al., 1991; Levy, Dumyahn, & Spengler, 2002), Buenos Aires (Murruni et al., 2009), Los Angeles (Kam, Cheung, Daher, & Sioutas, 2011; Kam, Delfino, Schauer, & Sioutas, 2013; Kam, Ning, Shafer, Schauer, & Sioutas, 2011; Sioutas, 2011), Mexico City (Gómez-Perales et al., 2007, 2004; Hernández-Castillo et al., 2014; Múgica-Alvarez, Figueroa-Lara, Romero-Romo, Sepúlveda-Sánchez, & López-Moreno, 2012); New York (Chillrud et al., 2004; Grass et al., 2010; Trattner et al., 1975; Wang & Gao, 2011), Santiago de Chile (Suárez et al., 2014), São Paulo (Silva et al., 2012), and Washington (Birenzvige et al., 2003). Elsewhere in the world there have been few studies so far (e.g., that in Cairo by Awad, 2002 and Tehran by Hoseini et al., 2013).

In this chapter we provide a summary of key points concerning what is known about subway air quality. We look first at ambient PM levels and size ranges of inhalable airborne particles present on platforms and in trains, then at their physicochemical characteristics and common sources. Following this, consideration is given to bioaerosols in the underground transport environment. Finally, we overview what has been published regarding potential health effects related to breathing subway air and comment on recent developments concerning mitigation, focusing especially on recent results obtained from the ongoing European project IMPROVE LIFE (http://improve-life.eu/en/).

1. PARTICULATE MATTER CONCENTRATIONS AND SIZE RANGES

1.1 On Platforms

$PM_{2.5}$ (particulate matter below 2.5 microns in size) concentrations recorded on subway platforms are variable, with published data ranging from extremes of $16-480 \, \mu g/m^3$ measured in Taipei and London, respectively (Cheng & Lin, 2010; Seaton et al., 2005), although values more commonly lie within the range of $40-65 \, \mu g/m^3$ (average values measured under different time resolutions, Tables 13.1 and 13.2). There is also temporal variation in air quality underground, as demonstrated, for example, by the "weekend effect" whereby weekday $PM_{2.5}$ concentrations on the station platforms tend to be considerably higher (typically in the range of 20%–50%) than those measured during weekends when there are fewer commuters and trains (Aarnio et al., 2005; Johansson & Johansson, 2003; Martins, Moreno et al., 2015; Múgica-Alvarez et al., 2012; Raut et al., 2009). Several studies have demonstrated that mean PM concentrations measured on subway platforms can be higher (in exceptional cases up to nearly seven times) than those simultaneously recorded in outdoor ambient air (Abbasi, Jansson, Sellgren, & Olofsson, 2013; Braniš, 2006; Cheng et al., 2008; Colombi et al., 2013; Cusack et al., 2015; Martins, Moreno et al., 2015; Nieuwenhuijsen et al., 2007; Salma, 2009; Wang & Gao, 2011). Other published measurements, however, have shown the opposite result can be the case (see, for example, Chan et al., 2002 in Hong Kong and Gómez-Perales et al., 2007 in Mexico City). Given the high variability of personal commuter exposure to outdoor air pollutants depending on the route taken through the city, transport mode chosen, and daily meteorological variations, it is difficult to arrive at a meaningful global average for conditions above and below ground. More specific comparisons between city commuting routes can, however, yield more useful data. Comparing air quality on subway, bus, tram, and walking journeys from the same origin to the same destination in Barcelona, for example, revealed that subway air had higher $PM_{2.5}$ concentrations ($43 \, \mu g/m^3$: range, $37-49 \, \mu g/m^3$) than in the tram or walking outdoors ($29 \, \mu g/m^3$: range, $23-35 \, \mu g/m^3$), but were slightly lower than in the bus ($45 \, \mu g/m^3$: range, $39-49 \, \mu g/m^3$, Moreno, Reche et al., 2015). Similar lower values for subway environments compared with other public transport modes have been demonstrated by commuting studies in Hong Kong (Chan et al., 2002), Mexico City (Gómez-Perales et al., 2007), Istanbul (Onat & Stakeeva, 2013), and Santiago de Chile (Suárez et al., 2014).

The transient burden of suspended PM in platform air is in constant flux due to the movement of trains through the system. High-resolution measurements have demonstrated how the "piston effect" of train arrival typically increases PM_X concentrations on the platform, followed by a corresponding decrease on departure when cleaner air is sucked into the station (Fig. 13.1). The importance of this "piston effect" is strongly influenced by station design, being most obvious on narrow platforms serving just a single rail track. Higher PM_X mass concentrations, especially of coarse particles, commonly occur at the train entry points and in the areas closer to the commuters access to the platforms, where air turbulence is greatest (e.g., Martins, Moreno et al., 2015; Moreno et al., 2014; Salma et al., 2007). Other important influences on platform air quality is the presence or absence of platform screen door (PSD)

TABLE 13.2 Mean PM$_{2.5}$ and Elemental Mass Concentrations on Platforms and Particulate Matter (PM) Mass Concentrations Inside Trains in Various Underground Subway Systems

City (Country)	On Platforms																							Inside Trains	Study Reference
	PM$_{2.5}$ (μg m^{-3})	Fe	Al	Ca	Mg	Ti	P	K	Cl	Ba (ng m^{-3})	Cu	Mn	Zn	Cr	Sb	Sr	Zr	Ni	Pb	Sn	V	As	Co	PM$_{2.5}$ (μg m^{-3})	
Amsterdam (Netherlands)																								75	Loxham et al. (2013)
Athens (Greece)	68	29	0.6	1.2	0.2	<0.1	<0.1	0.2	0.3	86	59	249	148	134	3	4	8	16	6	9	7	2	2	83–125	Martins et al. (2016)
Barcelona (Spain)	46[a]–125	32–56	0.4–0.7	1.2–1.8	0.1–0.9	0.02–0.05		0.08–0.10	0.4–0.7	20–2200	100–600	300–500	100–500	40–60	27–30	3–44	8–42	6–16	3–11	5–18	4–7	1.4–13	1.3–3.2	17–32	Querol et al. (2012)
Barcelona (Spain)	21[a]–93	7–52	0.1–0.8	0.2–1.8	0.1–0.3	<0.1	0.1	0.1–0.6	0.1–0.5	7–838	21–465	47–365	37–266	7.5–45	3–95	1.2–19	6–14	3–11	3–13	2–10	3–6	0.4–2.6	0.2–1.9	11–99	Martins, Moreno et al. (2015) and Martins et al., 2016
Boston (USA)	67																								Levy et al. (2002)
Budapest (Hungary)[b]	33	15.5	0.09	0.41	0.13	0.025		0.13	0.104		190	148	50	15				8	21						Salma et al. (2007, 2009)
Helsinki (Finland)	47–60	21–29	0.27–0.28	0.14, 0.33		0.02–0.04		0.04, 0.15–0.23	0.09, 0.105		117–173	234–311	34–124	42–59				23–34	10–13		5–8			21	Aarnio et al. (2005)
Hong Kong (China)																								21–48	Chan et al. (2002)

City																									Reference
Istanbul (Turkey)																								40–45	Onat and Stakeeva. (2013)
London (UK)																								157–247	Adams et al. (2001)
London (UK)	270–480																							130–200	Seaton et al. (2005)
Los Angeles (USA)	57[c]	10.6	0.15	0.19	0.06	0.012	0.06			216	65	85	30	23			12						1.2	24[c]	Kam, Cheung et al. (2011), Kam, Ning et al. (2011), 2013
Mexico city (Mexico)	61[d]																								Gómez-Perales et al. (2004)
Mexico city (Mexico)	48	2.6	0.6	0.9	0.13					30	220	26	131.2	17			6	60	17				1.6		Múgica-Alvarez et al. (2012)
Naples (Italy)	45–60																							18–36	Cartení et al. (2015)
New York (USA)	62	26								240		84													Chillrud et al. (2004)
Oporto (Portugal)	84	33	1	0.4	0.2	<0.1	<0.1	0.6	2.4	53	405	287	87	21	38	2	12	17	11	8	3	1.4	1.1	45–54	Martins et al. (2016)
Paris (France)	35–93																								Raut et al. (2009)

(Continued)

TABLE 13.2 Mean PM$_{2.5}$ and Elemental Mass Concentrations on Platforms and Particulate Matter (PM) Mass Concentrations Inside Trains in Various Underground Subway Systems—cont'd

City (Country)	On Platforms																								Inside Trains	Study Reference
	PM$_{2.5}$	Fe	Al	Ca	Mg	Ti	P	K	Cl	Ba	Cu	Mn	Zn	Cr	Sb	Sr	Zr	Ni	Pb	Sn	V	As	Co	PM$_{2.5}$	
	μg m^{-3}									ng m^{-3}														μg m^{-3}	
Prague (Czech Republic)	94	26	0.2	0.3	0.1	0.01		0.4		70	70	270	110	30	3.4	2		5	23	4	1	9	1.8		Cusack et al. (2015)
Seoul (Korea)	129																							126	Kim et al. (2008)
Seoul (Korea)	58[e]–66[f]																								Kim et al. (2012)
Seoul (Korea)																								58[g]–81[h]	Kim et al. (2014)
Seoul (Korea)	129[i]–219[j]																								Son, Dinh et al. (2014) and Son, Jeon et al. (2014)
Seoul (Korea)	52–83																								Kwon et al. (2015)
Shanghai (China)	287																								Ye et al. (2010)
Shanghai (China)	82–178	1–48	0.7–1.3	1.2–3.8	0.1–0.3	0.02–0.05		0.3–1.4		50–1947	38–240	36–420	20–400	24–570		0.9–69		6–30	1.1–37		4–10				Guo et al. (2014)
Shanghai (China)	49–66	6.6	1.6	0.24	0.25			0.45		119	48	84	127	26		10		34	21		5	2.0	1.3		Lu et al. (2015)

Location			Reference
Stockholm (Sweden)	258		Johansson and Johansson (2003)
Stockholm (Sweden)	60		Midander et al. (2012)
Taipei (Taiwan)	35	28[l]–32	Cheng et al. (2008, 2012)
Taipei (Taiwan)	25–40[k]		Cheng and Yan (2011)
Taipei (Taiwan)	16		Cheng and Lin (2010)

[a]PSDs station.
[b]PM$_2$.
[c]Only underground subway line data.
[d]Commuting.
[e]After PSDs.
[f]Before PSDs.
[g]After subway cabin air purifier.
[h]Before subway cabin air purifier.
[i]After using magnetic filters (tunnel samples).
[j]Before using magnetic filters (tunnel samples).
[k]Underground ticket hall and platform.
[l]Underground route.

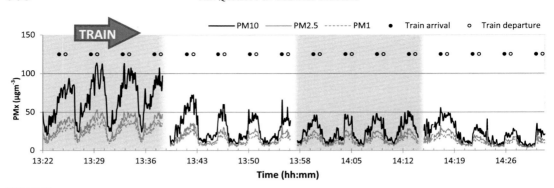

FIGURE 13.1 PMx variations along a subway platform showing the piston effect with an increase in particulate matter (PM) values on train arrival (especially at the platform entry point) and subsequent decrease on departure of each train.

systems, which can be highly effective in inhibiting tunnel air from entering and contaminating the platform, and the operating conditions of the ventilation system (discussed below).

The most extensive measurement program on subway platforms to date has been carried out in the Barcelona subway system where 30 stations with differing designs were studied, revealing substantial variations in $PM_{2.5}$ concentrations (hourly average ranging from 13 to 154 µg/m^3, Martins, Moreno et al., 2015). The stations with just a single tunnel with one rail track separated from the platform by PSDs showed on average around 50% lower $PM_{2.5}$ concentrations in comparison with the old conventional stations (Fig. 13.2). This significant improvement in air quality is attributed to a combination of PSDs preventing the air from the tunnel entering the platform, the more advanced ventilation system in the newer stations, as well as a lower train frequency. Among the conventional stations (no PSDs), those with single narrow tunnel and one rail track showed on average $PM_{2.5}$ concentrations higher

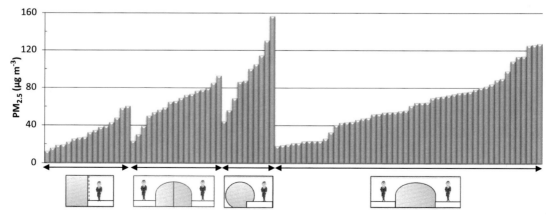

FIGURE 13.2 Average $PM_{2.5}$ concentrations measured at 24 stations with four different designs. *Modified from Martins, Moreno et al. (2015).*

than those with one wider tunnel and two rail tracks separated by a middle wall, most probably due to the less efficient dispersal of air pollutants enhancing the accumulation of PM. In the stations with one wide tunnel and two rail tracks without a middle wall, $PM_{2.5}$ concentrations were much more variable. Similarly, Jung et al. (2010) reported that on narrow platforms there is a larger dependence on strong ventilation to maintain lower PM concentrations.

As with PM mass, particle number concentrations and size ranges in subway air are also variable, but they tend to be different from those in outdoor air. In the comparative Barcelona commuting study mentioned previously (Moreno, Reche et al., 2015), average numbers (N) of ultrafine particles (UFPs) in the subway were 2.3×10^4 cm^{-3}, which is lower than in the other transport modes studied (especially in the bus: 5.4×10^4 cm^{-3}). With regard to particle size, the reverse is the case, with the highest mode values occurring in the metro (90 nm instead of 64−66 nm in vehicles above ground). In general, however, information on subway particle size is relatively scarce and comparisons are hindered by the fact that different studies use different particle size ranges. UFP number concentrations underground are usually found to be lower or similar to those reported both in ambient air (Cheng, Liu, & Lin, 2009; Midander et al., 2012) and in other transport modes (Moreno, Reche et al., 2015; Suárez et al., 2014). As with coarser particles, UFP number concentrations vary largely among subway systems and between stations, with higher values being correlated with greater depth of the platform in some cases (Reche et al., 2017), and number of passengers (Zheng et al., 2017). Another influence on UFP concentrations in the subway is the amount of contamination from polluted air above ground driven underground by mechanical ventilation (Kwon et al., 2016; Reche et al., 2017).

1.2 Inside Trains

Published measurements of air quality parameters inside trains are less common than on platforms, although once again a wide range has been observed, with recorded mean $PM_{2.5}$ concentrations between 11 and 250 $\mu g/m^3$ (Martins, Moreno et al., 2015 and Adams et al., 2001, for Barcelona and London, respectively). This variability presumably reflects the range of conditions encountered in subway lines with different design characteristics, with some systems using air-conditioning, others allowing open windows, and many newer lines operating PSDs. Excluding the extremes measured on some lines, $PM_{2.5}$ concentrations inside train carriages commonly lie in the range of 20−45 $\mu g/m^3$ (Table 13.2) and are usually lower than levels observed on platforms (Aarnio et al., 2005; Cartenì et al., 2015; Kam, Cheung et al., 2011; Querol et al., 2012; Seaton et al., 2005). In the case of Barcelona, where trains use air-conditioning, $PM_{2.5}$ levels inside trains are typically around 15% lower than on platforms (Martins, Moreno et al., 2015). Interestingly, in the case of the Seoul subway it has been noted that the introduction of PSDs into the platform, installed for safety reasons, produced a 30% increase in PM_{10} concentrations inside the trains (Son, Jeon, Lee, Ryu, & Kim, 2014), thus being attributed to reduced mixing of air between tunnel and platforms, enhanced by tunnel depth and length of the underground segments. In the absence of PSDs, interaction between platform air and interior train carriage air has also been demonstrated by other publications, including those from Barcelona (Martins, Moreno et al., 2015; Querol et al., 2012), Prague

(Braniš, 2006), and Taipei (Cheng et al., 2008), where $PM_{2.5}$ concentrations inside the trains can be correlated with the $PM_{2.5}$ concentrations on the corresponding platforms ($R^2 = 0.75$).

The use of air-conditioning has been proven to produce a clear abatement of PM concentrations inside train carriages, resulting in lower PM_X concentrations (by around 30% for $PM_{2.5}$), finer particles (PM_1/PM_{10} was around 15% higher), as well as reducing the variability of PM_X concentrations (Cartenì et al., 2015; Chan et al., 2002; Martins, Moreno et al., 2015; Salma, 2009; Tsai et al., 2008). PM_X concentrations inside trains tend to be relatively constant, although transient peaks can be observed after the train doors close, probably due to turbulence and consequent PM resuspension produced by the movement of passengers inside the trains (Martins, Moreno et al., 2015; Zheng et al., 2017). In trains where it is possible to open the windows, such as in Athens, PM_X concentrations can be shown generally to increase inside the train when passing through tunnel sections (Martins et al., 2016) and more specifically when the train enters the tunnel at high speed (Zheng et al., 2017).

2. AIRBORNE PARTICLE CHARACTERISTICS, SOURCES, AND CHEMICAL COMPOSITION

Inhalable particles in the subway are kept airborne by the motion of trains, turbulence from impulsion and extraction ventilation fans, and movement of commuters (Fig. 13.3, Table 13.3). As previously noted, a key characteristic of subway PM is that a majority of the particles are produced within the underground system itself. Thus ambient subway PM are mostly generated by mechanical wear and friction processes at the rail–wheel–brake interfaces, at the interface between power conductive materials providing electricity and the current collectors attached to trains, by the erosion and degradation of construction material,

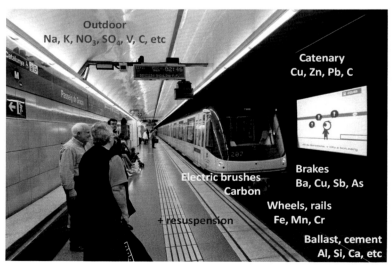

FIGURE 13.3   Main airborne particulate matter sources in a subway station. *Photo courtesy of TMB.*

TABLE 13.3   Summary of Subway Particulate Matter Sources

Sources	Subcategory	Examples
Trains	Non-exhaust emissions	Wheel-rail contact Braking process Interaction between pantographs and catenary (overhead wire) or between third rail and contact shoe Erosion by air turbulence caused by the subway motion (piston effect)
	Indirect	Volatilization of oil and other lubricants
Stationary processes (maintenance and construction)	Direct	Cleaning (e.g., sweeping) Rail cutting Rail welding Ballast changing
	Indirect	Volatilization of cleaning material
	Exhaust emissions	Diesel exhaust (maintenance machinery engines)
Air ventilation	Natural and/or forced	Moving and transferring particle emissions from outdoor Natural erosion of masonry structure
Commuters and subway staff	Human activities	Smoking on platforms Resuspension of deposited particles
	Other	Particles shed by commuters' clothes Degraded perishable materials and garbage

as well as by dust generation during tunnel maintenance works (Gustafsson et al., 2016; Johansson & Johansson, 2003; Jung et al., 2010; Kam et al., 2013; Loxham et al., 2013; Querol et al., 2012; Sundh et al., 2009). A railway is generally powered either by an overhead catenary with the current drawn through the contact material of the pantograph or by a third rail with the current drawn through the current-collecting component (contact shoe) on the train. In addition to the coarse and fine particles produced by abrasive shearing between wheels, rails, and brakes, UFPs can be generated via the high friction temperatures at interfaces between these components, with subsequent vaporization of the substrate (Gustafsson et al., 2016; Sundh et al., 2009; Zimmer & Maynard, 2002).

Given the fact that most subway PM are sourced from the trains and tracks, the resulting mix of inhalable particles underground is radically different from that breathed in outdoor city air. The most obvious difference is that subway particles are mostly highly ferruginous (FePM) and commonly accompanied by trace metals, such as Mn, Cr, Cu, Sb, Ba, and Zn (Table 13.2) (e.g., Aarnio et al., 2005; Braniš, 2006; Cui et al., 2016; Eom et al., 2013; Guo et al., 2014; Johansson & Johansson, 2003; Jung et al., 2010; Kang et al., 2008; Loxham et al., 2013; Lu et al., 2015; Ma et al., 2012; Martins et al., 2016; Moreno, Martins et al., 2015; Moreno, Reche et al., 2015; Múgica-Alvarez et al., 2012; D. Park et al., 2014; J. H. Park et al., 2014; Perrino et al., 2015; Querol et al., 2012; Salma et al., 2007; Zhang et al., 2011). Thus Fe can comprise over 75% of $PM_{2.5}$ mass concentration in subway air

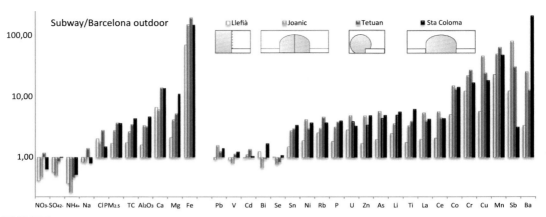

FIGURE 13.4 Chemical composition of PM$_{2.5}$ samples from four different stations in the Barcelona subway system normalized to corresponding ambient outdoor samples to better observe enrichment values for each element/ compound, separating major (left) from trace (right) components.

(Fig. 13.4) (e.g., Kang et al., 2008; C. Kim et al., 2010; Martins et al., 2016). These FePM particles are mostly generated by wear and friction processes as angular splintery shards and rough-surfaced flakes (Fig. 13.5A—D), with the dominant ferruginous component typically oxidized to magnetite, maghemite, and hematite (Cui et al., 2016; Eom et al., 2013; Jung et al., 2010; Moreno, Martins et al., 2015).

Although the trace metals represent less than 2% of the total PM$_{2.5}$ in the subway system, they provide a distinctive chemical signature that is important for source identification (Fig. 13.4 & Fig. 13.5E—F), with Mn and Cr, for example, mostly originating from the steel rails, and Ba and Sb from brake pads. Differences in PM trace metal concentrations observed between stations and subway systems can be up to one order of magnitude and in some cases linked to differences in chemical composition of brakes (Ba, Sb, Cu, Zn, Pb, Ni, Sr) and power supply materials (e.g., Cu-rich catenaries and Cu vs. C pantographs) (Moreno, Martins et al., 2015). Like the FePM in which they are typically embedded, these trace metals mostly originate from mechanical wear and friction processes degrading the composite manufactured materials (Abbasi et al., 2013; Bukowiecki et al., 2007; Cheng et al., 2008; Furuya et al., 2001; Gustafsson et al., 2012; Midander et al., 2012; Namgung et al., 2017; D. Park et al., 2014; Querol et al., 2012; Raut et al., 2009; Sundh et al., 2009; Zimmer & Maynard, 2002).

Carbonaceous particles are also very common in subway air, representing generally the second largest component of ambient PM$_{2.5}$ (Fig. 13.5G). Although some of these aerosols source from hydrocarbon combustion (infiltration of traffic-contaminated outdoor air as well as emissions from diesel-powered trains used for night tunnel maintenance activities), most of them derive from the abrasion of C-bearing brake pads and catenary power supply materials (Gustafsson et al., 2016; Midander et al., 2012; Moreno, Martins et al., 2015). In some cases total carbon has been found to be the most abundant component in subway PM, such as in Mexico City where an additional carbon source is provided by the use of rubber wheels (Gómez-Perales et al., 2004).

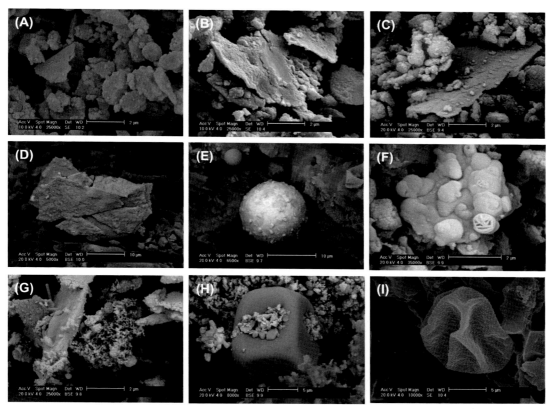

FIGURE 13.5 SEM images of subway airborne particles. (A) General view of Fe particles. (B–D) Close-up of Fe oxide flakes. (E) Fe–Mn sphere. (F) Sb-rich film. (G) Carbonaceous outdoor traffic-related soot aggregate (right) attached to Fe-rich flake (left). (H) Sea salt cube. (I) Pollen grain.

Particles of crustal origin (mostly silicates) found in subway $PM_{2.5}$ samples derive both from mineral dust in infiltrated outdoor air, as well as internal sources within the tunnel system, such as the weathering of wall and floor materials and maintenance activities. Of particular relevance here is dust generation during night tunnel works, especially when replacing rock ballast (these airborne particles are still present in the platform the following day). With regard to other pollutants common in outdoor air, secondary inorganic compounds (water-soluble nitrate, sulfate, and ammonium) and halite are rare in subway air (Fig. 13.5H) and, when present, are usually attributed to infiltration from outside (Kang et al., 2008) by both air and water infiltration, the latter related to the evaporation of water, condensation, and subsequent resuspension of mineral salts.

Airborne organic compounds have also been analyzed in subway platforms and are variable in their concentrations, presumably again due to the influence of different station designs and ventilation systems. H. Jung et al. (2012), for example, reported levels of polycyclic aromatic hydrocarbons (PAHs) in a subway tunnel in Seoul that were lower than that of PM_{10} in ambient air, whereas Zhang et al. (2012) found mean subway levels of major aromatic and chlorinated hydrocarbons higher than those outdoors in Shanghai. In general, however, subway PAH concentrations appear to lie within the range of those observed in urban road traffic sites above ground (Furuya et al., 2001; Levy et al., 2002; Reche et al., 2012). The presence of hopanes is probably related to the influence of nighttime diesel trains for maintenance activities and road traffic emission infiltration (Rogge, Hildemann, Mazurek, Cass, & Simoneit, 1993; Schauer et al., 2007). A measure of the amount of infiltration of particles derived from outdoor air is offered by the presence and abundance of tracer compounds for biomass burning (Martins et al., 2016). Levoglucosan concentrations measured on Barcelona subway platforms for example (up to 132 ng/m^3) are in the order of those observed in the outdoor atmosphere in the city (average around 100 ng/m^3; van Drooge et al., 2014). Chemical tracers for cigarette smoke also provide a potential, although less reliable, measure of outdoor air entry underground, as this activity is not allowed in the subway (Martins et al., 2016). In the case of Barcelona, nicotine levels of 1.3−12 ng/m^3 (van Drooge et al., 2014) may source not only from outside air but also from the transport of nicotine on passenger cloths, skin, and hair, as well as in some cases illegal smoking underground. Other organic compounds, such as aromatic musk compounds (methyl-dihydrojasmonate and galaxolide), are widely used as fragrance in cleaning agents, personal care, and consumer products (Matamoros & Bayona, 2006) and contribute to subway PM after desorption from commuters and subway train cleaning operations.

3. BIOAEROSOLS IN THE SUBWAY SYSTEM

The concentrations of culturable airborne bacteria and fungi colony-forming units (CFU) that are present in subway air have been investigated in several studies. Cho et al. (2006) reported that the highest average concentration of fungi (1574 cfu/m^3) was observed during the morning commute hour from 8:00 to 9:00, associated with the highest number of passengers, with fungi concentrations in subway stations being higher than outdoor concentrations, a fact also found in the Tehran and Mexico systems (Hernández-Castillo et al., 2014; Hoseini et al., 2013). In Saint Petersburg the additional influence of station design on fungal air particle number has been demonstrated, with more open and well-ventilated stations having lower CFU numbers than in closed stations with restricted airflow. This, together with higher CFU numbers in busy stations with transfer junctions, indicates that ventilation, seasonal time of the year, and the number of passengers are the main factors controlling fungi levels (Bogomolova & Kirtsideli, 2009; Heo & Lee, 2016). Hwang et al. (2010) detected concentrations of total airborne bacteria ranging from not detected to 4997 cfu/m^3, with an overall geometric mean of 191 cfu/m^3, and concluded that the presence of PSDs seemed to reduce concentrations. At four of the studied stations, concentrations were found to exceed the Korean indoor bioaerosol guideline. A later study of culturable airborne bacteria concentrations on platform floors also concluded that concentrations were lower in stations with PSDs and in which gram-negative bacteria were not detected (Hwang & Park, 2014). Microbial counts have been also performed in the subway system of Washington, D.C., Cairo, and Hong Kong, where authors detected *Bacillus* spores and concluded that less than 1% of the aerosols present were candidates for biological origin (Birenzvige et al., 2003), that different communities were found within each subway line (Leung et al., 2014), and that higher average concentrations were recorded at the surface than in tunnel stations (Awad, 2002). Low bioaerosol concentrations were measured in the Barcelona subway system (Triadó-Margarit et al., 2016), although the bacterial bioburden was rather high (10^4 bacteria/m^3). Airborne bacterial communities presented a high degree of overlap among the different subway environments sampled (inside trains, platforms, and lobbies) and were dominated by a few widespread taxa, with *Methylobacterium* being the most abundant genus.

4. HEALTH EFFECTS, MITIGATION, AND FUTURE TRENDS

The primary reason for studying the quality of subway air is to provide information and generate awareness that will in turn encourage the development of protocols designed to promote a healthier underground environment for commuters and workers. There are by now many published studies demonstrating that inhaled PM can cause cellular inflammation and the generation of reactive oxygen species and that they can induce both acute and chronic health effects in humans. Subway particles provide no exception to this rule, with several publications reporting on various toxic responses produced both in vitro and in vivo (Bachoual et al., 2007; Bigert et al., 2008; Chillrud et al., 2004; Grass et al., 2010; Janssen et al., 2014; M. H. Jung et al., 2012; Kam et al., 2013; Karlsson et al., 2008, 2006, 2005; Moreno, Kelly et al., 2017; Salma et al., 2009; Seaton et al., 2005; Sioutas, 2011; Spagnolo, Ottria, Perdelli, & Cristina, 2015; Steenhof et al., 2011).

The generally accepted maxim with regard to particulate air pollutants is that higher mass concentrations have greater negative health effects, thus providing the basis for legal standards and recommendations. However, much less is understood about which specific physical properties or chemical compositions of particles are more implicated in cell damage and health effects. Given the fact that subway particles are chemically so different from most outdoor PM, the obvious question arises: are they more toxic than other commonly inhaled particles in the city, for example, those characterizing traffic-polluted outdoor air? Some studies have concluded that subway PM are indeed relatively more toxic (e.g., Karlsson et al., 2006, 2005), whereas others have failed to detect any difference between the bioreactivity of outdoor and subway air (e.g., Spagnolo et al., 2015), while still others have reported higher oxidative potential (OP) of traffic PM as opposed to subway PM (e.g., Janssen et al., 2014). One reason for this apparent confusion is the fact that differing toxicological methods offer contrasting insights into the bioreactive behavior of PM samples. For example, Steenhof et al. (2011) investigated the in vitro toxicity of PM collected at different sites in the Netherlands and, using the MTT assay developed by Mosmann (1983), concluded that subway samples, characterized by a high content of transition metals, showed the largest decrease of metabolic activity in macrophages compared with traffic and urban background sites. However, the same study also concluded that fine PM collected from traffic-polluted outdoor sites produced higher proinflammatory activity (production of cytokines and chemokines) than subway PM. In addition, in the Netherlands, the more recent study on PM-related oxidative stress by Janssen et al. (2014) compared particulates from subway, traffic, urban background, and rural sites and concluded that "OP has several dimensions and that different measures for OP provide different results."

Some of the earlier studies on subway PM bioreactivity have implicated the unusually fer-ruginous nature of the material as primarily responsible for oxidative effects. In particular, the iron oxide magnetite has been singled out as a likely candidate for causing oxidative stress resulting from the production of hydroxyl radicals in the presence of hydrogen peroxide and Fe^{2+} via the Fenton reaction (e.g., Karlsson et al., 2005). However, further testing has failed to substantiate this hypothesis, with neither Fe in general (Bachoual et al., 2007) nor magnetite in particular (Karlsson, Holgersson, & Möller, 2008) being obvi-ously more genotoxic. The study by Karlsson et al. (2008) further ruled out water-soluble metals and intracellularly mobilized iron as responsible for subway PM toxicity, suggesting instead that the likely cause was "highly reactive surfaces giving rise to oxidative stress." A recent study of OP in samples collected from Barcelona similarly concluded that Fe in subway PM, despite its dominance, was not responsible for oxidative damage but that more likely candidates were the trace metals As, Mn, Zn, Ba, and especially Cu (Moreno, Kelly et al., 2017). From this perspective it is perhaps the particles derived from copper catenary materials and, especially, polymetallic aerosol flakes released by brake wear that may be primarily responsible for the bioreactive effects of subway PM. This same study also observed that the subway samples with lowest OP coincided with the new stations fitted with PSDs, again suggesting the likely health benefits to be derived from reducing the entry of train-generated PM into the platform area. Finally, an additional perspective on subway PM toxicity is offered by the work of H. J. Jung et al. (2012) and M. H. Jung et al. (2012) who demonstrated genotoxic effects and oxidative stress caused by the organic components in subway air, indi-cating that metals are not the only possible culprits.

Despite the strikingly metalliferous character of subway air particles, however, there is to date no obvious epidemiological indication of abnormal health effects on underground workers and commuters. Grass et al. (2010) calculated that the subway worker's mean time-weighted $PM_{2.5}$ exposure was $52 \, \mu g/m^3$, with the observed concentrations of iron ($7 \, \mu g/m^3$), manganese ($75 \, ng/m^3$), and chromium ($27 \, ng/m^3$) in $PM_{2.5}$ falling well below occupational standards. Gustavsson, Bigert, and Pollan (2008) concluded that no increased lung cancer risk was found among subway train drivers in the Stockholm subway system. Similarly, a study in New York by Grass et al. (2010) found no significant differences in health biomarkers (metal exposure, oxidative stress, DNA damage) between subway workers, bus drivers, and office employees. In addition, in New York, a study of the exposure to metals by subway workers and commuters concluded that there were no known health effects at the levels being inhaled (Chillrud et al., 2004). Even in more vulnerable individuals such as asthmatics, although some differences in airway inflammatory responses can be detected compared with healthy volunteers, a study by Klepczyńska-Nyström et al. (2012) concluded that "overall the acute health effects after exposure to the subway environment are few." The evidence thus far, therefore, suggests that subway commuters are not being exposed to a more toxic atmospheric environment underground than when traveling through the traffic-polluted city above. However, a note of caution is struck by the observations of Bigert et al. (2008) who found that subway employees working on the platforms, where PM concen-trations were greatest, tended to have higher levels of risk markers for cardiovascular disease than ticket sellers and train drivers. We return, therefore, to the general rule that higher mass concentrations of airborne PM have greater negative health effects. Irrespective of the fact that as yet there are no existing legal controls on air quality in the subway environment, the case for mitigating exposure to subway PM is already irrefutable.

Our summary provided in Table 13.1 demonstrates that there are by now abundant data available on concentrations of inhalable airborne particles in the subway environment, with wheels, rails, and, especially, brakes identified as the main sources of FePM (e.g., Gustafsson et al., 2016). However, as yet there are few publications dealing specifically with how subway air quality can be improved, although there are encouraging signs that this will be an increasingly important research focus. Recent results from the ongoing IMPROVE LIFE project (http://improve-life.eu/en) in the Barcelona metro system are focusing on realistic assessments of how mitigation strategies might be applied to reduce the contribution of specific sources, such as nighttime activities involving ballast replacement and rail welding. There is some evidence, for example, that in some stations applying dust suppressants to ballast can reduce ambient PM levels on platforms the day after night tunnel works (http://improve-life.eu/en/reports/).

It is already clear that more intelligent use of existing ventilation systems has the potential for significant improvements in subway air quality (e.g., Moreno, Reche et al., 2017; J. H. Park et al., 2014; Song et al., 2008), especially when used together with filters designed to dilute FePM concentrations (e.g., the magnetic filters tested by Son, Dinh, Chung, Lee, & Kim, 2014 in Seoul), and in addition to installing air purifiers, electrostatic precipitators (Tokarek & Bernis, 2006), or PSDs (Kim et al., 2012; Martins, Moreno et al., 2015; Querol et al., 2012). The importance of tunnel airflow is demonstrated, for example, by the observation that entirely eliminating any mechanical ventilation of the tunnel, leaving only the train piston effect to keep the air moving, can produce a marked increase in ambient FePM (in the case of Barcelona, platform $PM_{2.5}$ concentrations in non-PSD stations were on average around 50% higher: Martins, Moreno et al., 2015). Similarly, notable differences in air quality can be induced by changing the airflow direction: for example, during a series of IMPROVE LIFE project ventilation experiments, it has been observed that introducing outdoor air by mechanical impulsion can be more successful in diluting underground FePM than the fan extraction of indoor air (Martins, Moreno et al., 2015; Moreno, Reche et al., 2017).

Research on subway air quality appears, therefore, to be moving away from generalized monitoring and source apportionment studies and more toward realistic methods of mitigating ambient PM levels. Our experience in the Barcelona Metro, with its considerable range of different station designs and operating ventilation systems, is that each platform has its own specific atmospheric microenvironment. Further complexity is added by the fact that different catenary systems and train models are used on different lines, as are brake pads of differing chemical compositions. Applying effective protocols designed to reduce subway PM will ideally therefore need to take into account local conditions in each station, starting with a platform-specific air quality environmental audit before moving on to assess the influences of more general controls such as PM generation from moving train parts and their resuspension by piston effects and ventilation systems. Although improving air quality on underground platforms and in subway trains traditionally has not been given a high priority, this is likely to change as operating companies respond to the increasing awareness of how breathing cleaner air leads directly to a healthier city commuter lifestyle.

Acknowledgments

This work was supported by the IMPROVE LIFE project (LIFE13 ENV/ES/000263), the Spanish Ministry of Economy and Competitiveness with FEDER funds (METRO CGL2012-33066), the European Union Seventh Framework Programme (FP7/2007–2013) under grant agreement no. 315760 HEXACOMM, and the Generalitat de Catalunya 2014 SGR33. Wes Gibbons is thanked for improving the English.

References

Aarnio, P., Yli-Tuomi, T., Kousa, A., Makela, T., Hirsikko, A., Hammeri, K., et al. (2005). The concentrations and composition of and exposure to fine particles ($PM_{2.5}$) in the Helsinki subway system. *Atmospheric Environment, 39*(28), 5059–5066.

Abbasi, S., Jansson, A., Sellgren, U., & Olofsson, U. (2013). Particle emissions from rail traffic: A literature review. *Critical Reviews in Environmental Science and Technology, 43*, 2511–2544.

Adams, H. S., Nieuwenhuijsen, M. J., Colvile, R. N., McMullen, M. A. S., & Khandelwal, P. (2001). Fine particle ($PM_{2.5}$) personal exposure levels in transport microenvironments, London, UK. *The Science of the Total Environment, 279*, 29–44.

Awad, A. H. A. (2002). Environmental study in subway metro stations in Cairo. *Egyptian Journal of Occupational Health, 44*, 112–118.

Bachoual, R., Boczkowski, J., Goven, D., Amara, N., Tabet, L., On, D., et al. (2007). Biological effects of particles from the Paris subway system. *Chemical Research in Toxicology, 20*, 1426–1433.

Bigert, C., Alderling, M., Svartengren, M., Plato, N., de Faire, U., & Gustavsson, P. (2008). Blood markers of inflammation and coagulation and exposure to airborne particles in employees in the Stockholm underground. *Occupational and Environmental Medicine, 65*, 655–658.

Birenzvige, A., Eversole, J., Seaver, M., Francesconi, S., Valdes, E., & Kulaga, H. (2003). Aerosol characteristics in a subway environment. *Aerosol Science and Technology, 37*, 210–220.

Bogomolova, E., & Kirtsideli, I. (2009). Airborne fungi in four stations of the St. Petersburg Underground railway system. *International Biodeterioration and Biodegradation, 63*, 156–160.

Braniš, M. (2006). The contribution of ambient sources to particulate pollution in spaces and trains of the Prague underground transport system. *Atmospheric Environment, 40*(2), 348–356.

Bukowiecki, N., Gehrig, R., Hill, M., Lienemann, P., Zwicky, C. N., Buchmann, B., et al. (2007). Iron, manganese and copper emitted by cargo and passenger trains in Zürich (Switzerland): Size-segregated mass concentrations in ambient air. *Atmospheric Environment, 41*, 878–889.

Cartenì, A., Cascetta, F., & Campana, S. (2015). Underground and ground-level particulate matter concentrations in an Italian metro system. *Atmospheric Environment, 101*, 328–337.

Chan, C. C., Spengler, J. D., Özkaynak, H., & Lefkopoulou, M. (1991). Commuter exposures to VOCs in Boston, Massachusetts. *Journal of the Air & Waste Management Association, 41*, 1594–1600.

Chan, L., Lau, W., Lee, S., & Chan, C. (2002). Commuter exposure to particulate matter in public transportation modes in Hong Kong. *Atmospheric Environment, 36*(21), 3363–3373.

Chen, Y. Y., Sung, F. C., Chen, M. L., Mao, I. F., & Lu, C. L. (2016). Indoor air quality in the metro system in North Taiwan. *International Journal of Environmental Research and Public Health, 13*, 1200. http://dx.doi.org/10.3390/ijerph13121200.

Cheng, Y. H., & Lin, Y. L. (2010). Measurement of particle mass concentrations and size distributions in an underground station. *Aerosol and Air Quality Research, 22*–29. http://dx.doi.org/10.4209/aaqr.2009.05.0037.

Cheng, Y. H., Lin, Y. L., & Liu, C. C. (2008). Levels of PM_{10} and $PM_{2.5}$ in Taipei rapid transit system. *Atmospheric Environment, 42*, 7242–7249.

Cheng, Y. H., Liu, C. C., & Lin, Y. L. (2009). Levels of ultrafine particles in the Taipei rapid transit system. *Transportation Research Part D: Transport and Environment, 14*(7), 479–486. http://dx.doi.org/10.1016/j.trd.2009.06.002.

Cheng, Y. H., Liu, Z. S., & Yan, J. W. (2012). Comparisons of PM_{10}, $PM_{2.5}$, particle number, and CO_2 levels inside metro trains between traveling in underground tunnels and on elevated tracks. *Aerosol and Air Quality Research, 12*, 879–891.

Cheng, Y. H., & Yan, J. W. (2011). Comparisons of particulate matter, CO, and CO_2 levels in underground and ground-level stations in the Taipei mass rapid transit system. *Atmospheric Environment, 45*(28), 4882–4891.

Chillrud, S. N., Epstein, D., Ross, J. M., Sax, S. N., Pederson, D., Spengler, J. D., et al. (2004). Elevated airborne exposures of teenagers to manganese, chromium, and steel dust and New York City's subway system. *Environmental Science & Technology, 38*, 732–737.

Cho, J. H., Hee Min, K., & Paik, N. W. (2006). Temporal variation of airborne fungi concentrations and related factors in subway stations in Seoul, Korea. *International Journal of Hygiene and Environmental Health, 209*, 249–255.

Colombi, C., Angius, S., Gianelle, V., & Lazzarini, M. (2013). Particulate matter concentrations, physical characteristics and elemental composition in the Milan underground transport system. *Atmospheric Environment, 70*, 166–178.

Cui, G., Zhou, L., & Dearing, J. (2016). Granulometric and magnetic properties of deposited particles in the Beijing subway and the implications for air quality management. *Science of the Total Environment, 568*, 1059–1068.

Cusack, M., Talbot, N., Ondráček, J., Minguillón, M. C., Martins, V., Klouda, K., et al. (2015). Variability of aerosols and chemical composition of PM_{10}, $PM_{2.5}$ and PM_1 on a platform of the Prague underground metro. *Atmospheric Environment, 118*, 176–183.

van Drooge, B. L., Fontal, M., Bravo, N., Fernández, P., Fernández, M. A., Muñoz-Arnanz, J., et al. (2014). Seasonal and spatial variation of organic tracers for biomass burning in PM_1 aerosols from highly insulated urban areas. *Environmental Science and Pollution Research, 21*, 11661–11670.

Eom, H. Y., Jung, H. J., Sobanska, S., Chung, S. G., Son, Y. S., Kim, J. C., et al. (2013). Iron speciation of airborne subway particles by the combined use of energy dispersive electron probe X-ray microanalysis and Raman microspectrometry. *Analytical Chemistry, 85*, 10424–10431.

Fromme, H., Oddoy, A., Piloty, M., Krause, M., & Lahrz, T. (1998). Polycyclic aromatic hydrocarbons (PAH) and diesel engine emission (elemental carbon) inside a car and a subway train. *The Science of the Total Environment, 217*, 165–173.

Furuya, K., Kudo, Y., Okinaga, K., Yamuki, M., Takahashi, S., Araki, Y., et al.Hisamatsu, Y. (2001). Seasonal variation and their characterization of suspended particulate matter in the air of subway stations. *Journal of Trace and Microprobe Techniques, 19*(4), 469–485.

Gómez-Perales, J. E., Colvile, R. N., Fernández-Bremauntz, A. A., Gutiérrez-Avedoy, V., Páramo-Figueroa, V. H., Blanco-Jiménez, S., et al. (2007). Bus, minibus, metro inter-comparison of commuters' exposure to air pollution in Mexico City. *Atmospheric Environment, 41*, 890–901.

Gómez-Perales, J. E., Colvile, R. N., Nieuwenhuijsen, M. J., Fernández-Bremauntz, A., Gutiérrez-Avedoy, V. J., Páramo-Figueroa, V. H., et al. (2004). Commuters' exposure to $PM_{2.5}$, CO, and benzene in public transport in the metropolitan area of Mexico City. *Atmospheric Environment, 38*, 1219–1229.

Gong, Y., Wei, Y., Cheng, J., Jiang, T., Chen, L., & Xu, B. (2017). Health risk assessment and personal exposure to Volatile Organic Compounds (VOCs) in metro carriages - a case study in Shanghai, China. *Science of the Total Environment, 574*, 1432–1438.

Grass, D. S., Ross, J. M., Family, F., Barbour, J., James Simpson, H., Coulibaly, D., et al. (2010). Airborne particulate metals in the New York city subway: A pilot study to assess the potential for health impacts. *Environmental Research, 110*, 1–11.

Guo, L., Hu, Y., Hu, Q., Lin, J., Li, C., Chen, J., et al. (2014). Characteristics and chemical compositions of particulate matter collected at the selected metro stations of Shanghai, China. *The Science of the Total Environment, 496*, 443–452.

Gustafsson, M., Abbasi, S., Blomqvist, G., Gudmundsson, A., Janhäll, S., Johansson, C., et al. (2016). *Particles in road and railroad tunnel air. Sources, properties and abatement measures.* Swedish National Road and Transport Research Institute. VTI Report 917.

Gustafsson, M., Blomqvist, G., Swietlicki, E., Dahl, A., & Gudmundsson, A. (2012). Inhalable railroad particles at ground level and subterranean stations — physical and chemical properties and relation to train traffic. *Transportation Research Part D: Transport and Environment, 17*, 277–285.

Gustavsson, P., Bigert, C., & Pollan, M. (2008). Incidence of lung cancer among subway drivers in Stockholm. *American Journal of Industrial Medicine, 51*, 545–547.

Han, J., Kwon, S. B., & Chun, C. (2016). Indoor environment and passengers' comfort in subway stations in Seoul. *Building and Environment, 104*, 221–231.

Heo, K. J., & Lee, B. U. (2016). Seasonal variation in the concentrations of culturable bacterial and fungal aerosols in underground subway systems. *Journal of Aerosol Science, 92*, 122–129.

Hernández-Castillo, O., Mugica-Álvarez, V., Castañeda-Briones, M. T., Murcia, J. M., García-Franco, F., & Falcón Briseño, Y. (2014). Aerobiological study in the Mexico City subway system. *Aerobiologia, 30*, 357–367.

Hoseini, M., Jabbari, H., Naddafi, K., Nabizadeh, R., Rahbar, M., Yunesian, M., et al. (2013). Concentration and distribution characteristics of airborne fungi in indoor and outdoor air of Tehran subway stations. *Aerobiologia, 29,* 355–363.

Hwang, S. H., & Park, J. B. (2014). Comparison of culturable airborne bacteria and related environmental factors at underground subway stations between 2006 and 2013. *Atmospheric Environment, 84,* 289–293.

Hwang, S. H., Yoon, C. S., Ryu, K. N., Paik, S. Y., & Cho, J. H. (2010). Assessment of airborne environmental bacteria and related factors in 25 underground railway stations in Seoul, Korea. *Atmospheric Environment, 44,* 1658–1662.

Janssen, N., Yang, A., Strak, M., Steenhof, M., Hellack, B., Gerlofs-Nijland, M., et al. (2014). Oxidative potential of particulate matter collected at sites with different source characteristics. *Science of the Total Environment, 472,* 572–581.

Johansson, C., & Johansson, P. A. (2003). Particulate matter in the underground of Stockholm. *Atmospheric Environment, 37,* 3–9.

Jung, H. J., Kim, B., Malek, M., Koo, Y., Jung, J., Son, Y. S., et al. (2012). Chemical speciation of size segregated floor dusts and airborne magnetic particles collected at underground subway stations in Seoul, Korea. *Journal of Hazardous Materials, 213–214,* 331–340.

Jung, H. J., Kim, B., Ryu, J., Maskey, S., Kim, J. C., Sohn, J., et al. (2010). Source identification of particulate matter collected at underground subway stations in Seoul, Korea using quantitative single-particle analysis. *Atmospheric Environment, 44,* 2287–2293.

Jung, M. H., Kim, H. R., Park, Y. J., Park, D. S., Chung, K. H., & Oh, S. M. (2012). Genotoxic effects and oxidative stress induced by organic extracts of particulate matter (PM_{10}) collected from a subway tunnel in Seoul, Korea. *Mutation Research/Genetic Toxicology and Environmental Mutagenesis, 749,* 39–47.

Kam, W., Cheung, K., Daher, N., & Sioutas, C. (2011). Particulate matter concentrations in underground and ground-level rail systems of the Los Angeles Metro. *Atmospheric Environment, 45,* 1506–1516.

Kam, W., Delfino, R. J., Schauer, J. J., & Sioutas, C. (2013). A comparative assessment of $PM_{2.5}$ exposures in light-rail, subway, freeway, and surface street environments in Los Angeles and estimated lung cancer risk. *Environmental Science: Processes & Impacts, 15,* 234–243.

Kam, W., Ning, Z., Shafer, M., Schauer, J., & Sioutas, C. (2011). Chemical characterization and redox potential of coarse and fine particulate matter (PM) in underground and ground-level rail systems of the Los Angeles metro. *Environmental Science & Technology, 45,* 6769–6776.

Kang, S., Hwang, H., Park, Y., Kim, H., & Ro, C. U. (2008). Chemical compositions of subway particles in Seoul, Korea determined by a quantitative single particle analysis. *Environmental Science & Technology, 42,* 9051–9057.

Karlsson, H. L., Holgersson, A., & Möller, L. (2008). Mechanisms related to the genotoxicity of particles in the subway and from other sources. *Chemical Research in Toxicology, 21,* 726–731.

Karlsson, H. L., Ljungman, A. G., Lindbom, J., & Möller, L. (2006). Comparison of genotoxic and inflammatory effects of particles generated by wood combustion, a road simulator and collected from street and subway. *Toxicology Letters, 165,* 203–211.

Karlsson, H. L., Nilsson, L., & Möller, L. (2005). Subway particles are more genotoxic than street particles and induce oxidative stress in cultured human lung cells. *Chemical Research in Toxicology, 18,* 19–23.

Kim, B. W., Jung, H. J., Song, Y. C., Lee, M. J., Kim, H. K., Kim, J. C., et al. (2010). Characterization of summertime aerosol particles collected at subway stations in Seoul, Korea using low-Z particle electron probe X-ray microanalysis. *Asian Journal of Atmospheric Environment, 4,* 97–105.

Kim, C. H., Yoo, D. C., Kwon, Y. M., Han, W. S., Kim, G. S., Park, M. J., et al. (2010). A study on characteristics of atmospheric heavy metals in subway station. *Toxicology Research, 26,* 157–162.

Kim, J. B., Kim, S., Lee, G. J., Bae, G. N., Cho, Y., Park, D., et al.Kwon, S. B. (2014). Status of PM in Seoul metropolitan subway cabins and effectiveness of subway cabin air purifier (SCAP). *Clean Technologies and Environmental Policy, 16*(6), 1193–1200.

Kim, K. H., Ho, D. X., Jeon, J. S., & Kim, J. C. (2012). A noticeable shift in particulate matter levels after platform screen door installation in a Korean subway station. *Atmospheric Environment, 49,* 219–223.

Kim, K. Y., Kim, Y. S., Roh, Y. M., Lee, C. M., & Kim, C. N. (2008). Spatial distribution of PM_{10} and $PM_{2.5}$ in Seoul metropolitan subway stations. *Journal of Hazardous Materials, 154,* 440–443.

Kim, M. J., Braatz, R., Kim, J. T., & Yoo, C. K. (2015). Indoor air quality control for improving passenger health in subway platforms using an outdoor air quality dependent ventilation system. *Building and Environment, 92,* 407–417.

Kim, Y., Kim, M., Lim, J., Kim, J. T., & Yoo, C. (2010). Predictive monitoring and diagnosis of periodic air pollution in a subway station. *Journal of Hazardous Materials, 183*(1—3), 448—459.

Klepczyńska-Nyström, A., Larsson, B.-M., Grunewald, J., Pousette, C., Lundin, A., Eklund, A., Svartengren, M., et al. (2012). Health effects of a subway environment in mild asthmatic volunteers. *Respiratory Medicine, 106*(1), 25—33.

Kwon, S. B., Jeong, W., Park, D., Kim, K. T., & Cho, K. H. (2015). A multivariate study for characterizing particulate matter (PM_{10}, $PM_{2.5}$, and PM_1) in Seoul metropolitan subway stations, Korea. *Journal of Hazardous Materials, 297*, 295—303.

Kwon, S. B., Namgung, H. G., Jeong, W., Park, D., & Eom, J. K. (2016). Transient variation of aerosol size distribution in an underground subway station. *Environmental Monitoring and Assessment, 188*, 362. http://dx.doi.org/10.1007/s10661-016-5373-5.

Lee, K. R., Kim, W. G., Woo, S. H., Kim, J. B., Bae, G. N., Park, H. K., et al. (2016). Investigation of airflow and particle behaviour around a subway train running in the underground tunnel. *Aerosol Science and Technology, 50*(7), 669—678.

Leung, M. H. Y., Wilkins, D., Li, E. K. T., Kong, F. K. F., & Lee, P. K. H. (2014). Document Indoor-air microbiome in an urban subway network: Diversity and dynamics. *Applied and Environmental Microbiology, 80*, 6760—6770.

Levy, J. I., Dumyahn, T., & Spengler, J. D. (2002). Particulate matter and polycyclic aromatic hydrocarbon concentrations in indoor and outdoor microenvironments in Boston, Massachusetts. *Journal of Exposure Analysis and Environmental, 12*, 104—114.

Loxham, M., Cooper, M. J., Gerlofs-Nijland, M. E., Cassee, F., Davies, D. E., Palmer, M. R., et al. (2013). Physicochemical characterization of airborne particulate matter at a mainline underground railway station. *Environmental Science & Technology, 47*, 3614—3622.

Lu, S., Liu, D., Zhang, W., Liu, P., Fei, Y., Gu, Y., et al. (2015). Physico-chemical characterization of $PM_{2.5}$ in the microenvironment of Shanghai subway. *Atmospheric Research, 153*, 543—552.

Ma, C., Matuyama, S., Sera, K., & Kim, S. (2012). Physicochemical properties of indoor particulate matter collected on subway platforms in Japan. *Asian Journal of Atmospheric Environment, 6*, 73—82.

Martins, V., Minguillón, M. C., Moreno, T., Querol, X., de Miguel, E., Capdevila, M., et al. (2015). Deposition of aerosol particles from a subway microenvironment in the human respiratory tract. *Journal of Aerosol Science, 90*, 103—113.

Martins, V., Moreno, T., Mendes, L., Eleftheriadis, K., Diapouli, E., Alves, C., et al. (2016). Factors controlling air quality in different European subway systems. *Environmental Research, 146*, 35—46.

Martins, V., Moreno, T., Minguillón, M. C., Amato, F., De Miguel, E., Capdevila, M., et al.Querol, X. (2015). Exposure to airborne particulate matter in the subway system. *Science of the Total Environment, 511*, 711—722.

Martins, V., Moreno, T., Minguillón, M. C., van Drooge, B. L., Reche, C., Amato, F., et al. (2016). Origin of inorganic and organic components of $PM_{2.5}$ in subway stations of Barcelona, Spain. *Environmental Pollution, 208*, 125—136.

Matamoros, V., & Bayona, J. M. (2006). Elimination of pharmaceuticals and personal care products in subsurface flow constructed wetlands. *Environmental Science & Technology, 40*, 5811—5816.

Midander, K., Elihn, K., Wallén, A., Belova, L., Borg Karlsson, A., & Wallinder, I. (2012). Characterisation of nano- and micron-sized airborne and collected subway particles, a multi-analytical approach. *Science of the Total Environment, 427*, 390—400.

Morabia, A., Amstislavski, P. N., Mirer, F. E., Amstislavski, T. M., Eisl, H., Wolff, M. S, et al. (2009). Air pollution and activity during transportation by car, subway, and walking. *American Journal of Preventive Medicine, 37*, 72—77.

Moreno, T., Kelly, F., Dunster, C., Oliete, A., Martins, V., Minguillon, M. C., et al. (2017). Oxidative potential of subway $PM_{2.5}$. *Atmospheric Environment, 148*, 230—238.

Moreno, T., Martins, V., Querol, X., Jones, T., BéruBé, K., Minguillón, M. C., et al. (2015). A new look at inhalable metalliferous airborne particles on rail subway platforms. *Science of the Total Environment, 505*, 367—375.

Moreno, T., Pérez, N., Reche, C., Martins, V., de Miguel, E., Capdevila, M., et al. (2014). Subway platform air quality: Assessing the influences of tunnel ventilation, train piston effect and station design. *Atmospheric Environment, 92*, 461—468.

Moreno, T., Reche, C., Minguillón, M. C., Capdevila, M., de Miguel, E., & Querol, X. (2017). The effect of ventilation protocols on subway system air quality. *Science of the Total Environment, 584—585*, 1317—1323.

Moreno, T., Reche, C., Rivas, I., Minguillón, M. C., Martins, V., Vargas, C., et al. (2015). Urban air quality comparison for bus, tram, subway and pedestrian commutes in Barcelona. *Environmental Research, 142*, 495—510.

Mosmann, T. (1983). Rapid colorimetric assay for cellular growth and survival: Application to proliferation and cytotoxicity assays. *Journal of Immunological Methods, 65*(1—2), 55—63.

Múgica-Alvarez, V., Figueroa-Lara, J., Romero-Romo, M., Sepúlveda-Sánchez, J., & López-Moreno, T. (2012). Concentrations and properties of airborne particles in the Mexico City subway system. *Atmospheric Environment, 49*, 284−293.

Murruni, L. G., Solanes, V., Debray, M., Kreiner, A. J., Davidson, J., Davidson, M., et al. (2009). Concentrations and elemental composition of particulate matter in the Buenos Aires underground system. *Atmospheric Environment, 43*, 4577−4583.

Namgung, H. G., Kim, J. B., Kim, M. S., Kim, M., Park, S., Woo, S. H., et al. (2017). Size distribution analysis of airborne wear particles released by subway brake system. *Wear, 372−373*, 169−176.

Nieuwenhuijsen, M. J., Gómez-Perales, E., & Colvile, R. N. (2007). Levels of particulate air pollution, its elemental composition, determinants and health effects in metro systems. *Atmospheric Environment, 41*, 7995−8006.

Onat, B., & Stakeeva, B. (2013). Personal exposure of commuters in public transport to $PM_{2.5}$ and fine particle counts. *Atmospheric Pollution Research, 4*, 329−335.

Park, D., & Ha, K. (2008). Characteristics of PM_{10}, $PM_{2.5}$, CO_2 and CO monitored in interiors and platform of subway train in Seoul, Korea. *Environment International, 34*, 629−634.

Park, D., Lee, T., Hwang, D., Jung, W., Lee, Y., Cho, K., et al. (2014). Identification of the sources of PM_{10} in a subway tunnel using positive matrix factorization. *Journal of the Air & Waste Management Association, 64*, 1361−1368.

Park, D., Oh, M., Yoon, Y., Park, E., & Lee, K. (2012). Source identification of PM_{10} pollution in subway passenger cabins using positive matrix factorization. *Atmospheric Environment, 49*, 180−185.

Park, J. H., Woo, H., & Park, J. (2014). Major factors affecting the aerosol particulate concentration in the underground stations. *Indoor and Built Environment, 23*, 629−639.

Perrino, C., Marcovecchio, F., Tofful, L., & Canepari, S. (2015). Particulate matter concentration and chemical composition in the metro system of Rome, Italy. *Environmental Science and Pollution Research, 22*, 9204−9214.

Querol, X., Moreno, T., Karanasiou, A., Reche, C., Alastuey, A., Viana, M., et al. (2012). Variability of levels and composition of PM_{10} and $PM_{2.5}$ in the Barcelona metro system. *Atmospheric Chemistry and Physics, 12*(11), 5055−5076.

Raut, J. C., Chazette, P., & Fortain, A. (2009). Link between aerosol optical, microphysical and chemical measurements in an underground railway station in Paris. *Atmospheric Environment, 43*, 860−868.

Reche, C., Moreno, T., Amato, F., Viana, M., van Drooge, B. L., Chuang, H.-C., et al. (2012). A multidisciplinary approach to characterise exposure risk and toxicological effects of PM_{10} and $PM_{2.5}$ samples in urban environments. *Ecotoxicology and Environmental Safety, 78*, 327−335.

Reche, R., Moreno, T., Martins, V., Minguillón, M. C., Jones, T., de Miguel, E., et al. (2017). Factors controlling particle number concentration and size at metro stations. *Atmospheric Environment* (in press).

Ripanucci, G., Grana, M., Vicentini, L., Magrini, A., & Bergamaschi, A. (2006). Dust in the underground railway tunnels of an Italian Town. *Journal of Occupational and Environmental Hygiene, 3*(1), 16−25.

Rogge, W. F., Hildemann, L. M., Mazurek, M. A., Cass, G. R., & Simoneit, B. R. T. (1993). Sources of fine organic aerosol. 2. Noncatalyst and catalyst-equipped automobiles and heavy-duty diesel trucks. *Environmental Science & Technology, 27*, 636−651.

Şahin, Ü. A., Onat, B., Stakeeva, B., Ceran, T., & Karim, P. (2012). PM_{10} concentrations and the size distribution of Cu and Fe-containing particles in Istanbul's subway system. *Transportation Research Part D: Transport and Environment, 17*, 48−53.

Salma, I. (2009). Air pollution in underground railway systems. In R. E. Hester, & R. M. Harrison (Eds.), *Air quality in urban environments* (pp. 65−84). Royal Society of Chemistry.

Salma, I., Posfai, M., Kovacs, K., Kuzmann, E., Homonnay, Z., & Posta, J. (2009). Properties and sources of individual particles and some chemical species in the aerosol of a metropolitan underground railway station. *Atmospheric Environment, 43*, 3460−3466.

Salma, I., Weidinger, T., & Maenhaut, W. (2007). Time-resolved mass concentration, composition and sources of aerosol particles in a metropolitan underground railway station. *Atmospheric Environment, 41*, 8391−8405.

Schauer, J., Rogge, W., Hildemann, L., Mazurek, M., Cass, G., & Simoneit, B. (2007). Source apportionment of airborne particulate matter using organic compounds as tracers. *Atmospheric Environment, 41*, 241−259.

Seaton, A., Cherrie, J., Dennekamp, M., Donaldson, K., Hurley, J. F., & Tran, C. L. (2005). The London underground: Dust and hazards to health. *Occupational and Environmental Medicine, 62*(6), 355−362.

Silva, C. B. P., Saldiva, P. H. N., Amato-Lourenço, L. F., Rodrigues-Silva, F., & Miraglia, S. G. (2012). Evaluation of the air quality benefits of the subway system in São Paulo, Brazil. *Journal of Environmental Management, 101*, 191−196.

Sioutas, C. (2011). *Physical and chemical characterization of personal exposure to airborne particulate matter (PM) in the Los Angeles subways and light-rail trains METRANS final report*. http://www.metrans.org/sites/default/files/research-project/10-07_Sioutas_final_0_0.pdf.

Son, Y., Dinh, T., Chung, S., Lee, J., & Kim, J. (2014). Removal of particulate matter emitted from a subway tunnel using magnetic filters. *Environmental Science & Technology, 48*, 2870−2876.

Son, Y., Jeon, J. S., Lee, H. J., Ryu, I. C., & Kim, J. C. (2014). Installation of platform screen doors and their impact on indoor air quality: Seoul subway trains. *Journal of the Air & Waste Management Association, 64*, 1054−1061.

Song, J., Lee, H., Kim, S. D., & Kim, D. S. (2008). How about the IAQ in subway environment and its management? *Asian Journal of Atmospheric Environment, 2*, 60−67.

Spagnolo, A. M., Ottria, G., Perdelli, F., & Cristina, M. L. (2015). Chemical characterisation of the coarse and fine PM in the environment of an underground railway system: Cytotoxic effects and oxidative stress-a preliminary study. *International Journal of Environmental Research and Public Health, 12*(4), 4031−4046.

Steenhof, M., Gosens, I., Strak, M., Godri, K. J., Hoek, G., Cassee, F. R., et al. (2011). In vitro toxicity of particulate matter collected at different sites in The Netherlands is associated with PM composition, size fraction and oxidative potential − the RAPTES project. *Fibre Toxicology, 8*, 1−15.

Suárez, L., Mesías, S., Iglesias, V., Silva, C., Cácers, D. D., & Ruiz-Rudolph, P. (2014). Personal exposure to particulate matter in commuters using different transport modes (bus, bicycle, car and subway) in an assigned route in downtown Santiago, Chile. *Environmental Science: Processes & Impacts, 16*, 1309−1317.

Sundh, J., Olofsson, U., Olander, L., & Jansson, A. (2009). Wear rate testing in relation to airborne particles generated in a wheel − rail contact. *Lubrication Science, 21*, 135−150.

Sysalova, J., & Szakova, J. (2006). Mobility assessment and validation of toxic elements in tunnel dust samples—subway and road using sequential chemical extraction and ICP-OES/GF AAS measurements. *Environmental Research, 101*, 287−293.

Tokarek, S., & Bernis, A. (2006). An example of particle concentration reduction in Parisian subway stations by electrostatic precipitation. *Environmental Technology, 27*, 1279−1287.

Trattner, R. B., Perna, A. J., Kimmel, H. S., & Birch, R. (1975). Respirable dust content of subway air. *Environmental Letters, 10*(3), 247−252.

Triadó-Margarit, X., Veillette, M., Duchaine, C., Talbot, M., Amato, F., Minguillón, M. C., et al. (2016). Bioaerosols in the Barcelona subway system. *Indoor Air*. http://dx.doi.org/10.1111/ina.12343.

Tsai, D. H., Wu, Y. H., & Chan, C. C. (2008). Comparisons of commuter's exposure to particulate matters while using different transportation modes. *Science of the Total Environment, 405*(1−3), 71−77.

Vasconcellos, E. A. (2001). *Urban transport, environment and equity: The case for developing countries*. London: Earthscan.

Wang, B. Q., Liu, J. F., Ren, Z. H., & Chen, R. H. (2016). Concentrations, properties, and health risk of $PM_{2.5}$ in the Tianjin City subway system. *Environmental Science and Pollution Research, 23*, 22647−22657.

Wang, J., Zhao, L., Zhu, D., Gao, H. O., Xie, Y., Li, H., et al. (2016). Characteristics of particulate matter (PM) concentrations influenced by piston wind and train door opening in the Shanghai subway system. *Transportation Research D, 47*, 77−88.

Wang, X. R., & Gao, H. O. (2011). Exposure to fine particle mass and number concentrations in urban transportation environments of New York City. *Transportation Research Part D: Transport and Environment, 16*, 384−391.

Xu, B., Yu, X., Gu, H., Miao, B., Wang, M., & Huang, H. (2016). Commuters' exposure to $PM_{2.5}$ and CO_2 in metro carriages of Shanghai metro system. *Transportation Research Part D, 47*, 162−170.

Ye, X., Lian, Z., Jiang, C., Zhou, Z., & Chen, H. (2010). Investigation of indoor environmental quality in Shanghai metro stations, China. *Environmental Monitoring and Assessment, 167*, 643−651.

Zhang, W., Jiang, H., Dong, C., Yan, Q., Yu, L., & Yu, Y. (2011). Magnetic and geochemical characterization of iron pollution in subway dusts in Shanghai, China. *Geochemistry, Geophysics, Geosystems, 12*. http://dx.doi.org/10.1029/2011GC003524.

Zhang, Y., Li, C., Wang, X., Guo, H., Feng, Y., & Chen, J. (2012). Rush-hour aromatic and chlorinated hydrocarbons in selected subway stations of Shanghai, China. *Journal of Environmental Sciences, 24*, 131−141.

Zheng, H. L., Deng, W. J., Cheng, Y., & Guo, W. (2017). Characteristics of $PM_{2.5}$, CO_2 and particle-number concentration in mass transit railway carriages in Hong Kong. *Environmental Geochemistry and Health*. http://dx.doi.org/10.1007/s10653-016-9844-y.

Zimmer, A. T., & Maynard, A. D. (2002). Investigation of the aerosols produced by a high-speed, hand-held grinder using various substrates. *The Annals of Occupational Hygiene, 46*, 663−672.

Index

'*Note:* Page numbers followed by "f" indicate figures, "t" indicate tables.'

A

Access restriction measures, 244
Adaptive cruise control (ACC) systems, 224
Aerodynamic particle sizer (APS), 152—153
Air pollution
  AQPs and source attribution
    EU emission inventories, 233
    health protection thresholds, 233—234
    NOx source contributions, 231—232
    particulate matter source contributions, 232—233
    road traffic, air quality measures for, 234
    source contributions, 233—234
  critical urban pollutant, 230
  urban areas, nontechnological road traffic measures
    access restriction measures, 244
    chronogram, 235, 235f
    City Hall, 244
    city logistics, 241, 243t
    city planners, 243
    classification, 234, 235f
    cleaner fleet transformation, 239
    commerce, 243
    fuels and vehicles, taxation of, 240—241
    identify groups, 234
    information-assisted measures, 244
    infrastructural measures, 245
    in-use vehicles, 236—238
    lawyers, 244
    legislative and organizational measure, 244
    LEZs, 239—240, 241t
    manufacturers, 243
    neighbors and street users, 244
    police, 243—244
    policies, 239—241, 244
    public transport, 236
    reducing non-exhaust PM emissions, 254
    remediation measures, 247—253
    speed limit, 253
    territorial management measures, 244
    traffic engineers, 243
    traffic speed and road dust emission, 253—254,
      254f
    traffic speed management, 253

    transportation operators, 244
    urban transformation, 245—246
    vehicle manufacturers, 244
    vehicle speed reduction, 253—254
    WHO air quality guidelines, 231
Air Quality (AQ), 5
Air Quality Guidelines for Europe, 2—4
Air quality plans (AQPs), 231
Air quality standards
  PM10, 10, 12
  PM2.5, 10, 12
  spatial distribution, 11
Air quality. *See* Vehicle non-exhaust emissions, air
    quality
Airborne brake wear emissions, 123—125, 143
Aluminum brake discs
  Al-MMC, 210—212
  brake pad wear, 211—212
  lining material, 211
  spray-forming process, 210
  tribological system, 211
Aluminum metal matrix compound (Al-MMC), 210
American Thoracic Society (ATS), 71
Antimony (Sb), 23, 48, 69, 91, 93, 125—127, 129—130,
    136—137, 140—141, 216—217, 309—310, 311f
Asbestos-related regulation, 89
Asphalt pavements
  Nordic abrasion test, 168—170
  Nordic ball mill test, 170
  PM10 production, 168
  sandpaper effect, 167—168
Automatic number plate recognition (ANPR), 236—237
Automotive friction brake materials
  binders, 126
  brake lining material, 125
  brake pad formulations, 125, 126t
  characteristic brake pad material, 126
  design principles, 125
  disc brakes, 125
  fillers, 126
  friction modifiers, 126
  reinforcing constituents, 125
Average exposure indicator (AEI), 5

B

Bimodal particle distribution, 54–55
Brake linings
  brake pad concept, 215, 216f
  copper-free composition, 217
  friction layer, 217, 217f
  natural material, 217
  phenolic resin, 215
Brake particle emissions (BPEs), 205–206
  collection emitted brake particles, 220, 225
    abrasion powder, 218
    air flow, 218
    autonomous suction device, 218
    baseline test, 219–220
    brake dynamometer tests, 219
    brake pad waste collection system, 219,
      220f
  emitted brake particles, reduction measures of
    aluminum brake discs, 210–212
    brake linings, 214–218
    ceramic brake discs, 212–213
    gray cast iron brake discs, 206–210
    titanium brake discs, 213–214
  prevention
    driver assistance systems, role of, 223–225
    regenerative braking, role of, 221–223
Brake wear contribution, range, 69
Brake wear emissions
  airborne brake wear emissions, 123–125, 143
  automotive friction brake materials
    binders, 126
    brake lining material, 125
    brake pad formulations, 125, 126t
    characteristic brake pad material, 126
    design principles, 125
    disc brakes, 125
    fillers, 126
    friction modifiers, 126
    reinforcing constituents, 125
  brake dynamometer studies
    airborne wear particles, 136–137
    airflow, 134
    chemical composition, 132
    direct wear emissions, 132
    finest fractions, 134–136
    LM pads, 138
    NAO and LM materials, 134
    number *vs.* mass distributions, 132–133, 133f
    on-road particles, 134
    overview, 138
    oxygenated carbonaceous components, 136–137
    particle mass distribution and particle number
      distribution, 132

    particle size distributions, 134–136
    PM10 and PM2.5, TC emissions of, 136–137
    size fractions, 132–133, 136–137
    size-resolved sampling, 131–132
    submicron and nanofractions, 134–136
    TEM analysis, 134–136, 136f
    vehicle field tests, 138
    wear particles, online monitoring of, 134–136
    wear test procedure, 131–132
  composition-related issue, 124–125
  critical issues, 141–142
  exhaust and non-exhaust sources, 123–124
  field studies
    barium sulfate, 140–141
    categories, 138
    dynamometer tests, 140
    Fe-rich nanosized particles, 140–141
    filters, 140
    particle size distributions, 140
    PM, sampling and characterization of, 140–141
    wind tunnel measurements, 140
  fine fraction (PM2.5) and coarse fraction
    (PMc; PM2.5–10), 113
  friction brake materials
    friction layer, formation of, 126–127
    friction process, 126
    wear and wear products, 127–129
  friction industry, 142–143
  pin-on-disc studies
    airborne wear particles, 130
    fine particles, 130–131
    nonairborne fractions, 130
    submicron particles, 129–130
    ultrafine particles, 130–131
  regenerative braking technology, 142
  tracers, 116–118
    Cu, 116–118, 120
    PM10 and PM2.5, implied emission factors for,
      111t
    PM10, 117–118, 117t, 118f
Brake wear particles
  size distribution, 50–51
  unimodal brake wear mass distributions, 50–51
Brakes regulation
  asbestos-related regulation
    Directive 2003/18/EC, 90
    fibrous silicates, 90
    friction material, 89
    Japan, asbestos ban, 91
    motor vehicle brake friction materials, 91
    repair operations, 90
    Washington (Senate Bill (SB) 6557), 91
  CLP regulation, 96

REACH and REACH-Like regulations
aims, 94
Automotive Industry Guideline, 94
CCA, 95
concept, 94
CSCL, 95—96
general principle, 94
IECSC, 95
Korea REACH, 95
MEP, 95
processes, 94
progressive substitution, 94
proposals, 94
registration, 94
SVHC, 94
trace elements and heavy metals
brake pad partnership, 93
certification, marks used in, 93, 93f
third-party labs, 93
weight percentage, 92

C
Calcium magnesium acetate (CMA), 251—252
California (Senate Bill (SB) 346), 91—92
Cement concrete pavements, 171—172
Centre for Emission Inventories and Projections
(CEIP), 108
Ceramic brake discs
carbon/carbon composite, 213
manufacturing process, 212
spherical carbon particles, 212
Chemical Substances Control Law (CSCL), 95—96
Chemicals Control Act (CCA), 95
China, 95
City logistics
City Hall, 244
city planners, 243
commerce, 243
lawyers, 244
manufacturers, 243
neighbors and street users, 244
police, 243—244
policies, 244
access restriction measures, 244
information-assisted measures, 244
infrastructural measures, 245
legislative and organizational measure, 244
territorial management measures, 244
traffic engineers, 243
transportation operators, 244
vehicle manufacturers, 244
Clean Water Act, 93
CLP regulation, 96

Coarse fraction, 233
Colony-forming units (CFU), 312—313
Comparing emissions
implied emission factors/vehicle kilometer
country grouping, 110
Group I and Group II, 112
Groups III and IV, 112—113
HDV, category of, 109
road abrasion emissions, 110, 112—113
reported wear particles, size distribution of,
113—114
Convention on Long-Range Transboundary Air
Pollution (CLRTAP), 275
Copper (Cu), 93, 116—117, 120

D
Directive 2000/53/EC, 92
Directive 2002/24/EC of the European Parliament
and of the Council of March 18, 2002, 98
Directive 2003/37/EC of the European Parliament
and of the Council of May 26, 2003, 98
Directive 2007/46/EC of the European Parliament
and of the Council of September 5, 2007, 97
Driver assistance systems (DAS), 205—206, 223—225

E
Electric vehicle initiative (EVI), 263
Electric vehicles (EVs)
comparing ICEVs and, 278
electric vehicle market growth, 262—263
non-exhaust emissions and ICEVs, 282
Hooftman et al., 274—277
Platform for Electro-mobility, 273—274
Soret, Guevara, and Baldasano, 271
Timmers and Achten, 272—273
Van Zeebroek and De Ceuster, 269—271
passenger cars, 262
resuspension, 279—280
secondary particulate matter, contribution of,
278—279, 281
uncertainty, source of, 281
weight and non-exhaust PM emissions
brake wear, 264
electric and conventional passenger cars, weight
comparison of, 266—269
emission inventories, 266
experimental data and models, 264—266
qualitative evidence, 264
resuspension, 264
road surface wear, 264
Emission inventory
air quality management, 102
bottom-up approach, 102

Emission inventory (*Continued*)
  comparing emissions
    implied emission factors/vehicle kilometer,
      109–113
    reported wear particles, size distribution of,
      113–114
  emission data, 104
  Europe, emission reporting in, 108–109
    CEIP, 108
    NFR source categorization, 108–109
    resuspension emission, 108–109
    traffic-induced resuspension, 108–109
  exhaust emissions, 119
  light-duty vehicle tires, wear factors for, 102
  NFR14, 103t
  normal driving, 104
  road transport causes resuspension, 114–116
  road transport exhaust and wear emissions,
    102
    combustion emissions, 104
    EU15, emission data for, 108
    EU-NMS, 104
    European urban environment, 108
    PM10 emissions, 104–106, 105f
    PM2.5 emissions, 104–106, 106f
    traffic flow patterns, 106
    wear processes, emission factors of,
      107–108
  tracers, road transport wear emissions, 116–118
EMMA, 172
Europe
  CLP Regulation (Regulation 1272/2008), 96
  emission inventory. *See* Emission inventory
  emission reporting, 108–109
  End of Life Vehicle directive (Directive 2000/53/
    CEE), 92
  heavy metals restricted, 93
  PM
    air quality standards, 10–12
    European Union Limit Values, 5–6
    MK method, 7–8
    PM10, 9
    PM10, coarse particles in, 6–7
    PM2.5, 8–9
    PM2.5e10, 9–10
  REACH regulation (Regulation 1907/2006), 94, 97
  Regulation 1272/2008, 96
European Chemicals Agency (ECHA), 94
European Environmental Agency (EEA), 266

F

Ferritic nitrocarburizing (FNC) method, 209
Friction brake materials

friction layer, formation of, 126–127
friction process, 126
wear and wear products, 127–129

G

German Informative Inventory Report's (GIIR),
  275
Global Automotive Declarable Substance List
  (GADSL), 99
Global Automotive Stakeholders Group (GASG), 99,
  100t
Gray cast iron brake discs
  brake disc concepts, 208–209
  brake lining concept, 209
  coating material, 207
  diffusion layer, 209
  FNC, 209
  molybdenum, 206
  protective layer, 206–207
  rubbing ring, 207
  sealing layer, 206–207
  thermal spray coating method, 208–209
  titanium, 206
  wear-reducing surface coating, 207

H

Heavy-duty vehicles (HDV), 109, 165–166,
  196

I

Information-assisted measures, 244
Infrastructural measures, 245
Internal combustion engine vehicles (ICEVs),
  262
International Atomic Energy Agency (IAEA),
  14–15
International Material Data System (IMDS), 99
Inventory of Existing Chemical Substances Produced
  or Imported in China (IECSC), 95

J

Japan, 91
  CSCL, 95–96

K

Korea REACH, 95

L

Legislative and organizational measure, 244
Life cycle assessment (LCA), 274
Light-duty vehicles (LDV), 109, 165–166, 196
Lithuania's traffic sites, 8–9
Low emission zones (LEZs), 239–240

M

Mann–Kendall (MK) method, 5–6
Mechanical wear, 128
Ministry of Environment (MoE), 95
Ministry of Environmental Protection (MEP), 95

N

National Ambient Air Quality Standard (NAAQS), 12
New York State (Senate Bill (SB) A10871), 92
Noise vibration harshness (NVH), 220
Nomenclature for reporting (NFR), 102
Nonasbestos organic (NAO), 123–124
Non-exhaust particles
   epidemiological evidence
      air pollution, health effects of, 71
      biological effects, 72–73
      chemical profile, 72
      components, 77–80
      long-term effects, 77
      methodological progresses, 71–72
      physiologic or pathologic changes, 71
      short-term effects, 73–75
      size-fractioned particles, 72
   main sources and components, 68–69
   non-exhaust particulate matter
      long-term exposure, 70–71, 80
      receptor models, 69–70
      short-term exposure, 70, 80
Nonstudded tires, road wear particle emissions
   controlled studies, 165–166
   source apportionment studies
      ambient air, PM10 in, 163
      road dust, 163

O

Ontario's Occupational Health and Safety
      Act, 90
Oxidative wear, 128

P

Particulate air quality
   air quality data
      Europe, 5–12
      United States, 12–14
   Air Quality guidelines
      pollutants, 2–4
      WHO, 2–4, 2t–3t
Particulate matter (PM), 262
   Air Quality guidelines
      nonthreshold behavior, 2–4
      pollutants, 2–4
      WHO, 2–4, 2t–3t

air quality data
   Europe. See Europe
   South and Southeastern Asia, 14–17
   United States, 12–14
components, 77–80
health effects, 69
long-term effects, non-exhaust particles, 77
natural sources of, 69
non-exhaust. See Non-exhaust particles
short-term effects, non-exhaust particles
   cardiovascular diseases, 74
   morbidity outcomes, 74
   sources, 74
   traffic and daily health outcomes, 74
   traffic non-exhaust PM, 75, 75t–76t
subway systems. See Subway systems
Piston effect, 301–306
Platform screen door (PSD), 301–306
PM10
   air quality standards, 10, 12
   ambient air, 163
   asphalt pavements, 168
   fine and coarse modes, 69
   implied emission factors, brake and tire wear, 112, 112f
   International Standards for, 156–157, 157t
   road transport exhaust and wear emissions, 104–106, 105f
   TC emissions of, 136–137
   tire wear, emission factors for, 154, 156t
PM2.5
   air quality standards, 10, 12
   implied emission factors, brake and tire wear, 112, 112f
   International Standards for, 156–157, 157t
   road transport exhaust and wear emissions, 104–106, 106f
   TC emissions of, 136–137
Pollutant Release and Transfer Register (PRTR), 266
Polycyclic aromatic hydrocarbons (PAHs), 97, 98t, 149, 311–312
Positive matrix factorization (PMF), 48, 48f, 163

R

REACH regulation (Regulation 1907/2006), 94, 97
REACH-like regulations, 94
Regenerative braking, 225
   brake application scenario, 221
   brake blending, 221–222
   electric motor, 221
   front disc, temperature profiles, 222, 222f
   phases, 221–222
   test vehicle, 222
   velocity–deceleration behavior, 223, 223f

Remediation measures
  photocatalytic products, 252–253
  road sweeping and washing
    chemical dust suppressants, 250–251
    CMA, 251–252
    dust suppressants, 251
    dust-free performance, 248
    emission levels, 249–250
    factors, 249
    inhibit road dust suspension, 247
    moisture evaporation, 248–249
    preventative strategies, 247
    street washing, 249–250
    sweeper technology, 247
    water flushing, 248–249
    water sprays, 248
Rhode Island State (Senate Bill (SB) H7997), 92
Road dust emissions
  air quality modeling, 198
  direct measurements, 184
  mobile platform measurements, 194–195
  NOx emissions, use of, 195
  Omstedt model, 198
  real process-based model, 198
  road dust loading
    deposition, 189
    distribution, 190–191
    dry vacuum method, 191–192
    fugitive sources, 189
    local spatial variability, 194
    retention, 190
    road salt, 190
    road surface texture, 190
    sampling, 191–192
    silt loading, 191–192
    sources, 185
    surface abrasion, 186
    suspension, 185–186
    traction sand, 189
    water-related processes, 190
    wet dust sampling technique, 193
  road dust suspension, emission factors for, 196
  sources, 184
  suspension rates, 195
  upwind–downwind measurements, 195
  wet removal processes, 184
Road dust loading
  deposition, 189
  distribution, 190–191
  dry vacuum method, 191–192
  fugitive sources, 189
  local spatial variability, 194
  retention, 190

road salt, 190
road surface texture, 190
sampling, 191–192
silt loading, 191–192
sources, 185
surface abrasion, 186
suspension, 185–186
traction sand, 189
water-related processes, 190
wet dust sampling technique, 193
Road dust particles, 23
  lowest contributions, 54
Road dust, 1–2, 163
Road particles (RP), 173
Road transport causes resuspension, 114–116
Road transport exhaust and wear emissions, 102
  combustion emissions, 104
  EU15, emission data for, 108
  EU-NMS, 104
  European urban environment, 108
  PM10 emissions, 104–106, 105f
  PM2.5 emissions, 104–106, 106f
  traffic flow patterns, 106
  wear processes, emission factors of, 107–108
Road wear emissions
  dust particles, 161–162, 178
  factors, 172–173
  full road simulator tests, 177–178
  nonstudded tires
    controlled studies, 165–166
    source apportionment studies, 162–163
  road wear particle properties
    mineral content, 174
    street environment, 173
    tungsten, 174
    wearing pavements, 173
    zinc, 174
  studded tires, 162
    controlled studies, 167–172
    mobile measurement studies, 172
    source apportionment studies, 167
Road wear particles (RWPs)
  ambient air particulate, 51
  bimodal distribution, 51
Road wear
  fine fraction (PM2.5) and coarse fraction (PMc;
      PM2.5–10), 113
  tracers, 116–118
Rubber mixed pavements, 171

S

Sandpaper effect, 186
Scanning mobility particle sizer (SMPS), 152–153

Sen's slope method, 7—8
Senate Bill (SB) regulations
    California (Senate Bill (SB) 346), 91—92, 100t
    New York State (Senate Bill (SB) A10871), 92, 100t
    Rhode Island State (Senate Bill (SB) H7997), 92, 100t
    Washington (Senate Bill (SB) 6557), 91—92, 100t
Studded tires, road wear particle emissions
    controlled studies, 167—172
    mobile measurement studies, 172
    source apportionment studies
        alternative pavement materials and constructions,
            170—172
        asphalt pavements, 167—170
Substances of Very High Concern (SVHC), 94
Subway systems
    airborne particulate matter (PM), 289—290
        airborne organic compounds, 311—312
        carbonaceous particles, 310
        crustal origin, 311
        organic compounds, 311—312
        sources, 308—309, 308f, 309t
        trace metals, 309—310
    Barcelona metro system, 315
    bioaerosols, 312—313
    bioreactive behavior, 313
    cell damage and health effects, 313
    inside trains, 307—308
    literature review, 290, 291t—299t
    oxidative stress, 314
    platforms
        higher PMX mass concentrations, 301—306, 306f
        outdoor air pollutants, 301
        piston effect, 301—306
        PM2.5 concentrations, 301, 302t—305t, 306—307,
            306f
        PSDs, 306—307
        subway environments, lower values for, 301
        UFP, 307
    subway air quality, 300
    toxic atmospheric environment, 314—315
    tunnel airflow, 315
    ventilation systems, 315
Superblock system, 246, 247f

T

Tank-to-wheel (TTW), 276—277
Territorial management measures, 244
Testing Reentrained Aerosol Kinetic Emissions from
        Roads (TRAKER), 194
Tire wear
    fine fraction (PM2.5) and coarse fraction (PMc;
        PM2.5—10), 113
    NFR source categorization, 108—109

PM10 and PM2.5, implied emission factors for, 111t
range, 69
tracers, 116—118
Tire wear emissions
    characteristics, 147—148
        chemical composition, 149
        PAH concentrations, 149—150
        rubber-containing particles, 149
        TPs, 149—150, 151t—152t
        unreactive and reactive chemicals, 148—149
        volume size distribution, 152—153
    complex physicochemical process, 147—148
    data, 156
    non-exhaust vehicle particulate emissions, 148
    pyrolysis GC/MS method, 156
    road surface characteristics, 148
    tire wear emission rates, 154
    tire wear particle air concentrations, 153—154
    vehicle characteristics, 148
    vehicle operation, 148
Tire wear mass size distribution, 51
Tire wear particles (TWPs), 22
Tires regulation
    GADSL and IMDS, 99
    REACH regulation, 97—98
Titanium brake discs, 213—214
Total carbon (TC) emission, 132
Toxic Chemical Control Act, 95
Traffic-induced suspension, 185—186
Tread particles (TPs), 149

U

Ultrafine particles (UFPs), 307
United States, 91
Urban freight distribution (UFD), 241
US Environmental Protection Agency (EPA), 266

V

Vehicle non-exhaust emissions, air quality
    ambient air particulate
        atmospheric emissions, 49—50
        brake wear factor profiles, 48
        contributions, 25
        direct tire wear contribution, 48—49
        dispersion modeling/emission inventory, 25
        environment, types of, 25
        miscellaneous factors, 25
        PM chemical characterization, receptor modeling
            based, 24
        PM observations, 24
        PM10, 46
        PM2.5, 47
        PMF profiles, 48, 48f

Vehicle non-exhaust emissions, air quality (*Continued*)
    range and mean brake wear, 47, 47f
    road dust category, 25
    RWPs, 51
    sources, 25
    tire wear mass size distribution, 51
    vehicle exhaust contribution, 47, 47f
    brake pads and discs, 23
  brake wear particles, 22
  carbon-reinforcing filler material, 54—55
  emissions, 22
  microscale spatial variability
    brake wear, 53
    generic non-exhaust source, 53
    road dust suspension, 51—53
    tire wear, 53
  mineral oils, 54—55
  oxidative stress, 22
  receptor modeling technique, 24
  road dust particles, 23
  road surface wear, 23
  road traffic sector, 21—22
  submicron particles, 54
  TWPs composition, 22—23

W

Washington (Senate Bill (SB) 6557), 91—92
Wear emissions
  EU-NMS, 104
  NFR, 102
  road transport PM10 emissions, 118—119
  vehicle types and road types, 109
  wear processes, emission factors of, 107—108
Wet dust sampling technique, 193, 200
World Health Organization (WHO), 1—2
  Air Quality guidelines, 2—4, 2t—3t

Printed in the United States
By Bookmasters